Advances in Cryogenic Engineering - Materials: Proceedings of the International Cryogenic Materials Conference (ICMC 2019)

IOP Conference Series: Materials Science and Engineering
Volume 756

Hartford, Connecticut, USA
21 - 25 July 2019

ISBN: 978-1-7138-1970-7
ISSN: 1757-8981

Printed from e-media with permission by:

Curran Associates, Inc.
57 Morehouse Lane
Red Hook, NY 12571

CURRAN ASSOCIATES INC.
proceedings.com

Some format issues inherent in the e-media version may also appear in this print version.

Institute of Physics
Dirac House, Temple Back
Bristol BS1 6BE UK

Phone: 44 1 17 929 7481
Fax: 44 1 17 920 0979

techtracking@iop.org

Additional copies of this publication are available from:

Curran Associates, Inc.
57 Morehouse Lane
Red Hook, NY 12571 USA
Phone: 845-758-0400
Fax: 845-758-2633
Email: curran@proceedings.com
Web: www.proceedings.com

TABLE OF CONTENTS

CRYOGENIC PROPERTIES OF METALS, NON-METALS, AND RESINS

CRYOGENIC POWER ELECTRONICS AND APPLICATION TECHNOLOGIES

NB3SN AND NB3AL SUPERCONDUCTORS

BSCCO, MGB2, YBCO SUPERCONDUCTORS AND SUBSTRATES

FLUX PINNING, STABILITY, AND AC LOSS

CABLES, MAGNETS, AND ROTATING MACHINES

IOP Conf. Series: Materials Science and Engineering **756** (2020) 011001 doi:10.1088/1757-899X/756/1/011001

Preface

The 2019 Joint Cryogenic Engineering and International Cryogenic Materials Conferences were held from July 21 through July 25 at the Connecticut Convention Center in Hartford, Connecticut. As at past conferences, the international scope of these meetings was strongly maintained with 24 countries being represented by 613 attendees who gathered to enjoy the joint technical programs, industrial exhibits, and special events.

The program for the joint conferences included a total of 410 presentations in the plenary, oral, and poster sessions. Four plenary talks gave interesting in-depth overviews and discussions on the topics of the growing field of commercial application of hydrogen fuel-cells for specialized applications and large-scale semi-truck transportation, and recent 5x increase of liquid hydrogen production in the U.S.; Siemens' perspectives on evolving hybrid-electric propulsion technologies for large commercial aircraft; the world-leading research opportunities with rare isotopes in nuclear physics, nuclear chemistry, and the application of rare isotopes for society that the Facility for Rare Isotope Beams (FIRB) at Michigan State University will soon provide; and the impressive progress designing the next-gen SPARC fusion reactor 1st-ever utilizing 2nd-generation high-temperature-superconducting (HTS) wire. The attendees also convened for two Special Joint CEC-ICMC Symposia on i) Transportation, and ii) Quantum Computing, and seven Special-or-Focus Sessions were included in the programming emphasizing high-quality invited talks. Contributed papers covered a wide range of topics including many aspects in advances in cryogenics and superconductors, along with their applications. Both CEC and ICMC boards are encouraging student participation and career growth and provided registration support for 36 student applicants. In total, 207 papers were submitted for publication of which 190 are published in these conference proceedings.

The CEC/ICMC Cryo Industrial Expo displayed the products and services of 49 industrial exhibitors and provided a congenial venue for a reception and refreshments throughout the week as well as for the conference poster sessions.

The conference location at Hartford was pleasant and low-stress, with interesting local history to observe such as the Connecticut State Capitol, a beautiful-unique wrought-iron Railway station, and a few of the regional Tobacco producers in the "Tobacco Valley" running from Hartford to Springfield MA. Hartford was large enough to provide many high quality dining and entertaining options within walking distance, and small enough to allow easy transportation to most hotels, area attractions, and the regional airport. The venue at the Connecticut Convention Center was comfortably large and clean, with a beautiful vantage view of the Connecticut River always observable. During the week, participants enjoyed the Cryo Expo Exhibitor's Reception on Monday, the popular Fun Run/Walk along the river on Wednesday morning, and the Conference Banquet Event on Wednesday night. Entertaining music was provided at the banquet, with a buffet menu and sit-down seating available for everyone.

Conference Chairs for 2019 were Jennifer Marquardt from Ball Aerospace, for CEC and Timothy Haugan from U.S. Air Force Research Laboratory for ICMC. Program Chairs were Luisa Chiesa from Tufts University for CEC and Charles Rong from U.S. Army Research Laboratory for ICMC, assisted by the CEC Program Vice Chair, Jacob Leachman from Washington State University. The Exhibit Chair was Austin Capers of Scientific Instruments Inc. who also served as Publicity Chair. Finally, Awards Chairs for were Peter Kittel, Consultant, for CEC and Eric Hellstrom from the National High Magnetic Field Laboratory at Florida State University for ICMC.

On July 21, the 2nd ICMC short course was presented, which was seven lectures given by L. Cooley, D. Evans, R. Walsh, T. Shen, L. Graber, E. Hellstrom and T. Hartwig. This short course was fully attended with over 30 scientists and students, and every lecture was well received and appreciated.

IOP Conf. Series: Materials Science and Engineering **756** (2020) 011001 doi:10.1088/1757-899X/756/1/011001

Also on July 21, the Cryogenic Society of America presented three short courses: *Property and Cooler Considerations for Cryogenic Systems*, a full day course given by J. M. Pfotenhauer, F. K. Miller and R. C. Dhuley, *Cryocooler Fundamentals*, a half-day course by R. Radebaugh and M. Zagarola, and *Getting Stared with Cryogenic Fuels – Liquefied Hydrogen and Natural Gas*, a half-day course by J. Leachman. The courses were well attended and catered to both seasoned professionals and newcomers to cryogenics in engineering and materials.

The able leadership of Paula Pair with the hardworking, enthusiastic, and efficient staff from Centennial Conferences provided outstanding conference management and operations. Thank you, Paula and staff, for a truly outstanding and memorable conference experience! Also thanks to the CEC/ICMC board members who contributed many hours of excellent volunteer service for the programming and event. Finally, to the attendees, authors, editors, and reviewers whose combined efforts have contributed to these proceedings, thanks very much to you all.

We look forward to the next CEC/ICMC in Honolulu, Hawaii, May 16 – 20, 2021. Hawaii will offer attendees one of the most beautiful settings in the U.S., with paradise everywhere to see and experience. The conference will be held at the Hawai'i Convention Center, which is close to the Waikiki Beach strip and hotels, and only 5-10 minutes walking distance to ocean swimming, beachside parks and jogging, and the Ala Moana conference hotel. Many fast service restaurants and food courts are within a 5-10 minute walk, as well as the Ala Moana Mall which is the largest open-air shopping center in the world with almost 350 stores. The local food cuisine reflects the mix of many cultures in Hawaii, including Hawaii, Polynesia, Japan, Korea, Philippines, Puerto Rico, Portugal, China, American, and Fusion mixes. The Convention Center will provide comfortably large exhibit/poster and oral presentation rooms.

Many local sightseeing trips and tourist experiences will be available on Oahu, and one-day trips to any of the seven Hawaii islands are available with reasonable cost. Day trips anywhere on the entire island of Oahu can achieved using the all-day bus fee of $5.50, allowing easy access to all buses running in-and-around the entire island. The conference is working to arrange excursions to observatories on the Island of Hawaii, which incorporate many types of cryogenic-based technologies and detectors for astrophysics studies.

In addition to the rich spectrum of R&D and industrial cryogenic topics which will make up the conference, we hope you will be able to enjoy the comforts and attractions of Hawaii when you are attending. We wish you a very warm welcome and Aloha to Honolulu!

Jennifer Marquardt
2019 CEC Conference Chairman

Timothy Haugan
2019 ICMC Conference Chairman

IOP Conf. Series: Materials Science and Engineering **756** (2020) 011002 doi:10.1088/1757-899X/756/1/011002

ICMC Awards

ICMC LIFETIME ACHIEVEMENT AWARD 2019

Professor Teruo Matsushita

Kyushu Institute of Technology
Japan

ICMC has presented its Award for Lifetime Achievement biannually at the CEC/ICMC conference since its 30th anniversary conference in Keystone in 2005 to an individual to "recognize a lifetime's achievement in advancing the knowledge of cryogenic materials".

The ninth recipient of the ICMC Lifetime Achievement Award is Teruo Matsushita, who follows Edward Collings in 2005, David Larbalestier in 2007, Kyoji Tachikawa in 2009, René Flükiger in 2011, Harold Weber in 2013, Herbert Freyhardt in 2015, Kozo Osamura in 2017, and Archie Campbell in 2018.

Teruo Matsushita graduated from the Department of Electronics in the Faculty of Engineering of Kyushu University in 1970 and completed his Master thesis in the Graduate School of Engineering of Kyushu University in 1972. In 1973, after one year of his doctoral studies, he was appointed as a research associate at Kyushu University. He obtained his Doctor degree from Kyushu University in 1980 and was appointed as an Associate Professor at Kyushu University in the same year. He moved to Kyushu Institute of Technology as a Professor in 1990. He was a Vice President of the institute from 1999 to 2003 and retired from the institute in 2011. After retirement he continued to engage in research at the institute from 2011 to 2015.

His research area is wide, ranging from fundamental physics to applications associated with flux pinning phenomena. He has published 429 scientific papers. He studied flux pinning extensively. The elementary pinning force of thin normal conducting phases was theoretically calculated using the Ginzburg-Landau theory and he showed that the force is not reduced in spite of an appreciable proximity effect, which was contradictory with the theoretical prediction of Kramer and Freyhardt. The validity of his obtained result was proved by the strong pinning effect of thin α-Ti in Nb-Ti. He collaborated with researchers at a company to improve the critical current density of Nb-Ti by introducing artificial Nb pinning centers and explained the very high critical current density based on the kinetic energy interaction that was unknown at that time. He also challenged the mechanism of pinning force density that exceeds the maximum force, i.e., the shear force, in the Kramer model. The result he obtained is consistent with the pinning characteristics of Nb-Ti due to the mechanical treatment that changes the shape of α-Ti precipitates from cubic to ribbon-like.

Matsusita also contributed to the development of pinning theory. First, he solved the threshold problem for the elementary pinning force (f_p) in the summation theory of Labusch. He developed the coherent potential approximation theory, i.e., the mean field theory for random systems, based on the comment of Larkin and Ochinnikov that the pinning correlation lengths are finite. His result shows that there is no essential threshold problem, since the threshold value is always smaller than f_p. The theory explains the experimental results that the pinning force density (F_p) obeys the linear summation ($F_p \propto N_p f_p$) for strong pinning and the statistical summation ($F_p \propto N_p {f_p}^2$) for weak pinning, where N_p denotes the number density of pinning centers. Extending this theory to the dynamic state, he estimated the pinning loss power density, which is equal to the product of F_p and the mean velocity of flux lines as the critical state model predicts. That is, he proved that the microscopic summation theory and the macroscopic critical state model are consistent. Finally he proved the critical state model using first principles to minimize the free energy for the regime of the reversible flux motion and then, developed the irreversible case. As a result, the critical state model is no longer a phenomenological model but a rigorous theory. Thus, the pinning theory has been unified. Recently he calculated the maximum critical current density (J_{cm}) that can be achieved by the mechanism of flux pinning using the Ginzburg-Landau theory and found that it exceeded the depairing current density (j_d) predicted by Tinkham. Then, he reexamined the depairing current density and found that the correct value for j_d is twice Tinkham's predicted value, which solved the problem that J_{cm} exceeds j_d.

Matsushita also contributed to clarifying the mechanism of the longitudinal field effect, i.e., peculiar phenomena in a current-carrying superconductor in a parallel magnetic field. He found that the driving force that moves flux lines is not the Lorentz force but a torque to release the shearing deformation of flux lines and showed that the rotational flux motion by the torque exceeding the pinning torque explains the peculiar phenomena such as a break of Josephson's relationship for the induced electric field or appearance of negative electric field in the resistive state. This directly proves that the critical current density is determined by the flux pinning as well as in the usual transverse magnetic field. He invented an innovative new superconducting power cable, in which the current-carrying capacity can be enhanced using the back current in the outer shield layer. This is an application of the longitudinal field effect. He also developed the collective flux creep model to generally explain the electromagnetic properties of various superconductors with different dimensionality. This theory has been used to estimate the irreversibility field including its dependence on temperature, electric field, sample size and dimensionality of superconductor.

The citation on Matsushita's award plaque reads:

For his theoretical contributions to understanding vortex pinning in superconductors and flux-pinning characteristics of real materials, including the effects of nanometer size, geometry, normal-state properties, and collective flux creep.

IOP Conf. Series: Materials Science and Engineering **756** (2020) 011002 doi:10.1088/1757-899X/756/1/011002

ICMC CRYOGENIC MATERIALS AWARD FOR EXCELLENCE 2019

Professor Kohei Higashikawa

Department of Electrical Engineering
Graduate School of Information Science and Electrical Engineering
Kyushu University

The ICMC Cryogenic Materials Award for Excellence is awarded annually at the ICMC Conference, subject to the nomination of an appropriate candidate, to an individual, who is under 40 years of age by the application deadline, to "recognize excellence in advancing the knowledge of cryogenic materials over recent years". The 2019 award was given to Kohei Higashikawa. Previous awardees are Fumitake Kametani in 2014, Kazumasa Iida in 2016, Tengming Shen in 2017, and Anna Kario and Akiyasu Yamamoto in 2018.

Kohei Higashikawa received his doctorate in Electrical Engineering in 2007 from Kyoto University, Japan. In April 2007, he became a postdoctoral research fellow of the Japan Society for the Promotion of Science (JSPS), a position that is known to be very competitive (10 % success rate). In 2010, he became an assistant professor at Kyushu University and was promoted to associate professor in 2012. He was a visiting scientist/researcher in the Applied Superconductivity Center, National High Magnetic Field Laboratory, USA in 2013. He has received several awards, including from the International Energy Agency, MEXT, MRS-J, Cryogenics, the Superconductivity Society of Japan (CSSJ), and the Institute of Electrical Engineers of Japan (IEEJ). He has received research funding from JSPS "KAKENHI" and JST "A-STEP". He currently has significant research funding from NEDO in the field of superconductivity. He has published 60 peer-reviewed research papers and has given 7 invited talks at international conferences including at the Applied Superconductivity Conference, USA.

His research focuses on characterizing forefront superconducting wires/tapes and their application in high-field magnets, SMES systems, and novel superconducting cables. He developed several Scanning Hall-Probe Microscope (SHPM) systems to visualize local electromagnetic properties on the scale of several micrometers in forefront HTS tapes and MgB_2 wires at various operating conditions, which is not possible using conventional methods. He developed a method to investigate current blocking obstacles and track them to their microstructural origin at the nano-scale. He also built a facility integrating the SHPM with a long-wire traveling system, which allowed him to investigate two-dimensional critical current distribution in km-long HTS tapes. The study was essential to improve tape strand quality and to establish reliability of HTS coil windings. Furthermore, he applied these findings to the design of HTS coil applications such as for high-field magnets, energy storage, and power cables. His conceptual design work has been incorporated in the Japanese national project on HTS-SMES.

ICMC BEST PAPER AWARD

The ICMC Best Paper Awards provide an incentive for the production and presentation of high-quality papers at the International Cryogenic Materials Conferences and recognition of authors who, in the judgment of the ICMC Board of Directors, presented the best paper at the proceeding conference. Papers are nominated by the reviewers.

In 2019, two awards for the best papers delivered at the 2017 ICMC in Madison, Wisconsin, USA, and published in the *IOP Conference Series: Materials Science and Engineering, Vol. 279, 2017*, were presented at the 2019 Hartford Conference to the following:

Best Superconducting Materials Paper

Christopher Segal, Chiara Tarantini, Peter J Lee and David C Larbalestier

for their paper

"Improvement of small to large grain A15 ratio in Nb_3Sn PIT wires by inverted multistage heat treatments"

IOP Conference Series: Materials Science and Engineering, Vol. 279, 2017; 012019

Best Structural Materials Paper

Norimitsu Koga, W Zhang, O Umezawa, V Tschan, J Sas and K P Weiss

for their paper

"Temperature dependence on tensile properties of Cu-40mass%Fe dual phase alloy"

IOP Conference Series: Materials Science and Engineering, Vol. 279, 2017; 012004

ICMC STUDENT MERITORIOUS PAPER AWARD

The ICMC Board of Directors also recognizes students who write high quality papers. All students who applied for and participated in the ICMC Early Registration Fee Waiver program during the 2017 ICMC Madison Conference were automatically entered for consideration of the ICMC Student Meritorious Paper Award. The papers are ranked on the basis of research merit and quality of writing.

The recipient of the 2019 ICMC Student Meritorious Paper Award is:

Imam Hossain

for his paper

"Effect of sheath material and reaction overpressure on Ag protrusions into the TiO_2 insulation coating of Bi-2212 round wire"

I Hossain, J Jiang, M Matras, U P Trociewitz, J Lu, F Kametani, D Larbalestier, and E Hellstrom

IOP Conference Series: Materials Science and Engineering, Vol. 279, 2017; 012021

IOP Conf. Series: Materials Science and Engineering **756** (2020) 011003 doi:10.1088/1757-899X/756/1/011003

2019 International Cryogenic Materials Conference Board of Directors

CONFERENCE AND PROGRAM

Michael Sumption, *President* – The Ohio State University
Timothy Haugan, *Conference Chair* – U.S. Air Force Research Laboratory
Charles Rong, *Program Chair* – U.S. Army Research Laboratory

BOARD OF DIRECTORS

U. (Balu) Balachandran, *Chief Technical Editor* – Argonne National Laboratory
David A. Cardwell – University of Cambridge, United Kingdom
Lance D. Cooley – NHMFL, Florida State University
David Evans – Advanced Cryogenic Materials Ltd, United Kingdom
Herbert C. Freyhardt – TcSUH / University of Göttingen (retired) and ZFW gGmbH Göttingen, Germany
Wilfried O. Goldacker – Karlsruhe Institute of Technology, Germany
K. Theodore Hartwig, Jr. – Texas A&M University
Timothy J. Haugan – U.S. Air Force Research Laboratory
Eric E. Hellstrom – NHMFL, Florida State University
Juan Knaster – Fusion for Energy at ITER Organisation, France
Yanwei Ma – Institute of Electrical Engineering, Chinese Academy of Science, China
Thomas H. Nicol – Fermi National Accelerator Laboratory
Richard P. Reed – Cryogenic Materials, Inc.
Charles Rong – U.S. Army CCDC Army Research Laboratory
Kenichi Sato – Japan Science and Technology Agency, Japan
Jun-ichi Shimoyama – Aoyama Gakuin University, Japan
Hyung-Seop Shin – Andong National University, South Korea
Michael D. Sumption – The Ohio State University
Herman H. J. ten Kate – CERN, Switzerland
Marcel H. ter Brake – University of Twente, The Netherlands
Robert P. Walsh – NHMFL, Florida State University

AWARDS COMMITTEE

David A. Cardwell – University of Cambridge, United Kingdom
Lance D. Cooley – NHMFL, Florida State University
K. Theodore Hartwig, Jr. – Texas A&M University
Eric E. Hellstrom, *Chair* – NHMFL, Florida State University
Michael D. Sumption – The Ohio State University
Robert P. Walsh – NHMFL, Florida State University

ICMC Technical Editors

U. (Balu) Balachandran
Chief Technical Editor
Argonne National Laboratory
9700 S. Cass Avenue
Building #212
Argonne, IL 60439
USA
Email: balu@anl.gov

Shreyas Balachandran
NHMFL, Florida State University
2031 E. Paul Dirac Drive
Tallahassee, FL 32319
USA
Email: shreyasb@asc.magnet.fsu.edu

Emanuela Barzi
Fermi National Accelerator Laboratory
Pine and Kirk Roads
MS 315
Batavia, IL 60510
USA
Email: barzi@fnal.gov

David Cardwell
University of Cambridge
Department of Engineering
Trumpington Street
Cambridge, Cambridgeshire CB21PZ
United Kingdom
Email: dc135@cam.ac.uk

Timothy Haugan
U.S. Air Force Research Laboratory
AFRL/RQQM, 1950 Fifth Street
Bldg. 18
Wright-Patterson AFB, OH 45433
USA
Email: timothy.haugan@us.af.mil

Eric Hellstrom
NHMFL, Florida State University
2031 E. Paul Dirac Drive
Tallahassee, FL 32310
USA
Email: hellstrom@asc.magnet.fsu.edu

Jianyi Jiang
NHMFL, Florida State University
2031 E. Paul Dirac Drive
Tallahassee, FL 32319
USA
Email: jjiang@asc.magnet.fsu.edu

Swarn Kalsi
Kalsi Green Power Systems, LLC
46 Renfield Drive
Princeton, NJ 08540
USA
Email: swarnkalsi@yahoo.com

Takanobu Kiss
Kyushu University
Dept. of Electrical Engineering
744 Motooka, Nish-ku
Fukuoka, 819-0395
Japan
Email: kiss@sc.kyushu-u.ac.jp

Peter Lee
NHMFL, Florida State University
2031 E. Paul Dirac Drive
Tallahassee, FL 32310
USA
Email: lee@asc.magnet.fsu.edu

Tatiana Prikhna
V. Bakul Institute for Superhard Materials of the National Academy of Sciences of Ukraine
2, Avtozavodskaya Str.
Kiev, 04074
Ukraine
Email: prikhna@ukr.net

Chiara Tarantini
NHMFL, Florida State University
2031 E. Paul Dirac Drive
Tallahassee, FL 32310
USA
Email: tarantini@asc.magnet.fsu.edu

Wolfgang Stautner
GE Research
Biology & Applied Physics Electro-magnetics and Superconductivity
One Research Circle
EP 113
Niskayuna, NY 12309
USA
Email: stautner@research.ge.com
Email: stautner@ge.com

Editorial Management
Annett Cady and Carrie Lian
Centennial Conferences
908 Main Street, Suite 350
Louisville, CO 80027
USA
Email: cecicmc@centennialconferences.com

IOP Conf. Series: Materials Science and Engineering **756** (2020) 011005 doi:10.1088/1757-899X/756/1/011005

Acknowledgments

The CEC and ICMC Boards wish to thank our supporter and sponsors who have contributed to the 2019 CEC/ICMC Conference.

SUPPORTER

- **Fermilab**

SPONSORS & ADVERTISERS

- **Ability Engineering Technology, Inc.**
- **Aerospace Fabrication & Materials, LLC**
- **Beijing Sinoscience Fullcryo Technology Co., Ltd.**
- **Cryomech, Inc.**
- **Demaco Holland bv**
- **Fujikura Ltd.**
- **GE Global Research**
- **High Precision Devices**
- **Scientific Instruments, Inc.**

Their contributions helped ensure the success of the 2019 Conference.

Producing the **Open Access *IOP Conference Series: Materials Science and Engineering (MSE)*** requires the dedicated effort of many individuals. The quality of the publication relies on the dedication of those who took time from their busy schedules to review and edit papers. The Boards wish to specifically acknowledge the efforts of Centennial Conferences for their editorial support, and the chief technical editors, John Weisend (CEC) and Balu Balachandran (ICMC). Their contributions to ensure a quality publication in a timely manner are greatly appreciated. The chief technical editors are indebted to the individual technical sub-editors for their time spent with reviewers and authors to get the final manuscripts ready for publication. John Weisend is also indebted to Gunilla Jacobsson for her assistance with the cumulative CEC subject index. The Boards also wish to thank all students and post-doctoral researchers from a number of universities for their service with the initial format checking of the papers submitted for peer review.

CEC/ICMC 2019 was managed by Paula Pair and Centennial Conferences. Recognizing this is a tremendous team effort, the Boards would especially acknowledge, in addition to Paula, all her staff – Annett Cady, Troy Christensen, Brion Jacobs, Andrew Kahl, Carrie Lian, Diane Mehling, Sydney Pair, and Letitia Xu – for the fantastic job they did wonderfully.

Last, but not least, the Conference would not have been possible without our exhibitors and their sustained commitment and support, and the effort and dedication of the attendees in preparing their presentations and papers, traveling to Hartford, Connecticut, and showing the professionalism and enthusiasm that makes the CEC/ICMC such a great Conference time after time.

IOP Conf. Series: Materials Science and Engineering **756** (2020) 011006 doi:10.1088/1757-899X/756/1/011006

Materials Index

IOP Conf. Series: Materials Science and Engineering **756** (2020) 011006 doi:10.1088/1757-899X/756/1/011006

Subject Index

IOP Conf. Series: Materials Science and Engineering **756** (2020) 011008 doi:10.1088/1757-899X/756/1/011008

Peer review statement

All papers published in this volume of *IOP Conference Series: Materials Science and Engineering* have been peer reviewed through processes administered by the proceedings Editors. Reviews were conducted by expert referees to the professional and scientific standards expected of a proceedings journal published by IOP Publishing.

IOP Conf. Series: Materials Science and Engineering **756** (2020) 012001 doi:10.1088/1757-899X/756/1/012001

The low temperature mechanical properties of a Nitronic 40 forging

R P Walsh, K J Radcliff, J Lu and K Han
National High Magnetic Field Laboratory, Florida State University
800 E. Paul Dirac Dr.

Abstract. Nitronic 40 forged shells are typically used for structural reinforcement in high field pulse magnet design and applications. To better understand the mechanical performance of this versatile high strength austenitic steel a series of mechanical tests were conducted. Tensile were performed at 295 K, 77 K and 4 K, and cryogenic fracture mechanics tests were performed at 77 K and 4 K. The effect of temperature on strength, ductility, toughness and fatigue crack growth rate are evaluated. Microstructure and composition effects are also presented and discussed.

1. Introduction

Pulse magnet reinforcement shells are required to have good mechanical properties over the temperature range from 77 K to 350 K. Pulse magnets operate at temperatures ranging from 77 K to 350K with peak stress at the low end of the temperature range. The austenitic steel alloy Nitronic 40 is a material of choice for pulse magnet reinforcement shells and experience has shown that process and quality control of this complex alloy is necessary [1]. The mechanical properties of a Nitronic 40 forged reinforcement shell have been investigated at three test temperatures (295, 77, and 4 K) with tensile, fracture toughness, and fatigue crack growth rate (FCGR) tests. The fracture toughness and FCGR tests are focused on the 77 K test temperature while the 4 K tensile and fracture toughness tests were conducted to obtain additional low temperature data on this versatile alloy.

The test material was obtained from a LANL pulse magnet shell that was available after retirement from some years of pulse magnet service. We tested the forging in two different conditions, the as-received (AR) condition and a heat-treated condition (HT). The HT condition was achieved with a relatively short-time, low-temperature heat treatment (400C for 20 min) that simulates the proposed insulation curing process to be used on future pulse magnets. Fracture toughness measurements were made at 77 K to obtain design data to make sure the heat treatment did not compromise the materials properties. The magnetic permeability, before and after low temperature deformation, were measured and are reported here too.

2. Material Information

The chemistry of the forging was determined commercially at two locations within the forging and is given in Table 1. Nitronic 40 is a commercially available high-manganese nitrogen-strengthened austenitic steel that is often referred to as 21-6-9, due to the nominal percentages of Cr/Ni/Mn. It has extremely low magnetic permeability and good low temperature properties with higher yield and tensile strength at 295 K than typical of the 300 series austenitic steels.

Table 1: The Nitronic 40 forging chemistry at 2 locations within the forging is shown below.

Element (wt %)		C	Mn	P	S	Si	Cr	Ni	Mo	Cu	N
Required	Min	---	8.00	---	---	---	19.0	5.50	---	---	0.15
Specifications	Max	0.04	10.00	0.060	0.030	1.00	21.5	7.50	0.75	---	0.40
110 mm from end		0.05	9.39	0.024	0.00	0.47	19.67	7.28	0.25	0.20	0.34
560 mm from end		0.05	9.44	0.023	0.003	0.49	19.67	7.29	0.25	0.20	0.34

The heat treated (HT) material condition was achieved by heat treating (4) tensile and (3) CTs. The specimens were placed in the preheated furnace retort, with an Ar atmosphere and subjected to the thermal profile in Figure 1.

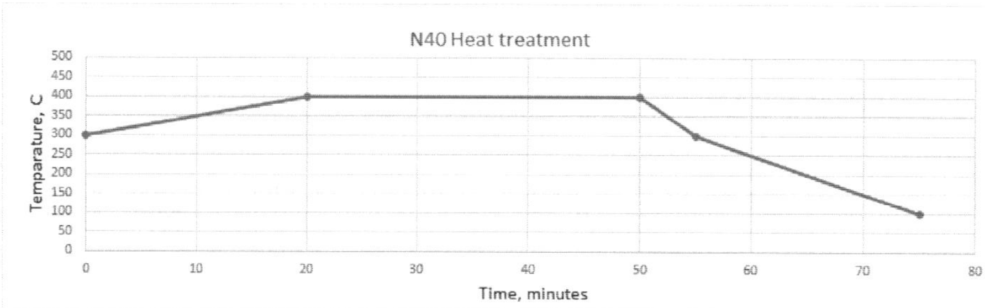

Figure 1: The nominal heat treatment temperature profile for HT condition specimens.

3. Procedures

The schematic in Figure 2 shows the ASTM compliant tensile and fracture toughness test specimen location and orientation as they were removed from the forged tube. All the tests were conducted on either a 100 kN or 250 kN capacity MTS machines that are equipped with a cryostat to enable testing at 77 K and 4 K with the test specimen and fixture immersed in liquid nitrogen or helium respectively.

Tensile tests are conducted, according to procedures in ASTM E8 and ASTM E1450, in displacement control at a rate = 0.5 mm/min (initial elastic strain rate = 2.2e-4 s^{-1}). The cylindrical tensile specimens (6.35 mm dia. X 38 mm long gage section) were oriented with their axes in the circumferential direction (C). Strain was measured using a set of (3)-25 mm gage length clip-on extensometers. Elongation is determined from scribe marks made 25 mm apart on the specimen gage

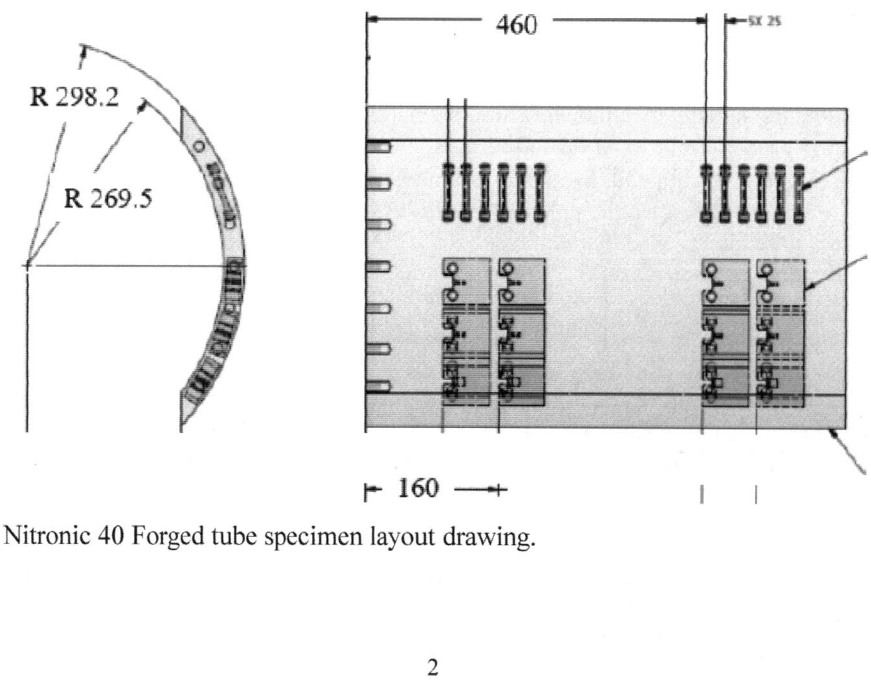

Figure 2: Nitronic 40 Forged tube specimen layout drawing.

IOP Conf. Series: Materials Science and Engineering **756** (2020) 012001 doi:10.1088/1757-899X/756/1/012001

section prior to testing. The 77 K FCGR and the 77 K and 4 K fracture toughness were performed according to the guidelines in ASTM E647 and ASTM E1820 respectively. The 1.0 CT specimens are oriented CL (circumferential load direction, longitudinal crack direction). The CT specimens were Electro-Discharge-Machined (EDM) with a short notch (15.5 mm) so they can be used for both the FCGR tests and the J_{IC} tests.

The specimens were initially pre-cracked ~2 mm at 77 K before FCGR testing at 77 K or 4 K. The FCGR test is conducted from the crack length (a) = 17.5 mm until (a) reaches ~ 30 mm for an (a/W) ratio of 0.6 that is desired for the J_{IC} test. The FCGR test was performed in constant load control (an increasing Δ K mode) and during the test the Δ K ranges from ~ 30 MPa√m to 70 MPa√m until the 30 mm crack length is reached. The fracture toughness test is performed next.

For ductile materials, the elastic-plastic fracture mechanics test method ASTM E1820 allows for the determination of $K_{IC}(J)$ in a J-integral (J_{IC}) test where K_{IC} is converted from J_{IC} using the equation:

$$K_{IC}(J) = (J_{IC} * E^1)\ \hat{}\ 1/2$$

Where; $E^1 = E/(1-\nu\hat{}2)$
E = Young's Modulus
ν = Poisson's Ratio

The values determined in 77 K and 4 K fracture toughness tests typically do not satisfy all the ASTM E1820 requirements and are considered estimates of the 4 K toughness.

Magnetic permeability measurements were made on test specimens using a Low-Mu Permeability Indicator according to ASTM A342 procedures. The indicator can sense the relative permeability (μ) of low μ materials (μ of 6.0 or less).

4. Results and Discussion

The tensile test results are shown in Table 2 and Figure 4 plot of strength vs test temperature. The 295 K tensile properties of both the AR and the HT material are similar, suggesting no effect from the HT process. The forging has a high yield strength and tensile strength (432 MPa and 791 MPa) at 295 K that increases almost linearly with decreasing temperature to approximately 1400 MPa and 1845 MPa respectively at 4 K. The tensile results for the forging compare favourably to previously published low

Table 2: Nitronic 40 tensile test results for circumferential direction.

Temp	Condition	Spec. No	Modulus of Elasticity GPa	Yield Strength MPa	Tensile Strength MPa	Elong. in 25 mm %	Reduction of Area %	Relative Permeability μ
295	AR	T1	199	430	784	63.0	76.2	<1.01
		T2	205	434	794	60.5	76.1	<1.01
		T3	208	433	795	64.1	75.0	<1.01
		AVG	204	432	791	63	76	
	HT	T3	204	440	790	69.0	79.2	<1.01
		T4	202	441	787	60.5	76.3	<1.01
		AVG	203	441	789	65	78	
77	AR	T4	NA	1125	1577	35.1	29.5	3.0
		T5	210	1133	1593	45.9	39.0	5.0
		T6	205	1129	1585	35.6	29.4	2
		AVG	208	1129	1585	39	33	
	HT	T5	204	1107	1490	35.9	28.9	2.0
		T6	204	1118	1434	42.5	33.9	3.0
		AVG	204	1113	1462	39	31	
4	AR	T1	209	1414	1855	19.1	24.3	2.0
		T2	205	1393	1836	15.9	17.1	1.2
		AVG	207	1404	1845.5	18	21	

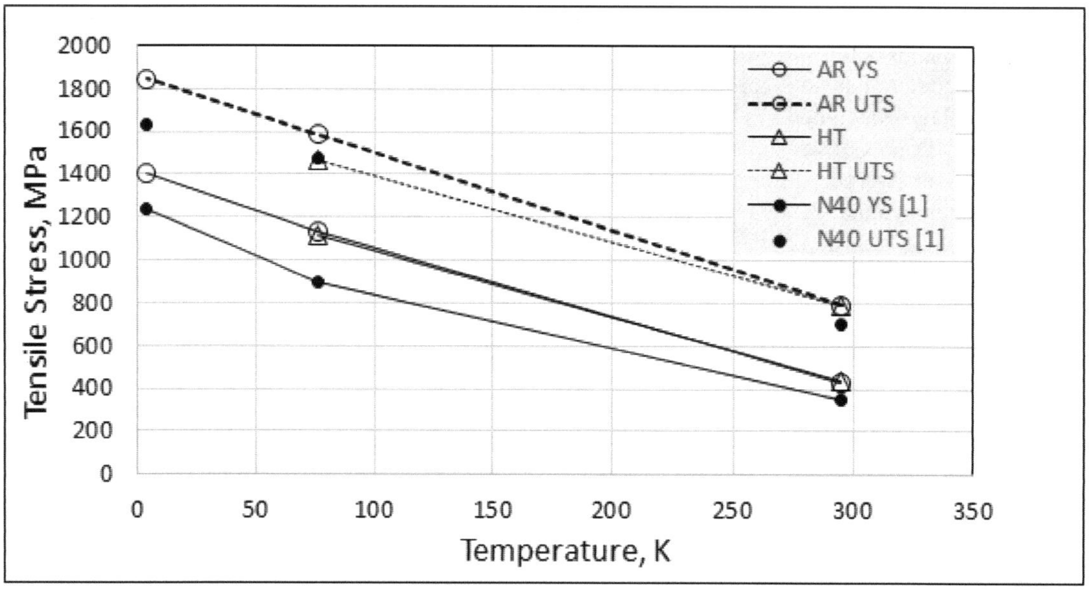

Figure 3: Plot shows an almost linear relationship between strength and temperature.

temperature data [2-4] for Nitronic 40 which is included on the graph in Figure 3 and this forging is somewhat stronger at all three temperatures probably since the reference data is for annealed plate. Both the AR and HT material exhibit good ductility at 77 K with almost 40% elongation and about 30 % reduction in area.

The fracture toughness and FCGR results are shown in Table 3. The 77 K fracture toughness of the forging is excellent with a slight difference in the average toughness (3 tests each) of the HT and AR

Table 3: Nitronic 40 fracture toughness and FCGR test results.

Temperature		Specimen	KIC(J)	Paris Law Constants (see note)		Delta K Range
K		Number	MPa√m	C	m	MPa√m
77	AR	110-CT-1	357	1.16E-08	2.63	32 to 66
		110-CT-2	353	5.85E-09	2.82	33 to 66
		110-CT-3	n/a	7.39E-09	2.78	34 to 67
		460-CT-1	348	n/a	n/a	n/a
		Average	352	7.49E-09	2.76	34 to 67
	HT	110-CT-4	318	7.49E-09	2.76	35 to 70
		110-CT-5	344	1.12E-09	3.27	28 to 55
		110-CT-6	366	2.07E-09	3.16	28 to 55
		Average	343	9.42E-10	3.36	28 to 70
4	AR	460-CT-4	144	3.53E-10	3.88	28 to 55
		460-CT-5	n/a	1.19E-08	3.00	27 to 40
		460-CT-6	164	7.99E-10	3.74	31 to 68
		460-CT-2	118	5.33E-09	3.23	27 to 57
		Average	142	1.54E-09	3.54	27 to 68
1. K_{IC}(J) values are estimates since validity requirements of ASTM E813 are not met						
2. Relevant units for Paris Law constants: ΔK = MPa√m, da/dn = mm/cycle						
3. Specimen Orientation is (CL)						

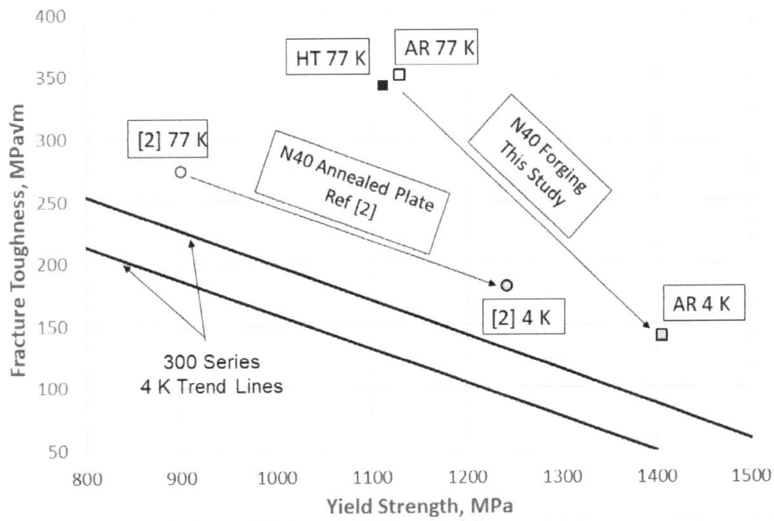

Figure 4: Fracture toughness vs yield strength comparison with the published data [3]. The NIST 4 K trend lines for 300 series stainless steel are also included for reference.

conditions, the $K_{IC}(J)$ values are 355 MPa√m and 343 MPa√m respectively. About a 60% decrease in toughness was observed for the AR material from 77 K to 4 K.

The 4 K toughness of the AR is still good with an average value = 142 MPa√m for 3 tests. Low temperature fracture toughness in austenitic steels is usually inversely dependent to the tensile yield strength. The Figure 4 graph of Yield Strength vs Fracture Toughness shows the 77 K and 4 K results for the Nitronic forging and compares them with published data [3]. The NIST trend lines show the scatter band typical for austenitic steels at 4 K. This Nitronic 40 forging has better performance than the annealed plate [3] at 77 K while the 4 K strength toughness relationship is similar.

FCGR tests have been conducted to determine the Stage II (Paris Regime) crack growth behaviour. In the Paris regime (FCGR from about 10^{-5} mm/ cycle to 10^{-3} mm/cycle) the crack growth rate obeys the simple power law;

$$da/dn = C^* \Delta K^m$$

where;
da/dn = crack growth increment/cycle
ΔK = stress intensity factor range (Kmax-Kmin)
C, m = constants related to material, test temperature, etc.

The plot of da/dn vs ΔK in Figure 5a is typical FCGR data from the test of one specimen (Specimen AR-CT2). A power law curve fit is applied to the data and the Paris law parameters (C and m) for this test are shown on the graph and in Table 3. The average Paris law parameters shown on Table 3 are obtained by combining the FCGR data specific to one material condition (AR or HT) and one test temperature (77 K or 4 K) from all the specimens tested for that condition and curve fitting the entire data set. The combined data set for the (3) 77 K tests of AR material are plotted in Figure 5b providing a visual representation of the FCGR data scatter.

The Paris Law parameters are used to present the average FCGR for the test and material conditions in Figure 6 and 7. The 77 K FCGR results for the AR and HT are very similar, and for example for a

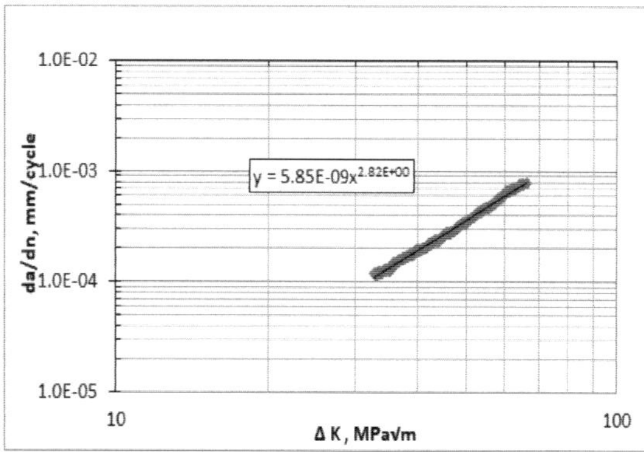

Figure 5a: Power law curve fit of FCGR test data for one specimen.

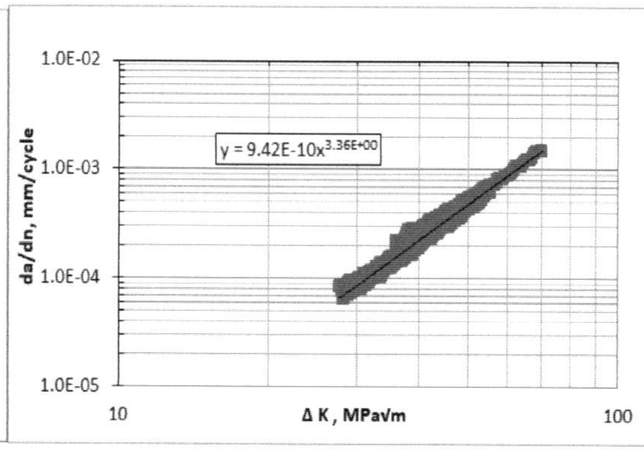

Figure 5b: Power law curve fit FCGR test data for three tests to get the average parameters.

ΔK of 30 MPa√m the crack growth rates are 8.84×10^{-5} and 8.65×10^{-5} mm/cycle respectively. This rate is comparable to the previously published 102 K data that is about 8.76×10^{-5} mm/cycle at 30 MPa√m. Figure 6 shows the 4 K FCGR is quite a bit higher than the 77 K rate, approximately 2 times higher at ΔK of 30 MPa√m and about 4 times higher at ΔK of 70 MPa√m. Although we could not find any published FCGR data for Nitronic 40 at 4 K, we compare the 4 K FCGR of Nitronic 40 to Nitronic 50 in Figure 7. This Nitronic 40 forging has considerably higher crack growth rates than published values for Nitronic 50.

More research is necessary to understand why Nitronic 40 FCGR is higher at 4 K than 77 K and why the FCGR is higher than Nitronic 50 at 4 K. Nitronic 50 is susceptible to low temperature strain induced martensitic transformation so we made magnetic measurements on the fractured Nitronic 40 tensile specimens.

The Nitronic 40 forging is non-magnetic due to its composition and metallurgical condition (austenitic phase with FCC crystal structure). Magnetic permeability (μ) measurements are made on the fractured tensile specimens as a measure of austenitic stability after low temperature deformation. The measurements are localized around the fracture region and the results are shown in the last column in

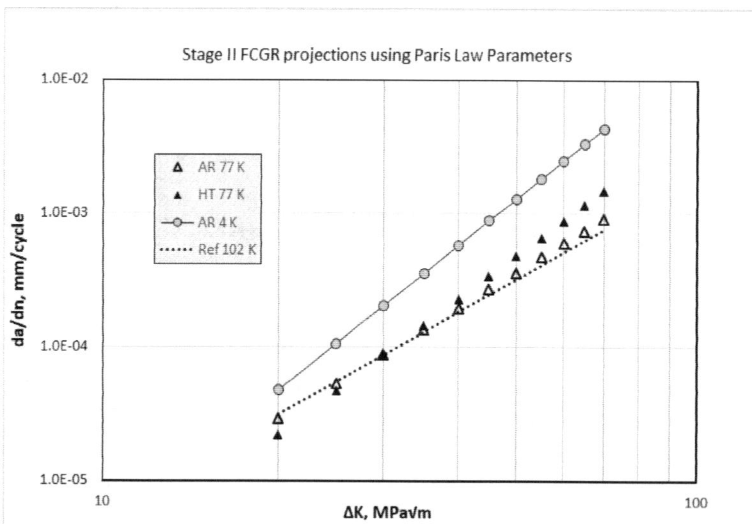

Figure 6: The 77 K projections compare well with the 102 K reference [2] data. The 4 K FCGR trendline shows a much higher crack growth rate than at 77 K.

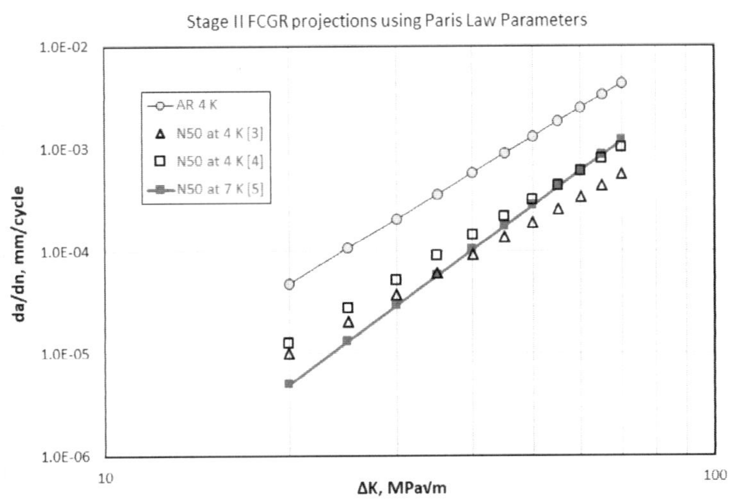

Figure 7: The 4 K Nitronic 40 data show a considerably higher crack growth rate than Nitronic 50 at 4 K [4-6].

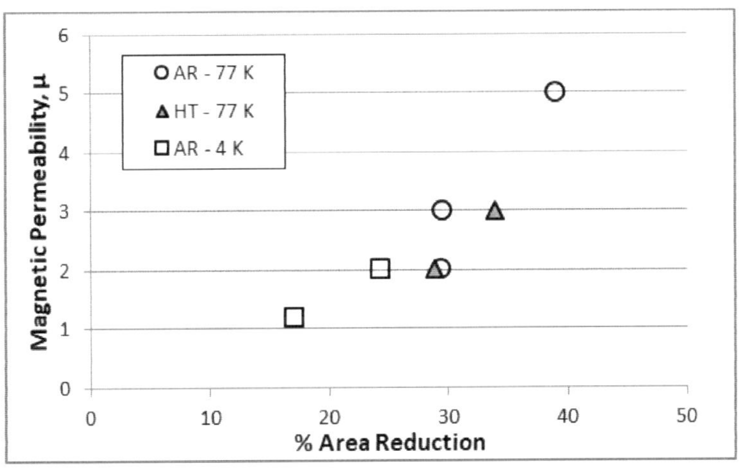

Figure 8: Relative permeability measurement taken at fracture locations on tensile specimens.

Table 2 and on the graph in Figure 8. While there was no indication of change in magnetic permeability after 295 K deformation, there was a noticeable increase in permeability after 77 K and 4 K deformation as was also observed in [3]. The increase in permeability is attributed to martensitic transformation where a percent of the FCC austenite crystal structure is converted to BCC martensite which may contribute to a reduction in toughness at cryogenic temperatures. Figure 8 also shows the magnetic transformation appears to be insensitive to the deformation temperature (77 K or 4 K) which is different than the findings in [3] where the amount of transformation at 4 K was supressed compared to 77 K.

5. Conclusions

The Nitronic 40 forged material researched here has excellent mechanical properties at both room and cryogenic temperatures that agree well with expectations and reference data in the literature. The mechanical properties are not degraded after exposure to a low-temperature short-time heat treatment that the material must undergo in future magnet construction procedures. Although the fracture toughness and FCGR are significantly reduced at 4 K compared to 77 K the retained values are considered excellent for cryogenic applications.

6. References

[1] JR Sims, J B Schillig, G S Boebinger, H Coe, A W Paris, M J Gordon, M D Pacheco, T G Abeln, R G Hoagland, M C Mataya, k Han, and A Ishmaku, The U. S. NHMFL 60 T Long Pulse Magnet Failure, IEEE Transactions on Applied Superconductivity, Vol. 12, No. 1, March 2002.

[2] Domack, M A, Fracture toughness and flaw growth in Nitronic 40 at cryogenic temperatures, Langley Research Center, Hampton, Va, NASA Technical Paper 2312, 1984.

[3] Tobler, R L and Reed, R P, Tensile and fracture behavior of a nitrogen-strengthened, chromium-nickel-manganese stainless steel at cryogenic temperatures, Elastic-Plastic Fracture, ASTM STP668,1979, pp. 537-552.

[4] D M McRae, R P Walsh, E N C Dalder, S Litherland, M Trosen and D J Kuhlman, Fatigue and fracture properties of a super-austenitic stainless steel at 295 K and 4 K, Advances in Cryogenic Engineering- Materials Vol. 60 (2013), pp. 59-66.

[5] R P Walsh and D M McRae, Report on CS tie plate base metal and weld characterization program, FSU/NHMFL Internal Engineering – Memo # MD&C-EM11-009E, 2013.

[6] A. Nyilas, Fatigue crack growth rate and fracture toughness properties at 7 K of forged Nitronic 50 material and its welding, Cryogenic Engineering & Materials Expertise (CEME) Inc., Germany, Document: 060741-14, CEME 2013.

Acknowledgments
This work was performed at the National High magnetic Field Laboratory which is supported by the National Science Foundation Cooperative Agreement DMR-1644779 and the State of Florida.

RRR measurements and tensile tests of high purity large grain ingot niobium

Z Zhao[1], C Owen[1], G G R Abdo[2], R M Ribas[2], G R Myneni[3], and L Chiesa[1]

[1] Tufts University, Mechanical Engineering Department,
Medford, MA 02155, USA

[2] Companhia Brasileira de Metalurgia e Mineração (CBMM),
Córrego da Mata, s/n Araxá, Minas Gerais 38183-903 Brazil

[3] BSCE, Systems, Inc.,
Yorktown, VA 23693, USA

Abstract. Residual resistivity ratio (RRR) measurements and tensile tests were performed on samples taken from seven high purity large grain niobium ingots produced by CBMM. Test specimens were taken from the top, middle and bottom regions of each ingot. The RRR measurements were performed with a GM cryocooler system and more than 60% of the samples exceeded RRR values of 250. Room temperature tensile tests were performed on samples from the same regions. 17 samples were machined according to ASTM standard test methods for tension testing of metallic material. Tests were performed with an Instron tensile/compression testing machine, and various mechanical properties such as Young's modulus, 0.2% proof stress, ultimate tensile strength and strain at fracture were measured. For all the samples, the fracture strain exceeds 35% while the ultimate tensile stress is close to 100 MPa. Contents of impurity (N, O, C, H, Ta) of each region of the seven ingots were also measured. These data provide insights for a possible correlation among RRR values, mechanical properties and impurity contents of niobium ingots.

1. Introduction

Superconducting radio-frequency (SRF) cavities are used in modern particle accelerators to efficiently accelerate a beam of charged particles. The power dissipated in the cavity wall is several orders of magnitude lower if superconductors are utilized compared to a normal conductor, thus making superconductors highly efficient in this application [1]. Niobium is the material of choice for making SRF cavities because of its relatively high superconducting transition temperature (~9.3 K), good electrical conductivity and thermal diffusivity, relative abundance and availability, and good formability [2]. Residual Resistivity Ratio (RRR) is a great indicator of the purity and thermal conductivity of a material and the typical RRR value of niobium for RF applications is 200~300 [3]. The RRR value of niobium is measured by comparing the resistivity of niobium at ~10 K (just above its transition temperature) to the resistivity at room temperature [4]. In addition, strength and formability is also essential for the fabrication of SRF cavity, and the improvement of those characteristics allows the cavities to be formed, transported and more easily assembled. The mechanical properties can be provided through tensile test. Those include elastic modulus (E), yield strength (Y), ultimate tensile strength (UTS), and elongation to failure. The yield strength is evaluated using a standard 0.2 % offset strain method to find the 0.2% proof stress ($R_{p0.2}$).

In the standard SRF fabrication, high purity niobium sheets with grain size over 50 μm (ASTM ≥5) are selected. The small and uniform grain size of the polycrystalline niobium indicates the good formability and good surface finish [5]. However, the production processes of fine-grain niobium sheet, including hot forging, rolling and annealing, are complicated and expensive. The high RRR material may also experience oxidization and its quality could degrade during the machining. In contrast, the major advantages of large grain ingots niobium are lower cost and simpler machining compared to the standard polycrystalline material [3,6]. Started around 2004, the collaboration between Jefferson Nation Lab and CBMM, a leading global supplier of high purity niobium, has been focusing on machining and testing SRF cavities produced with large diameter, large grain niobium ingot. The results showed that the performance of those cavities satisfy most accelerator requirements, which makes this material very promising for future use [7-9].

In this work, to further investigate the properties of this material, the quality among ingots and within the same ingot at different locations are investigated through measurements of mechanical properties and RRR measurements. Test specimens were taken from the top, middle and bottom regions of seven niobium ingots produced by CBMM. The quality and the mechanical properties of the material was characterized through RRR measurements and tensile tests. The RRR measurements were repeated 10 times for each sample (a total of 23 samples were tested). The samples were tested inside a Gifford-McMahon (GM) cryocooler system (1.5 W at 4.2 K). Tensile tests at room temperature were performed on samples from the same regions as the RRR samples. 17 tensile test specimens were machined according to ASTM E8/E8M standard test methods for tension testing of metallic material. Tests were performed with Instron tensile/compression testing machine, and various mechanical properties including Young's modulus (E), 0.2% proof stress ($R_{p0.2}$) and strain at fracture were measured. Contents of impurity (N, O, C, H, Ta, etc.) of regions of the seven ingots were also provided by CBMM to ensure the high purity of the ingots.

2. Sample preparation

Samples were taken from seven ingots with ID: 4162, 4163, 4164, 3782, 3773, 3769, and 3706. The seven ingots were machined by CBMM, and the fabrication process can be found in [8]. During the refining electron beam (EB) melting process in an EB furnace as illustrated in figure 1 (a), the bottom of the ingot is where the solidification starts, and the top of the ingot is where it ends. The RRR samples were removed by diamond disc cutting from the ingots, and chemical analysis was performed at exactly the same locations. The locations where tensile samples were removed are shown in figure 1 (b).

2.1. Samples for RRR measurement

The dimensions of the samples for RRR measurements are 45 mm in length, 2 mm in width and 0.5 mm in thickness. For ingots 4162, 4163 and 4164, four samples were taken from each ingot. Among those samples, 2 samples were taken at the top and bottom positions to ensure the homogeneity of each location. For the other 4 ingots, three samples were taken from top, middle and bottom regions of the

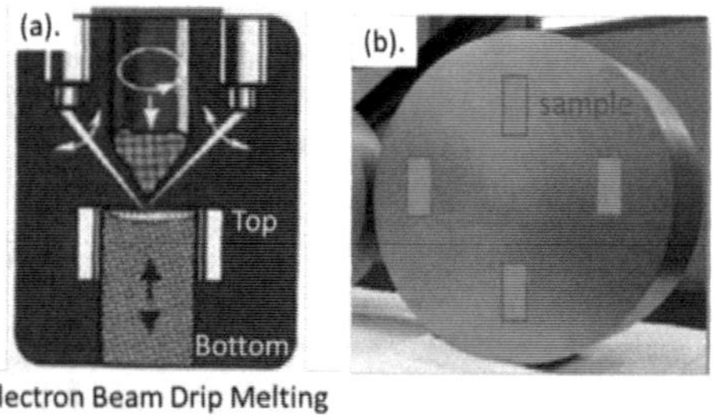

Figure 1. (a) Illustration of refining EB melting of large grain niobium ingot. Bottom of the ingot is where the solidification starts. (b) Locations at which tensile test samples were taken.

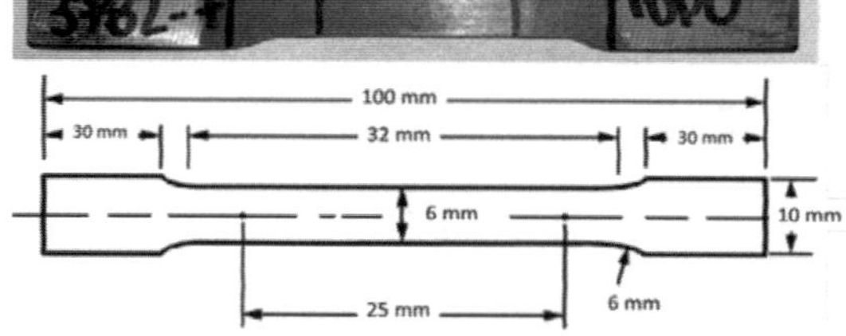

Figure 2. Machined tensile test sample and the illustration of the dimensions according to ASTM E8/E8M [11]. The thickness of the tensile sample is 3 mm.

ingot. No specific cleaning procedure (buffered chemical polishing or electropolishing) was applied onto the sample surface prior to the test, since work presented in [10] indicated that the standard BCP processing for cavity fabrication does not affect the RRR values.

2.2. Specimens for tensile test
The tensile test specimens were taken at the same locations as the RRR samples. A rectangular dog-bone tension test specimen was selected for the mechanical testing of niobium, since it requires less material and it is easier to machine. Standard dimensions for the samples were adopted according to the ASTM E8/E8M standard specifications [11]. The machined tensile test sample with relevant dimensions are shown in figure 2.

2.3. Chemical impurity content
The most important chemical impurity contents (C, N, O, H, Ta) at each location of the ingot where the samples were taken are listed in Table 1 as provided by CBMM. The chemical analysis presents very low level of interstitial impurities. Other low-level impurity contents are W < 50 ppm, Si < 20 ppm, S and Ti <10 ppm; Sn, Al, Fe, Hf, Mo and P < 5 ppm; V, B, Be, Cd, Co, Cr, Ni, U and Zr < 1 ppm. The Tantalum content varies from less than 100 ppm to almost 1700 ppm between ingots.

Table 1. Major Impurity Contents for 17 Samples

Ingot	4162		4163		4164		3706			3769			3773			3782		
Region	Top	Bot	Top	Bot	Top	Bot	Top	Mid	Bot	Top	Mid	Bot	Top	Mid	Bot	Top	Mid	Bot
H (ppm)	3	2	2	3	<2	<2	<2	<2	<2	2	3	<2		2	2	4	2	<2
N (ppm)	5	10	<5	8	<5	6	10	<5	5	<5	6	6		8	13	31	5	7
O (ppm)	10	13	14	6	5	10	15	6	10	13	11	6	N/A	17	11	13	20	22
C (ppm)									<30									
Ta (ppm)	81	255	918	1278	1697	1670	775	568	657	1161	1292	1223		1313	1501	85	93	158

3. RRR measurement
In this section, the measurement apparatus utilized for the RRR measurements and the results are discussed in detail. The RRR measurements were performed applying a DC-method, and the resistance of the sample was obtained by a standard 4-point measurement. The RRR value in this work is defined as the ratio of the sample resistance at room temperature (298 K), and the resistance at 9.5 K with no background magnetic field, as discussed in [3]. Note that another method defines RRR to be the ratio of resistance at room temperature and the one at 4.2 K. In this method, the resistance at 4.2 K is extrapolated from the one at 9.5 K [3]. The RRR values of the samples reported in this work are roughly 10% lower than the extrapolated RRR values. Considering niobium is not a solder-friendly material, electrical

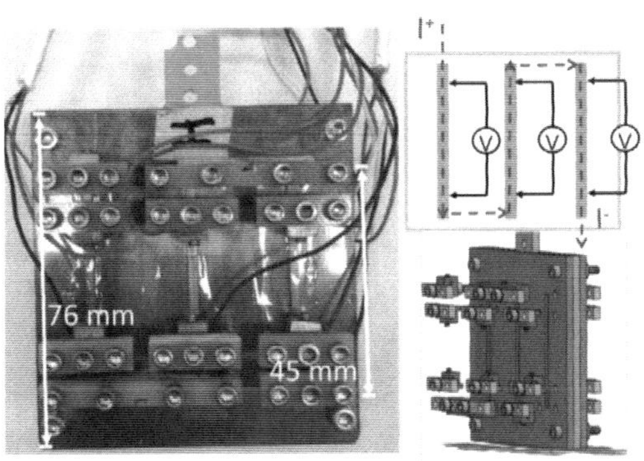

Figure 3. Sample holder with the samples mounted, and the current flow illustration. The locations of the voltage taps are also shown for illustration.

Figure 4. Sample holder mounted onto the probe and covered with mylar film, and ready to be cooled down.

contact between sample and wires to measure the voltage across the samples was achieved through mechanical contact instead of soldering. The measurements were performed on a custom designed sample holder, and a GM cryocooler was utilized as cooling system.

3.1. Sample holder design and procedures of RRR measurement

The illustration of the sample holder is shown in figure 3 and a picture of the fully assembled probe is shown in figure 4. The base of the sample holder was assembled from sandwiching a G10 plate with two identical copper plates. The copper plates were wrapped with a thin non-adhesive Kapton tape to be electrically insulated from the samples (figure 3). The copper plates, due to their excellent thermal conductivity, serve as heat sink to maintain a homogeneous temperature of the niobium samples. Current leads and voltage taps are secured through mechanical contacts on pieces of G10 and copper (exploded view in figure 3). The copper block was fixed on the G10 piece with screws. Good electrical and mechanical contact between the sample and the current leads and voltage taps was established by tightening bolts and nuts shown in the figure. Three samples were mounted on each side of the sample holder and wired in series so that six samples can be tested in one cool down. After the samples were mounted, and the sample holder was assembled onto the probe, the resistance was measured at room temperature. The probe was then placed inside the cryocooler to be cooled down to 9.5 K for another resistance measurement of the samples.

During each resistance measurement, a staircase current sweep from -200 mA to 200 mA with an increment of 50 mA was performed, with 20 voltage data points recorded at each step. Each resistance measurement was repeated 10 times. This method yields an uncertainty of less than 3% for each RRR measurement.

3.2. Results of RRR measurements

For ingots 4162, 4163 and 4164, 2 samples were taken at the bottom and the top of the ingots to investigate the homogeneity at the same location. As shown in figure 5, the two samples taken at the same at location have similar RRR values. The measurements also agree with the reference provided by CBMM. However, samples taken at different regions of the ingot have different RRR values. For example, sample *4162 Top* has RRR of >500, while the RRR of sample *4162 bottom* is close to 300.

This behaviour was also observed from the samples taken from ingots 3782, the RRR value at top is >500, and <300 at the bottom. This variation along the ingot is due to the fact that during the refining melting process in an EB furnace, as shown in figure 1 (a), the starting solidification point is on the bottom of the ingot and it ends at the top of the ingot. Normally the pressure inside the melting chamber

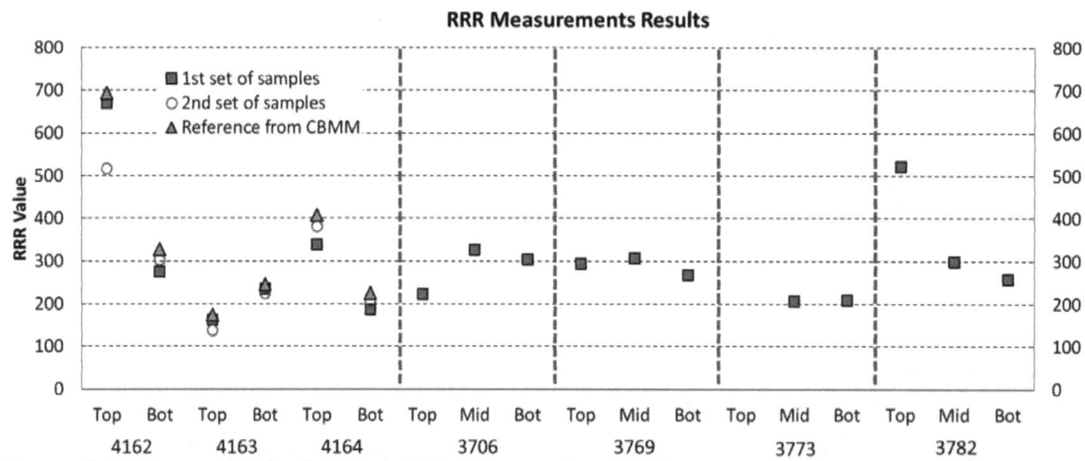

Figure 5. For ingots 4162, 4163 and 4164, two samples are taken at the same location, from top and bottom of the ingot. Three samples were taken for the rest of the ingots (top, mid and bottom). The results show that the samples at the same location have similar RRR values, but the RRR value varies through the ingot. The measured RRR values are listed in Table 2.

gets better (decrease) along the melting. At the beginning inside the furnace, some traces of gases (especially O, N and H) can be found, and they gradually decrease along the melting time. Generally, this allows better refining process of the ending point of process, which is at the top of ingots. This can also be observed from the impurities of Top and Bottom regions of ingot 4162 listed in Table 1.

The RRR values of the samples from the seven ingots are listed in Table 2. The RRR values of 11 out of 17 samples exceed 250, and only 2 samples' RRR values are lower than 200. This indicates that the overall quality of the ingot is good. The major difference impurity content between the ingots is the Ta content, and test result shows that with high Ta content (> 1000 ppm), the RRR can still exceed 200. Since the impurities of C, N and O vary significantly among those samples (from 2 to 20 ppm as shown in Table 1) and they have much more impact on the RRR values than Ta, it is hard to conclude the correlation between RRR and Ta contents from these data.

4. Tensile tests

In this section, the results of tensile tests are discussed in detail. 17 samples were taken from seven ingots, and a picture of a representative sample and its dimensions are shown in figure 2. Tensile tests were performed on an Instron 5500R tensile/compression testing machine, with a strain rate of 1% strain per minute. The strain was recorded by a pair of extensometers. The mechanical properties including Young's modulus (E), 0.2% proof stress ($R_{p0.2}$), UTS, and strain at fracture were measured through the tests. The fracture profiles of the samples are also presented and the potential correlation between strain at fracture and RRR values is discussed. According to the illustration on how the samples were cut shown in figure 1 (b), there is a great probability that these tensile samples contain some grain boundaries (samples were not cut trying to avoid grain boundary). Studying the crystal orientation was beyond the scope of this work but future work will be conducted to the tested samples to understand how the grain boundaries and crystal orientations might have affected the rupture.

4.1. Stress-strain behaviour

The work hardening behaviour and the elastic range between samples are very different at low strain, and 16 out of 17 samples started to yield under 0.2% strain. Five stress-strain curves are shown in figure 6 to illustrate the behaviour of the samples. The offset in stress at 0% strain is due to the residual stress when gripping the sample and it is small compared to the overall stress measured. 3 out of 17 samples started yielding at or slightly below 0.05% strain, with stresses less than 50 MPa. To compare all the samples, the elastic modules were defined as the slope of the linear fitting in the strain range from 0% to 0.05%. The Young's modulus of each sample was used to evaluate the 0.2% proof stress ($R_{p0.2}$) with an offset in strain of 0.2%.

For typical ductile materials, once the material is in the plastic region, the speed of the strain applied can be increased to speed up the test. A slower rate during the elastic phase is needed to record good values for the linear region, but a faster rate in the plastic region typically does not impact the material

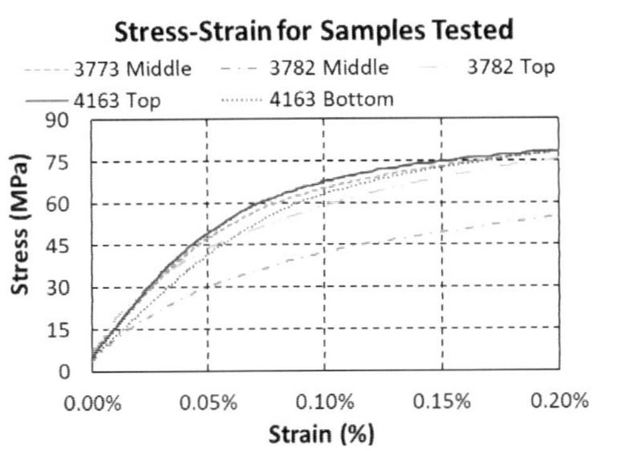

Figure 6. Stress-strain curves for 5 sample in strain range from 0% to 0.2%.

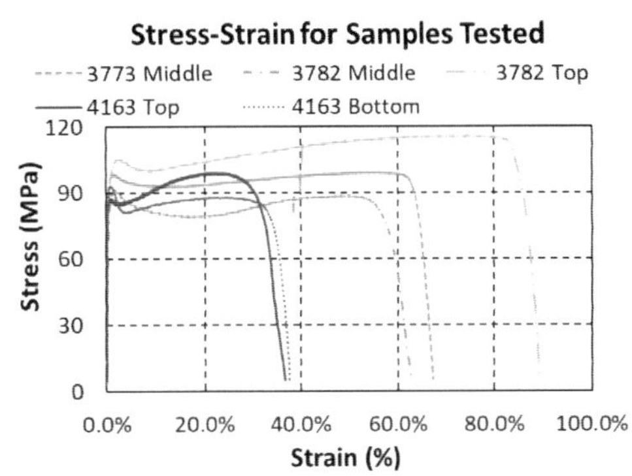

Figure 7. 5 typical stress-strain curves up to fracture of the measured niobium samples.

properties. While testing sample 4162 Top and Bottom, the strain rate was increased to 5% strain per minute when the strain reached 10% but a sudden increase in stress (~5 MPa) was observed when the strain rate was changed. This behaviour has been observed before as reported in [12] and it is considered detrimental in reporting the mechanical properties values. Therefore, following those first two samples, it was decided to apply a uniform strain rate of 1% per minute for the entire duration of the test and for all the samples despite significantly increasing the time to perform one single test. Five typical stress-strain curves up to fracture obtained for the niobium samples are shown in Figure 7. The behaviour observed fall within the category of single-crystal niobium as discussed in [13,14], with the most notable characteristic being the rapid increase and a peak in stress at less than 5% strain.

4.2. Mechanical properties and fracture profiles

The strain at fracture varied from less than 40% to close to 90% among the samples, and different fracture profiles were observed. Three typical fracture profiles are shown in Figure 8. The rupture of sample 4162 Top shown in figure 8.a has the longest %elongation at fracture, close to 90%. An angle close to 45° is observed at the location of fracture between two pieces. This shear fracture is very similar to the result of a mono-slip behaviour in single crystal [15]. The shear fracture in figure 8.b has lower %elongation, and the angle formed at the place it fractured is close to zero (flat line of separation). The third fracture profile shown in figure 8.c has one of the lowest %elongation of all the samples, and the chisel-edge fracture can be observed on the side. The fracture profiles reported are possibly affected by both grain boundaries and crystal orientations. At this point, the mechanism of the ruptures is not well understood, and it is beyond the scope of the work presented here but it will be investigated in the future through the tests of additional samples.

The detailed mechanical properties, including E, $R_{p0.2}$, UTS, and strain at fracture, combined with the RRR values and Tantalum impurity content are shown in Table 2. Note that the RRR value for samples from ingots 4162, 4163 and 4164 are the average of two measurements. As shown in figure 9, for the 6 samples taken from ingots 4162, 4163 and 4164, there is a rather strong correlation showing that higher RRR value samples experience higher elongation. Higher RRR niobium indicates lower content of interstitial elements (C, N, O and H). Consequently, the material will have fewer barrier to block the dislocations slipping in the material matrix and therefore higher elongation is expected. If all the 17 samples are considered (figure 10), the correlation becomes weaker. The weaker correlation could

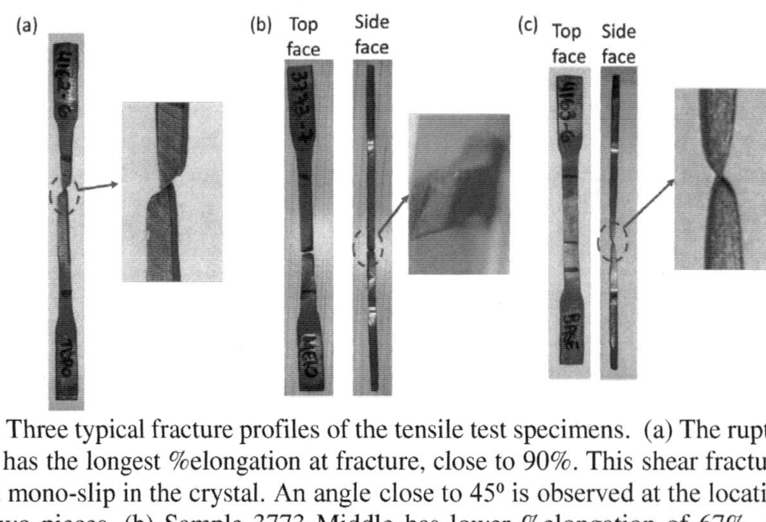

Figure 8. Three typical fracture profiles of the tensile test specimens. (a) The rupture of sample 4162 Top has the longest %elongation at fracture, close to 90%. This shear fracture is mainly a result of a mono-slip in the crystal. An angle close to 45° is observed at the location of fracture between two pieces. (b) Sample 3773 Middle has lower %elongation of 67%, and the angle formed at the place it fractured between two pieces is close to zero. The shear fracture can be observed on the cross section at fracture. (c) Sample 4163 Bottom has one of the lowest %elongation of 37.7%, and the chisel-edge fracture can be observed on the side face. [14]

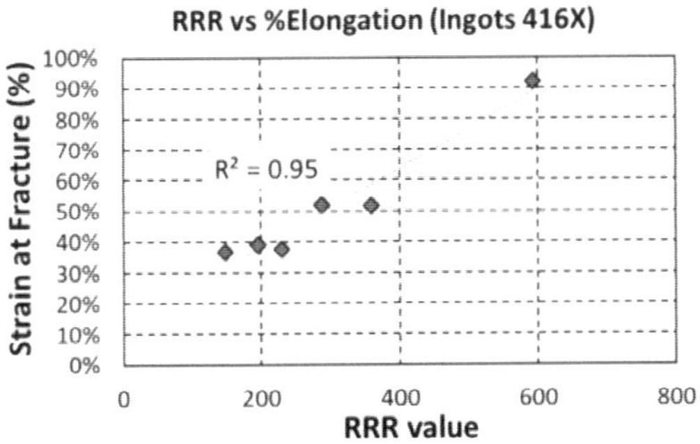

Figure 9. Correlation between RRR value and the %elongation at fracture six samples taken from ingots 4162, 4163 and 4164, a nice linear trend is observed (R^2 value of 0.95).

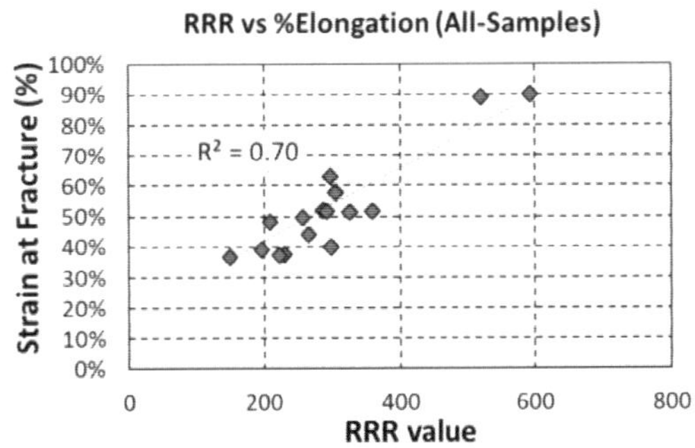

Figure 10. Correlation between RRR value and the %elongation at fracture for all the samples tests. The correlation becomes weaker considering all the samples (R^2 value of 0.70).

be caused by different grain orientation of the different samples. In addition, there is no obvious correlation observed between strength (yield stress, UTS) and Ta contents. More study on the crystal orientation will be needed to further interpolate these data.

Table 2. Summary of mechanical properties, RRR and Ta impurity for all the samples tested

Ingot	Region	UTS (MPa)	$R_{p0.2}$ (MPa)	E (GPa)	%Elongation	RRR	Ta (ppm)
4162	Top	115	93	65	92.0%	594	81
	Bottom	94	87	143	52.0%	289	255
4163	Top	99	82	73	36.8%	150	918
	Bottom	93	86	94	37.7%	231	1278
4164	Top	136	124	75	51.7%	360	1697
	Bottom	101	88	127	39.2%	197	1670
3706	Top	107	105	94	37.4%	223	775
	Middle	95	87	114	51.4%	327	568
	Bottom	95	90	83	40.0%	325	657
3769	Top	84	77	89	51.7%	294	1161
	Middle	116	71	87	57.8%	307	1292
	Bottom	104	73	67	43.9%	268	1223
3773	Top				N/A		
	Middle	99	85	86	67.5%	207	1313
	Bottom	96	74	72	48.0%	209	1501
3782	Top	99	66	86	89.2%	522	85
	Middle	91	66	59	62.9%	299	93
	Bottom	116	84	68	49.6%	258	158

5. Conclusions

RRR measurements and tensile tests were performed on samples taken from the top, middle, and bottom positions of seven high purity large grain niobium ingots produced by CBMM. High RRR values (15 out of 17 samples exceed 200) were obtained from the measurements indicating the good quality of the samples. The RRR values are similar at the same location, however vary along the ingot (from ~300 to >500 for ingot 4162 bottom to top). The variation can be explained by the change of impurities during the refining melting. The major different impurity content between the ingots is Tantalum content, however, no obvious correlation with RRR values or strength was observed. Mechanical properties including Young's modulus, 0.2% proof stress, UTS, and strain at fracture of the specimens taken at the same location as the RRR samples were obtained through standard tensile tests. The results show that the %elongation of all the samples exceed 35% and the greatest one is close to 90%. Three types of fracture profiles are observed from these measurements, and the profiles are similar to the rupture profiles of single crystal reported in literature. Samples with high RRR values experience higher elongation. The 6 samples from ingot 4162, 4163 and 4164 show a good correlation with a linear fit with R^2 value of 0.95. If all 17 samples are considered, the correlation becomes weaker (R^2 value of 0.70). The reason of this weaker correlation may due to the grain orientation of the samples. The effects of crystal orientation of the samples were beyond the scope of the work presented here but it will be investigated in the future through more tests and etching of the ruptured sample surfaces.

References

[1] Ciovati G, Dhakal P and Myneni G R 2016 Superconducting radio-frequency cavities made from medium and low-purity niobium ingots *Supercond. Sci. Technol.* **29** 064002

[2] Jana A R, Kumar A, Kumar V and Roy S B 2017 Influence of material parameters on the performance of niobium based superconducting RF cavities. (*preprint arXiv:1703.07985*.)

[3] Antoine C 2012 Materials and surface aspects in the development of SRF Niobium cavities 111

[4] Singer W, Ermakov A and Singer X 2010 RRR-measurement techniques on high purity niobium. *TTC report*

[5] Kneisel P, Ciovati G, Dhakal P, Saito K, Singer W, Singer X and Myneni G R 2015 Review of ingot niobium as a material for superconducting radiofrequency accelerating cavities *Nuclear Instruments and Methods in Physics Research Section A: Accelerators, Spectrometers, Detectors and Associated Equipment* **774** 133–50

[6] Kneisel P, Ciovati G, Dhakal P, Saito K, Singer W, Singer X and Myneni G R 2013 The Rise of Ingot Niobium as a Material for Superconducting Radiofrequency Accelerating Cavities (*preprint arXiv:1304.1722*)

[7] Ciovati G, Kneisel P and Myneni G R 2011 America's Overview of Superconducting Science and Technology of Ingot Niobium *AIP Conference Proceedings* **1352** 25-37

[8] Abdo G G, Sousa C A, Guimarães R C, Ribas R M, Vieira A S, Menezes A D, Fridman D P and Cruz E B 2015 Recent developments in high purity niobium metal production at CBMM *AIP Conference Proceedings* **1687** 020001

[9] Ciovati G, Dhakal P, Kneisel P and Myneni G R 2015 Summary of performance of superconducting radio-frequency cavities built from CBMM niobium ingots *AIP Conference Proceedings* **1687** 030001

[10] Bauer P, Berenc T, Boffo C, Foley M, Kuchnir M, Tereshkin Y and Wokas T 2003 RRR measurements on niobium for superconducting RF cavities at Fermilab *Proc. of 11th International Workshop on RF Superconductivity*

[11] E28 Committee Test Methods for Tension Testing of Metallic Materials (ASTM International)

[12] Myneni G R and Umezawa H 2003 Variation of Mechanical Properties of High RRR And Reactor Grade Niobium With Heat Treatments *Materiaux & Techniques* **91** 7-8-9 922

[13] Bieler T R, Wright N T, Pourboghrat F, Compton C, Hartwig K T, Baars D, Zamiri A, Chandrasekaran S, Darbandi P, Jiang H, Skoug E, Balachandran S, Ice G E and Liu W 2010 Physical and mechanical metallurgy of high purity Nb for accelerator cavities *Phys. Rev. ST Accel. Beams* **13** 031002

[14] Kneisel P. 2007 Progress on Large Grain and Single Grain Niobium – Ingots and Sheet and Review of Progress on Large Grain and Single Grain Niobium Cavities

[15] Abbaschian R, Abbaschian L and Reed-Hill R E 2009 *Physical metallurgy principles*. 4th ed. (Cengage Learning) chapter 21 pp 686-689

Resins for Superconducting Magnet Construction – An Overview of Requirements, Processing and Properties

David Evans

Advanced Cryogenic Materials Ltd
Brunel House
3, Norman Avenue, Abingdon,, Oxon, OX14 2HQ UK.

David Evans276@btinternet.com

Abstract. The resin used for bonding and insulating superconducting magnets can be a major factor in ensuring reliable and stable operation of the completed assembly. There are many resin systems and available to magnet designers and engineers. The selection of the correct material depends on the application technique selected, the processing requirements and the end properties required of the cured resin. Available techniques for magnet construction and bonding include vacuum impregnation (VPI), so called 'wet winding' and the use of pre-impregnated fabrics. The advantages and disadvantages of these techniques are considered and the processing requirements for each are discussed. Materials selection requires an understanding of the processing characteristics such as viscosity and 'useable lifetime' but these must be married with the properties required of the cured resin. A low viscosity and long useable life is a frequent requirement for VPI processing of large magnet systems and this may be difficult to match with high thermal shock resistance that may be required to minimize cracking possibilities in resin rich regions. Examples are presented of resin systems resistant to cracking and structural features that enhance this parameter. For many magnets that may operate in an ionising radiation environment, radiation stability is an important requirement. Radiation stable systems are described and structural features that promote such stability are considered, along with the difficulty of matching all competing requirements.

1. Introduction

Of the three possible techniques for magnet insulation / bonding, wet winding and use of pre-impregnated materials introduce the resin during the winding procedure. Vacuum impregnation applies the resin only after winding has been completed. All three techniques predominantly use epoxy resins. Whatever technique is selected for coil manufacture, the resin system selected for bonding must have an overall 'suitability for purpose' and have the characteristics believed to be necessary. The insulating / bonding materials may be required to withstand high mechanical loads, high electrical stresses and be subjected thermally induced stresses as a result of cooling and possible thermal cycling that could result in resin cracking or delamination.

Since the vacuum impregnation technique (VPI) is the dominant procedure selected for magnet construction, most reports have concentrated on this technique, and there have been reports on optimising the procedure [1].[2] but the limited use of the other techniques and the **infinite number** of variables, limits the opportunities for any form of 'normalising'.

The aim of any insulation / bonding technology is to ensure full "wetting" with resin of all surfaces to be bonded and to minimise, so far as is reasonably possible 'voids' within the insulation. It was reported [3] that the partial discharge signal in a composite manufactured in air was an order of magnitude higher than for composite insulation produced by vacuum impregnation. It was also claimed that the existence of voids in electrical insulation should not be tolerated unless the sole purpose of that insulation was mechanical support.

2. Available Techniques for Coil bonding / Insulation

There is not a single correct method for insulating and bonding a particular coil. Rather, the choice between vacuum impregnation, wet winding or pre-impregnated materials depends on many factors and the perceived advantages and disadvantages of the two techniques are set out below (Tables 1, 2 and 3.) the decision on which route to employ may depend on coil geometry, numbers to be built (and therefore amortised costs) and a number of local factors – but the final decision can often be a compromise..

2.1 Vacuum Impregnation

A liquid resin is infused between conductors that have been previously wrapped (or coated) with an insulating material to provide turn and layer insulation. The conductor may be wrapped with a glass tape (or a glass tape /polymer film combination), or coated with an inert (inorganic material). For the VPI process the coil is sealed within a vacuum tight enclosure, heated and the pressure reduced to allow a period of out-gassing, followed by filling with a fully de-gassed resin. After filling, the pressure on the resin / coil assembly is raised and the prevailing atmospheric pressure (or an over-pressure, if considered necessary) drives the resin into the structure.

2.2 Use Pre-pregs

Pre-pregs are fabrics or tapes that have been pre-impregnated with resin and in most cases, partially cured. These materials are used, under tension, to wrap or cover conductors and so provide the necessary insulation and the resin for bonding. Such materials may be applied only after reaction of niobium / tin coils. Pre-pregs need to be consolidated to encourage resin flow during winding and resin flow that occurs during the gel / cure phase after winding may lead to overall dimensional changes in the coil. tool to ensure all surfaces are bonded. Pre-pregs offer only limited resin formulating properties (hot curing only) but radiation stable materials are readily available.

2.3 Wet Winding

The insulation is applied to the pre-insulated conductor (after reaction in the case of niobium tin coils) or by wrapping (say) glass tape onto the conductor and at the same time wetting the tape with liquid resin. Adding a layer of dry glass fabric between layers of conductor and wetting the glass before winding the next conductor layer is another variation on the wet wining theme.

From the Discussion above and considering the vast majority of magnet coils are vacuum impregnated, the remainder of this paper is devoted to aspects of the VPI process

3. Techniques for VPI

Within the general vacuum impregnation process, there are three variations on a VPI theme that may be used: These three variations on an impregnation theme are outlined below.

3.1 Flood filling

The coil complete with some tooling to define dimensions etc, is placed within a tank to hold resin and the complete system placed within a vacuum chamber. After evacuation at elevated temperature for some period of time (See [1] and [2]) depending on the nature of the coil assembly, liquid resin is allowed to completely cover the tool and at this stage, atmospheric pressure may be applied and used to force the resin into the free space within the coil assembly. If a pressure greater than atmospheric is required, the 'vacuum' chamber has to be pressurised. .

3.2. Vacuum Impregnation with Vacuum Chamber

The coil is contained within a tool that is sufficiently well sealed to prevent resin leakage under the operating conditions (i.e. using atmospheric pressure or greater) but need not necessarily be vacuum tight. The coil and tooling assembly is placed inside a chamber and evacuated. Resin is introduced directly into the tool, with the valve on the evacuation line open. When resin appears in the 'riser', this outlet valve is closed and atmospheric pressure is then available to force the resin into the coil structure. Over-pressure may be applied if required. If resin is able to leak from the tool at this stage, either there will be insufficient pressure generated to ensure the structure is thoroughly impregnated, or drainage of resin could result in starvation of the coil. The vacuum in the chamber should be maintained until the coil is fully impregnated and resin is no longer flowing.

3.3 Vacuum Sealed Tool
Similar to the technique described in Section 3(b), the major difference being that the tool is vacuum tight and an external vacuum vessel is not used. Assuming the mould is sealed and remains so throughout the procedure, coils may be successfully impregnated using this technique.

4. Materials for The VPI Process
For coil impregnation, the ideal system would have the following characteristics:

(a) Adequate 'useable life - Low viscosity, to be maintained if possible for many hours
(b) Resistant to cracking when subjected to thermal shock
(c) Low curing temperature (<130C).
(d) Low shrinkage on cure
(e) Low thermal contraction, related to that of the conductors
(f) Adequate room temperature mechanical strength and resistance to creep under the loads from clamping forces
(g) Adequate electrical properties, reliably reproducible, to provide insulation
(h) Good adhesion to conductor and / or insulation – possibly also to tooling
(i) Resistance to ionising radiation up to the doses anticipated- if this is a design requirement

It is unlikely that a single resin system will meet all of these criteria and inevitably, a compromise is required. While it is necessary for designers of superconducting magnets to take into account the technology of impregnation and insulation materials as they are today, there are also opportunities to blend and formulate materials for specialist applications.

4.1 Viscosity, Useable life and Toughness
The 'useable life' of a mixed resin / hardener system is defined by many suppliers as the time for the viscosity to double, at the operational temperature, from that of the initially mixed value. This loose definition doesn't provide guidance for magnet engineers and the 'useable life' required would depend very much on the details of the magnet system that is to be vacuum impregnated.

There is not a simple definition of 'useable life' since this is usually defined by the needs of the particular application. Essentially, 'useable life' is the time for the viscosity to rise to a level at which it is no longer 'suitable for purpose' and this point will be different for different applications.

The viscosity achieved immediately after mixing resin and hardener is influenced by temperature, a higher temperature lowers this initial viscosity but also increases the reaction rate and so shortens the usable life. The 'useable life' is usually quoted 'at the processing temperature', which is an optimised temperature to maximise the working life while achieving a sufficiently low viscosity for the work involved.

The mass of resin involved may also have a significant influence on the working life. Preparing a large batch of mixed resin and hardener may reduce the working life of the material relative to resin prepared by automatic proportioning and mixing in a dedicated resin dispensing machine.

The viscosity of the resin system will determine the rate of flow of the resin into the magnet and the extent to which the resin will flow within the structure. The time for which the resin remains 'useable' is determined by the rate of change of viscosity at the processing temperature and the value that is defined as the 'upper limit' for the process

Hacker et al [4] described low viscosity as 20 - 100 mPa.s and the upper limit for flow was defined by Green et al [5] as 500 mPa.s

More generally, it is regarded that the upper limit of viscosity for impregnation is reached when the resin viscosity has risen to 1000 mPa.s providing the initial viscosity is less than 200MPa.s. Therefore the time for the viscosity to rise from its initial value (at the prevailing temperature) to 750 mPa.s may be termed the "useable life". However, there is not a fixed and immutable value for an upper limit to viscosity, since this will depend on largely on the nature of the coil,

The rate of change of viscosity, at a fixed temperature for an epoxide resin / hardener system (with accelerator if appropriate) can be determined and from this the 'useable life' may be estimated.

In general, for impregnation, high addition, low viscosity hardeners, such as some of the acid anhydrides, form the most efficient impregnating systems i.e. low initial viscosity and long 'useable life'. In order to minimize the gel / cure temperatures for these systems, an accelerator is generally used and the rate of addition of such a component will strongly influence the rate of viscosity change (Figure 1).

If the upper limit of viscosity is taken as a conservative 500 mPa.s, the highest level of added accelerator shows a useable life of ~17 hours and the lowest >35 hours. Figure 1 shows the rate of change of viscosity of diglycidyl ether of bisphenol F (DGEBF) resin with MTHPA acid anhydride hardner (typically, Huntsman HY 917) and with varying levels of added accelerator.

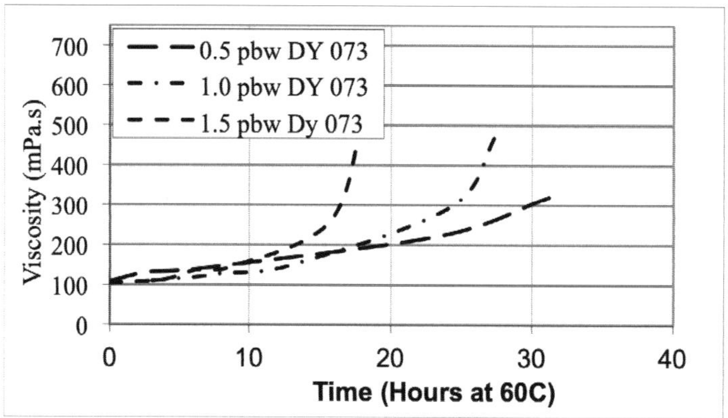

Figure1. Viscosity Change of DGEBF Resin with Acid Anhydride Hardener and Various Levels of Accelerator (Huntsman DY 073-1)

4. 2 Toughness

Most anhydride cured resins are inherently brittle – along with many other curing agents that have long 'useable life' and therefore require elevated temperature curing.

An aliphatic amine hardener, typically Huntsman HY 5922, generally known as "D400' or POPDA – (Polyoxypropylene diamine)) when combined with DGEBF resin, has outstanding thermal shock resistance and has found widespread use for the impregnation of superconducting magnets, notwithstanding the relative short 'useable life (Figure 2).

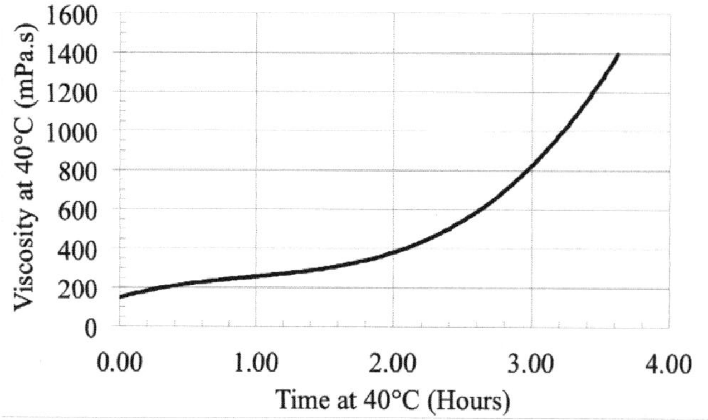

Figure 2. Change in Viscosity at 40°C of DGEBF Resin with Curing Agent POPDA

IOP Conf. Series: Materials Science and Engineering **756** (2020) 012003 doi:10.1088/1757-899X/756/1/012003

Many applications have used raised injection pressures to increase the speed of resin flow into the structure. Tensile and compressive properties of the brittle acid anhydride cured resins and the tough POPDA cured resin system are similar at 4K.

Colyer et al [6] measured the work of fracture of epoxy resins and reported that for the tough POPDA system the work of fracture was approximately a factor of five greater at 4K than the acid anhydride cured system and showed how this parameter could be useful in defining differences in the 'cracking potential' of resins.

The work of fracture was used by Evans [7] to estimate the 'allowable' thickness of resin, that would not crack due to differential thermal movements, when restrained between conductors in a superconducting coil.

The work of fracture of epoxy resins is an important parameter in assessing the cracking potential of resin in a given i.e. if the stored strain energy in the resin exceeds the work of fracture, then there is a possibility of forming cracks:

$$\frac{\sigma^2 t}{2E} \geq 2\gamma_p \tag{1}$$

where σ is the stress, t is the material thickness, γp is the work of fracture and E is Young's modulus.

Equation 1 may be re-written to define the minimum thickness of resin - at the given stress - which will have the necessary stored strain energy for crack formation

$$t \geq \frac{4E\gamma_P}{\sigma^2} \tag{2}$$

The stress in the resin when constrained between conductors in two dimensions nay be simplistically stimated from:

$$\text{Stress}\ (\sigma) = \frac{\int_{4K}^{300K} E \propto dT}{1\text{-}2\mu} \tag{3}$$

Where μ = Poisson's Ratio and is taken to have the value of *0.3*

A calculation based on equation (3) and using 'average' values for young's modulus can only approximate to the true condition because both E and α vary with temperature and the value for Poisson's ratio is uncertain. In addition, the calculation does not take into account strain induced in a resin due to the volume change during cure, which in itself could lead to damage during cooling from the cure temperature. However, even this simplistic estimation of stress is a useful starting point in trying to understand and predict the overall effect of cooling the resin/ conductor combination.

Using the above equations to calculate the induced stress and from that estimate the 'maximum allowable thickness between conductors for two resin systems:

(i) A 'brittle' acid anhydride cured DGEBA resin

(ii) The same resin cured with POPDA (D400)

Young's Modulus (E) is similar at 4 K for the two resins (3000 MPa at RT and 6000 MPa at 4K) - and is "averaged" over the temperature range. The contraction coefficient ($\propto$) is also taken to be "constant" and is derived from the total contraction (RT to 4K) divided by 290 and subtracting the contraction coefficient for copper (1.0 x 10-5/K) from that of the resin (3.8 x10-5/K). The temperature range (dT) is taken as the difference between the Tg and the operational temperature (4K).

Tg for Resin (i) is taken as 120°C , (dT ~390K) and for resin (ii) as 40°C (dT ~ 310K)

For resin (i):

$$\text{Stress}\ (\sigma) = 124\ \text{MPa}$$

and for resin (ii)

$$\text{Stress}\ (\sigma) = 90\ \text{MPa}$$

IOP Conf. Series: Materials Science and Engineering **756** (2020) 012003 doi:10.1088/1757-899X/756/1/012003

The work of fracture of resin (i) was esimated [6] as 72J/m^2 and for resin (ii) as 360 J/m^2
Using these values in equation 2 to estimate the maximum allowable thickness (E and σ in N/m^2):

Resin (i) $t = 8.4 \times 10^{-5}$m (0.084 mm)

Resin (ii) $t = 8.0 \times 10^{-4}$m (0.8 mm)

To increase toughness for low temperature applications, there is a range of additives that may be used with epoxies that fully react into the system. These materials are preferred to other diluents that literally just dilute the resin and 'plasticize' by their presence among the long chain molecules of the resin. Other so called flexibilizers have only one reactive group within the molecular structure and these function by the one reactive group combining with the growing epoxy chain, so preventing further growth of that chain.

i.e. these are "chain stoppers", limiting the molecular weight of the cured system and limiting cross-link density.

In Figure 3 details are presented of DGEBF resin but cured with an aromatic amine (DETD – typically Huntsman HY 5200). The unmodified resin / hardener system offers a high degree of radiation stability and shows a 'useable life' of ~7 hours (at 50°C), based on an 'upper limit of viscosity of 500mPa.s

When the DGEBF resin is blended with a flexibilising resin, polypropylene glycol diglycidyl ether (PPGDGE, typically Dow Resin DER 732 a di-epoxide), not only does the toughness increase dramatically but the rate of change of viscosity is reduced so that the 'useable life' of the widely used "60/40" blend of DGEBF / PPGDGE is > 30 hours [8].

The resin PPGDGE and curing agent POPDA both have the same repeating 'polyoxypropylene' unit in the structure and it is this feature that confers the toughness on resin systems.

Evans and Zhang [9] reported work of fracture on the resin systems defined in Figure 3 and these data are presented in Table 1. Results for DGEBF /POPDA at 4K differ by approximately a factor of two from that in [6] but the value for an an anhydride cured resin is similar. The blend of DGEBF with PPGDGE (60/40 is shown to have a work of fracture three times that of the DGEBF / MTHPA system at RT and nearly twice that at 4K.

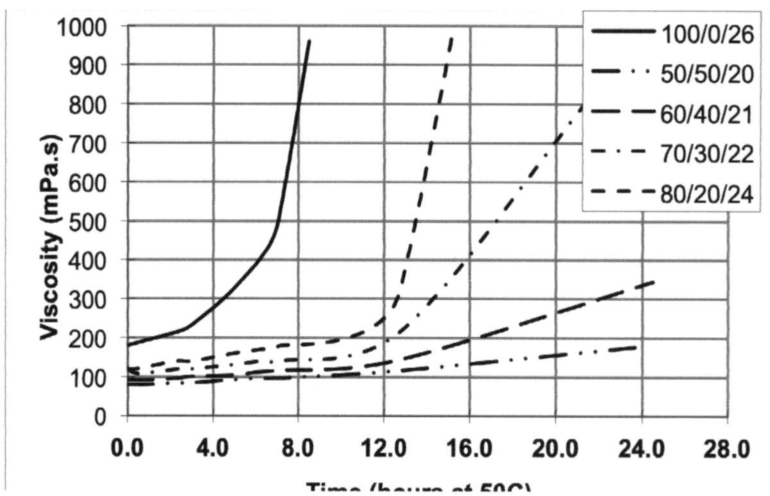

Figure 3. Effect of Added Flexibilising Resin (PPGDGE) on 'Useable life' of DGEBF Resin Cured with DETD (Ratios are in pbw of DGEBF/PPDGE/Hardener DETD)

Table 1. Work of Fracture of Resin Systems at Three Tempersatures [9]

	DGEBF/ MTHPA*	DGEBF POPDA	DGEBF/ PPGDGE 50/50	DGEBF/ PPGDGE 60/40	DGEBF/ PPGDGE 80/20	DGEBF/ PPGDGE 100/0
	Work (J/m^2)	Work (J/m^2)	Work (J/m^2)	Work (J/m^2)	Work (J/m^2)	Work (J/m^2)
RT	103	899**	2241**	353	310	205
	± 12			± 53	± 42	± 58
77K	199	525	396	359	249	110
	± 13	± 67	± 27	± 49	± 49	± 25
4.2K	61	178	128	100	93	63
	± 5	± 10	± 9	± 2	± 9	± 3

*MTHPA – methyl tetrahydrophthalic anhydride – typically Huntsman HY 917

**Results at RT are geometry dependent - no single value for the work of fracture

A 'low' Tg (but sufficiently raised to maintain acceptable room temperature properties) along with a high work of fracture, combine to increase the cracking index. Calculated values for the induced thermal stress (equation 3) and 'cracking index '(maximum allowable thickness without cracking – equation 2) are presented in Table 2.

Table 2. Calculated "Cracking Index'at 4K For Four Resin Sysatems

Resins		DGEBA / POPDA	DGEBA/ MTHPA	DGEBF/ PPGDGE 100/0	DGEBF/ PPGDGE 60/40
Thermal Stress (MPa)	4K	95	124	124	101
Cracking Index (mm)	4K	0.36	0.07	0.07	0.19

5. Outgassing and Resin Flow & Distribution

5.1 Coil Outgassing

The importance of resin de-gassing was explained in [1] and [2]. The only additional comment here is to emphasize that a coil structure is a very efficient de-gassing medium. When the advancing resin front is spread as a thin film, with many nucleation points, if the resin has not been fully out-gassed, any residual gas will certainly be release into the coil structure. The need to ensure complete de-gassing of resin prior to impregnation cannot be over-stated.

Another procedures include purging the evacuated coil structure with a dry gas as an aid to accelerating the removal of chemically absorbed water molecules. Frequently dry nitrogen is specified for this purpose but others prefer to use carbon dioxide on the basis of a claim that this gas is more soluble in resin than nitrogen. Increased solubility was said to be an advantage because pockets of the purge gas that may be trapped within the structure would more readily dissolve in resin and so minimise the risk of void formation. The comparative figures for the solubility of nitrogen and CO_2 on which the above conclusion was based were not available.

Adsorption of species onto a surface may take place as either: (a) Physical adsorption or (b) chemical adsorption. When physical adsorption occurs, gas molecules are attracted and held by weak van der Waals forces with low binding energies (<40 kJ/mol). Adsorption does not require energy for activation. When chemical adsorption occurs, there is a definite (exothermic) chemical reaction between the gas and the the absorbing surface

The binding energies for chemically absorbed species can vary anywhere between 80 kJ/mol and 800 kJ/mol. When the binding energies are low (physical adsorption) the rates of desorption are high and the desorbed molecule is readily removed by pumping. When binding energy is high, the rates of desorption are low.

Elsey [10] found that when pumping metal structures, de-gassing occurred quickly but once the pressure had been reduced to a low value, the major residual gases being removed were water vapour and carbon dioxide.

Work by Bakaeva et al [11], compared the adsorption of the technically no-polar CO_2 with non-polar argon on glass surfaces. It was found that the pattern of adsorption of CO_2 mirrored that of water vapour but there was no indication of adsorption of non-polar gases such argon or nitrogen.

Siperstein and Gubbins [12] measured the heat of adsorption of argon and krypton on glass at 14 – 18 kJ/mol at room temperature.

Baptist and Levy [13]confirmed a chemical reaction between a freshly broken glass surface and CO2.

Battjes et al [14] showed that when exposed to the gas, surfaces that were coated in amino-silane coupling agents, became coated with a strongly adhering layer of a white powder. This powder was shown to be an 'amine carbamate', the product of the interaction between the amine and CO2.

Details of the reaction between CO2 and Bisphenol A epoxy resin were documented by Khoshkish et al [15]. It was demonstrated that carbon dioxide reacted with epoxy groups to form 'cyclic carbonates'.

The apparent high solubility of carbon dioxide in water is well known and this is due to chemical action (to form carbonic acid) and not just simple solubility. Carbon monoxide is around a factor of 60 less soluble in water than carbon dioxide.

5.2 Resin Flow & Distribution

The rate of resin flow through packed glass fabric is governed by the permeability of the glass fabric, the resin viscosity and pressure difference. the volumetric flow rate, A the cross-sectional area, η the fluid viscosity, K the permeability, and ΔP the pressure difference over the length L. Permeability, in turn, is defined as that property of a porous material, which characterizes the ease of fluid to flow through the material by the applied pressure gradient

The complicated and non-uniform micro-structure of the compacted insulation, where the local permeability may vary by more than an order of magnitude, will lead to non-uniform flow of resin (race tracking) and the presence of un-wetted regions in the coil as the resin flows past.

The manner in which the permeability of compacted glass cloth varies with the degree of compaction was demonstrated by Hammond [16] and it was shown that permeability of glass fabric compacted to 60 volume % was a factor of approximately twenty less than the same fabric compacted to 40 volume %.

6. Closing Remarks

The various methods of coil bonding with epoxy resins, wet-winding, use of pre-pregs and vacuum impregnation VPI have been reviewed and the various options within the VPI process considered in detail. To complement these various procedures, the material requirements have been considered and discussed, including resin viscosity and the concept of 'useable life' outlined.

The resistance of resins to cracking at low temperatures has been discussed along with a summary of some published data, to show how the induced thermal stress may be minimized and toughness maximized by modification of resin formulation.

Factors influencing resin flow into compacted insulation structures have been outlined and evidence is presented that carbon dioxide reacts with resin and is chemically adsorbed onto most solid surfaces. It is confirmed that carbon dioxide is not more soluble in resin than dry nitrogen and apparent solubility of the gas has been confused with a chemical reaction.

7. References

[1] Canfer, S.J. & Evans, D. A definition of the parameters that influence and control the flow of resin during vacuum impregnation of magnets and other structures", Adv. Cry. Eng. (Materials) Vol. 48, 2002, pp 287 – 294 (Plenum Publishing)

[2] Evans, D. Knaster, J. & Rajainmäki, H. Vacuum Pressure Impregnation process in superconducting coils: best practice. Vacuum pressure impregnation process in superconducting coils: best practice, IEEE Transactions on Applied Superconductivity 22, 2012. Paper presented at MT -22, Marseille, France, 2011

[3] Kutil, A. and K. Frölich, Partial Discharge Phenomena in Composite Insulation Materials, IEEE 1995, Annual Report Conf. On Electrical Insulation and Dielectric Phenomena, 1995, pp 343-346,

[4] Hacker, H. Albrecht, C. Laupenmuhlen, H. Salzburger, H. & Ihlein, W. Epoxies for low temperature application impregnation technology. Adv. Cry. Eng. Vol 30, 1984, pp 51 - 60 (Plenum Publishing)

[5] Green, M.A., Coyle, D.E., Miller, P.B. & Wenzel, W.F. Vacuum impregnation with epoxy of large superconducting magnet structures Nonmetallic Materials and Composites at Low Temperatures, Vol. 1 - p409 (Plenum Publishing

[6] Colyer, B, Ball, M.T. and Campbell, M.P. The Work of Fracture of Epoxy Resins. Cryogenics, 1974, 281 – 283

[7] Evans D. Unpublished Rutherford Appleton Laboratory internal note.

[8] Canfer, S.J. and Evans, D. A new resin system for the impregnation and bonding of large magnet coils. Proceedings ICEC 17: Superconductivity, Section 12 IOP publishing (July 1998)

[9] Evans, D & Zhang, Z. The work of fracture of epoxide resins at temperatures to 4K Adv. Cry. Eng. Vol 46, 2000, 235 - 242 (Plenum Publishing)

[10] Elsey, R.J. Outgassing of Vacuum Materials, 2. Vacuum Volume: 25 Issue: 8 Pages: 347-361,

[11] Bakaev, V.A, et al Adsorption of CO2 and Ar on glass surfaces. Computer simulation and experimental study. Journal of Chemical Physics, Volume: 111, Issue: 21, Pages: 9813-9821

[12] Siperstein, F.R & Gubbin, K.E Studies in Surface Science and catalysis s. (Elesevier, 2002)

[13] Baptist, R. & Levy, F. Carbon Dioxide Adsorption on Glass. Vacuum, volume 43, No. 3, pp 213 – 214 (1992)

[14] Batrtjes, K.P. et al New Evidence Related to Reactions of Aminated Silane Coupling Agents with Carbon Dioxide. Adhesion Sci. Technol. Volume 5, No.10, pp785 – 799 (1991).

[15] Khoshkish, M. et al The Influence of CO2 Fixation Reaction on the Kinetic Curing Parameters of Bifunctional Resin. l Thermochima Acta, 617, (2015) 95-101

[16] Hammond, V.H. and Loos, A.C The effect of fluid type and Viscosity on the Steady-State and Advancing Front Permeability Behavior of Textile Preforms. J. Reinforced Plastics and Composites, Vol. 16, No. 1, (1997)

DC surface flashover characteristics of ZnO/Epoxy composites at room temperature and 77 K

Kaikai Zhang[1,2], Chuanjun Huang[1]*, Yongguang Wang[1,2], Hua Zhang[1], Chi Zhang[1,2], Zhixiong Wu[1], Huiming Liu[1], Dong Xu[1], Rongjin Huang[1,2] and Laifeng Li[1,2]

[1]Key Laboratory of Cryogenics, Technical Institute of Physics and Chemistry, Chinese Academy of Sciences, Beijing 100190, P. R. China.
[2]Center of Materials Science and Optoelectronics Engineering, University of Chinese Academy of Sciences, Beijing 100049, P. R. China

cjhuang@mail.ipc.ac.cn

Abstract. Epoxy (EP) resin based nano-composites are widely used in high voltage direct current (HVDC) high temperature superconducting (HTS) power cable. In this paper, the DC surface flashover characteristics of ZnO/EP composites were studied at both room temperature and 77 K. The samples were made by dispersing ZnO nano-particles into EP resin with various weight percentages of 0%, 1%, 3%, 6% and 10%, respectively. The experiments were carried out with a cryogenic system in which DC high voltages ranging from 0 kV to 100 kV were supplied. The results showed that the surface flashover voltages increased with the increase of ZnO content at both room temperature and 77 K, and the surface flashover voltages at 77 K were higher than that at room temperature for composites with the same ZnO content.

1. Introduction

High temperature superconducting (HTS) power cable is used in high voltage direct current (HVDC) transmission. The operating temperature of the cable termination ranges from room temperature to 77 K, and the termination is exposed to high electrical field which can cause surface flashover or breakdown of the insulating structure [1, 2]. For solid insulation, the surface flashover voltage is usually much lower than the breakdown voltage [3]. Therefore, it is of great significance to study the surface flashover characteristics of insulating materials at both room temperature and 77 K.

At present, researchers have studied surface flashover characteristics of various materials used in the insulating structure of the HVDC HTS cable. For example, Kim S H et al [4] have found that the surface flashover voltages of GFRP, Bakelite and Teflon were nearly equal at 77 K. In addition to these materials, EP resins doped with various inorganic fillers were widely used. Chen Y et al [5] have studied surface flashover characteristics of EP resins containing different kind of micro-inorganic oxide particles such as Al_2O_3, $BaTiO_3$ and TiO_2 under nanosecond pulse. The results showed that the surface flashover voltages of the composites were lower than that of the pure EP. Li T et al [6] have found that the DC surface flashover voltage of the TiO_2/EP nano-composites was higher than that of pure EP. The difference was due to the fact that the surface flashover voltages were related to the sizes of the particles. Moreover, fillers with non-linear conductivity such as SiC were used to reduce the surface charge accumulation thus increase the surface flashover voltage.

ZnO is a new type of filler. Researches have found that the inclusion of ZnO nano-particles in EP resin could produce non-linear conductivity behavior [7]. Zhang Z et al [8] have shown that the DC surface flashover voltage of the polyimide doped with 3 wt% nano-ZnO was higher than that of pure polyimide. For ZnO/EP nano-composites, surface flashover characteristics at cryogenic temperatures were rarely studied. Therefore, in this paper, the DC surface flashover characteristics of ZnO/EP nano-composites at 77 K were studied and compared with those at room temperature. The samples were made by dispersing ZnO nano-particles into EP resin with weight (wt) percentages of 0%, 1%, 3%, 6% and 10%, respectively. The experiments were carried out under a cryogenic system in which DC high voltages ranging from 0kV to 100kV were supplied.

2. Experiment

2.1 Surface modification of nano-ZnO

ZnO nano-particles (4g) and the ethanol/deionized water (50ml/100ml) solution were mixed by ultrasonic for 30 minutes (mixture I). KH560 silane coupling agent (5 wt% of ZnO), the main function of which was to enhance the adhesion between EP and ZnO particles, was dripped into the ethanol/deionized water (50ml/50ml) solution and the solution was then heated to 80°C (solution I) by a magnetic stirring device. After that, mixture I was slowly dripped into solution I and reacted at 80°C for 2 hours. Finally, ZnO nano-particles were filtered, dried and ground [9].

2.2 Preparation of the nano-composites

The modified nano-ZnO were dispersed into EP (bisphenol A, 128, South Asia epoxy Co., Ltd.) with weight percentages of 0%, 1%, 3%, 6% and 10% respectively and the mixture were mixed and stirred for 5 min at 800 rpm, 15 min at 1000 rpm by a planetary gravity mixer. Then the mixture was dispersed by ultrasonic for 30 minutes, and then curing agent (D230) was added. A rotary vacuum pump was used to degas the mixture. Finally, the mixture was poured into the mould and cured at 80°C for 3 hours. The cured samples were thin slices with the diameter of 55mm and the thickness of 1.5mm.

2.3 Test of the surface flashover voltage

The experimental setup consisted mainly of a cooling system and a DC high voltage power system. The cooling system was sealed in a cryostat cooled by a GM cryocooler (KDE 415, CSIC PengLi Cryogenic Technology Co., Ltd), including the external chamber, two radiation shields connected with the 1st and 2nd stages, respectively. The electrode system was encapsulated in the 2nd radiation shield chamber, as shown in figure 1. High voltage was supplied by a DC high voltage power supply (DW-P104-2AC2). The surface flashover voltage was measured by an oscilloscope, as shown in figure 2. The experiment was carried out under 10^{-4} Pa. Each sample was cleaned in ethanol, deionized water respectively with ultrasonic before test.

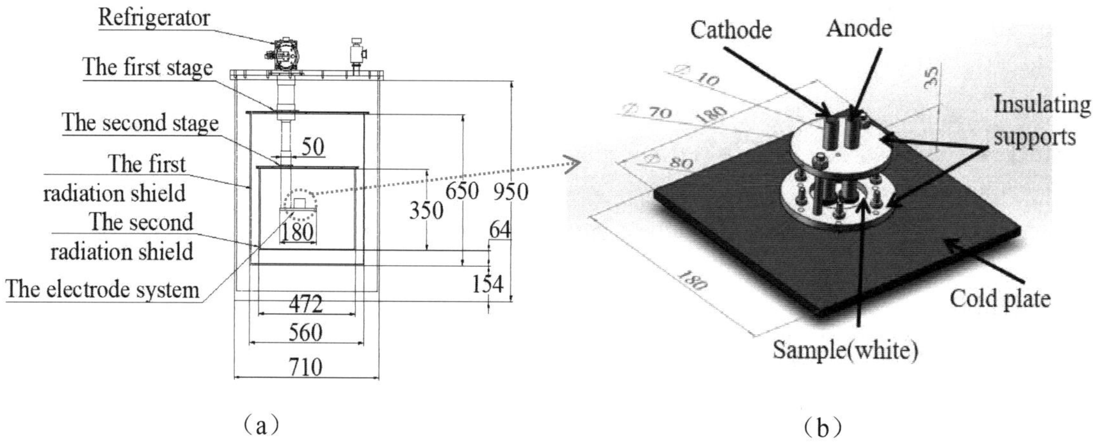

Figure 1. The cooling system (a) and the electrode system (b).

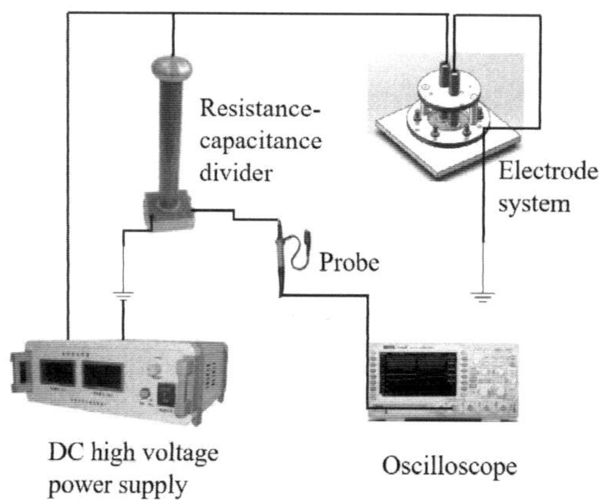

Figure 2. The schematic diagram of the DC high voltage power system.

3. Results and discussion

3.1 Results

The test results of each sample were described by Weibull distribution. The cumulative density function are shown in equation (3.1) :

$$F(U,\alpha,\beta)=1-\exp\left[-\left(\frac{U}{\alpha}\right)^{\beta}\right] \tag{3.1}$$

The equation (3.1) can be written in another form, i.e

$$\beta(\ln U-\ln\alpha)=\ln[-\ln(1-F)] \tag{3.2}$$

Where the scale parameter α represents the voltage for which the failure probability is 0.632 (that is

(1-1/e), where e is the exponential constant, when F=0.632, U=α). It is analogous to the mean of the Normal distribution [10]. Place the n flashover voltages in order from the smallest to the largest and assign them a rank from i = 1 to i = n, then F can be calculated by equation (3.3),

$$F = \frac{i - 0.44}{n + 0.25} \tag{3.3}$$

In order to get the Weibull distribution, Let

$$X_i = \ln U \tag{3.4}$$

$$Y_i = \ln[-\ln(1-F)] \tag{3.5}$$

X_i, Y_i were plotted in figure 3 and figure 4, respectively. In both figures, the left part (a) was the diagram of the Weibull distribution while the right part (b) was the enlarged diagram of the rectangular mark in the left diagram.

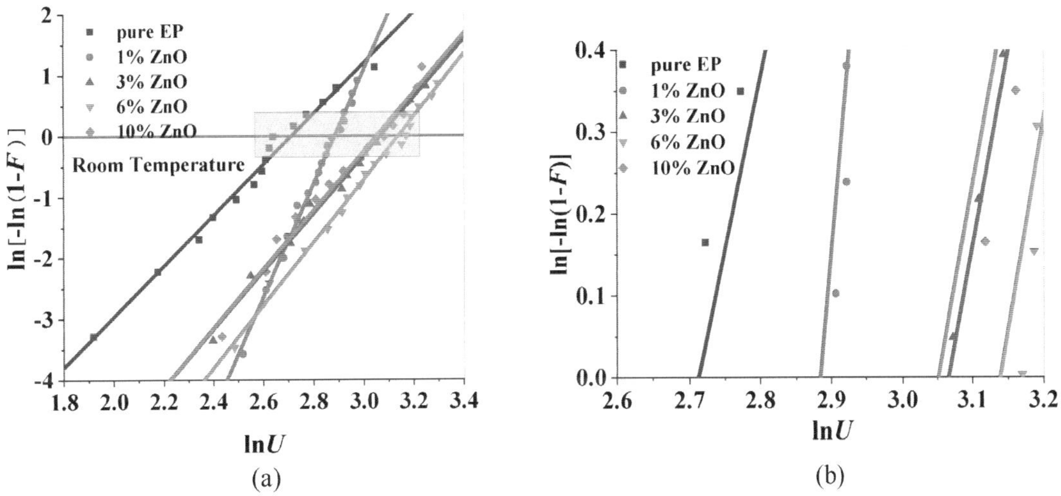

Figure 3. The surface flashover test results at room temperature.

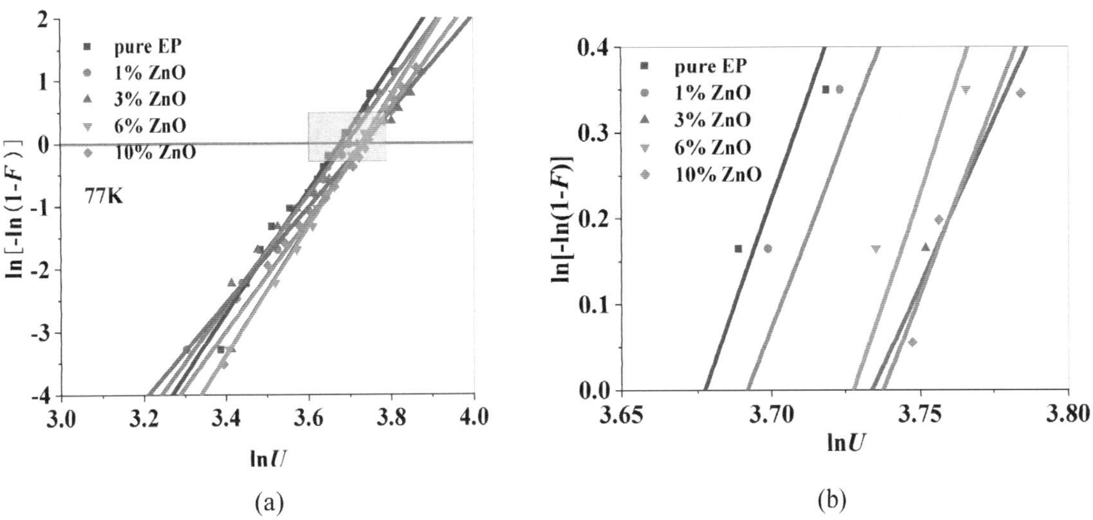

Figure 4. The surface flashover test results at 77 K.

The results obtained at room temperature were shown in figure 3. Compared with the pure EP , the surface flashover voltage of the sample containing 1 wt% ZnO increased obviously. When 3 wt%, 6 wt% and 10 wt% ZnO were added to EP, the surface flashover voltages had little difference, but were higher than that of the sample containing 1 wt% ZnO. The measurement uncertainties of X_i of each sample were 0.28, 0.12, 0.25, 0.23 and 0.24 respectively.

The results at 77K were shown in figure 4. The surface flashover voltages of each sample manifested little difference. To be specific, for the samples containing 0 wt%, 1 wt%, 3 wt% and 10 wt% ZnO, the surface flashover voltages increased with the increase of the content of ZnO. The flashover voltage of the sample containing 6 wt% ZnO was slightly lower than that of samples containing 3 wt% ZnO. The uncertainties of X_i of each sample were 0.12, 0.13,0.15, 0.12 and 0.16 respectively

The surface flashover voltages for which the failure probability were 0.632 were calculated and plotted in figure 5. It showed that the surface flashover voltages at 77 K were higher than that at room temperature for samples with the same weight percentages of ZnO .

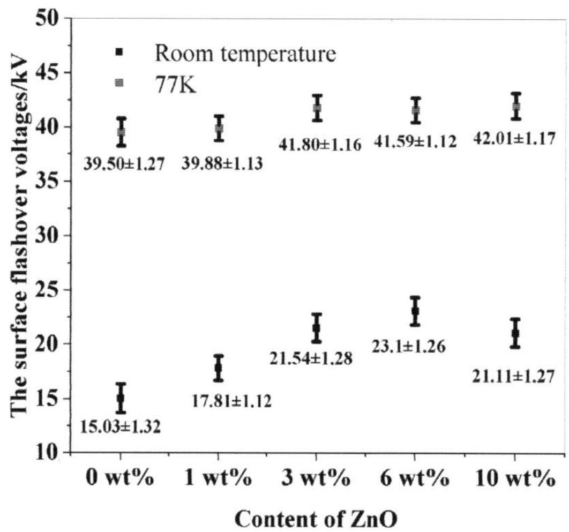

Figure 5. The voltages for which the failure probability were 0.632.

3.2 Discussion

The mechanism of the surface flashover phenomenon in vacuum can be explained by the SEEA (Secondary Electron Emission Avalanche) model and the ETPR (Electron Triggered Polarization Relaxation) model. According to the SEEA model, the initial electrons, emitted from the triple-junction of the material-vacuum-cathode due to the distortion of the electrical field, will hit the material and produce the secondary electrons. The initial electrons and the secondary electrons will continuously hit the material and lead to electron avalanche. Meanwhile, the adsorbed gas on the surface of the material will be removed. Oxygen released will capture the electrons and then carry negative charges, moving with the electrons towards the anode. When the electrons form a channel connecting the cathode and anode, the surface flashover occurs. According to the ETPR model proposed by Blaise G [11], trapping centers and space charges can be found inside the insulating material. External turbulence can lead to the depolarization of the material. The stored electrostatic energy released by charge relaxation can contribute to the surface flashover.

To study the influence of space charges and trap levels on the surface flashover voltage, TSDC (Thermally Stimulated Direct Current) were measured. Samples were polarized by electrical field of 5 kV/mm at 300 K for 20 min, and then were cooled to 250 K rapidly. Depolarization currents were measured while the samples were heated with the rate of 5 K/min. The results were shown in figure 6. The trapped charges Q and the activation energy E were calculated according to the method proposed by Singh S D et al [12]. The results were shown in table 1.

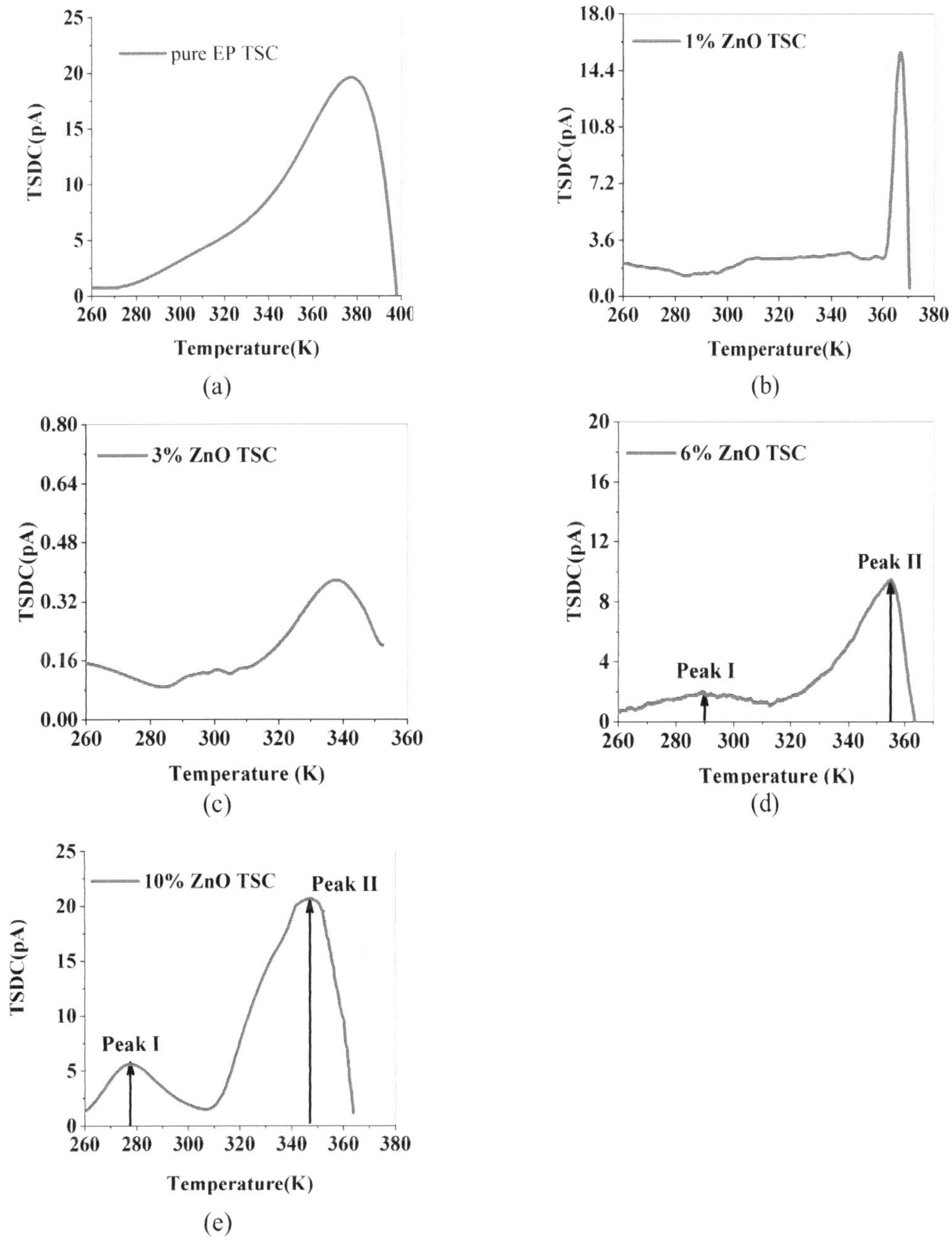

Figure 6. TSDC of each sample (a~e).

Table 1. The trapped charge Q and the activation energy E.

	Q /nC	E_I /eV	E_{II} /eV
Pure EP	13.35	—	0.28
1%ZnO	3.51	—	0.56
3%ZnO	0.22	—	0.69
6%ZnO	3.48	0.22	0.89
10%ZnO	10.22	0.65	0.89

As shown in table 1, the activation energy of each sample increased with the increase of the content of ZnO, meaning that the trap levels were deeper. The charge carriers trapped in the deep traps were more difficult to escape than those trapped in the shallow traps, thus the development of the electrons avalanche of the samples containing deep traps was not as rapid as that of the samples containing shallow traps. Moreover, samples containing more space charges would release more charges and energy in the process of depolarization, contributing to the development of the surface flashover.

At room temperature, it was assumed that the pure EP contained more space charges and the trap level was shallow, hence the surface flashover voltage was the smallest of all the samples. For the samples containing 1 wt%, 3 wt% and 6 wt% ZnO, especially for the sample containing 1 wt% ZnO, due to the fact that the trap levels were deeper and the amount of the space charges decreased obviously, the surface flashover voltages increased obviously. For sample containing 10 wt% ZnO, although the trap level of it was deeper than that of the other samples, the space charges of it were much more than that of the other samples (except the pure EP), either. Hence the surface flashover voltage of it was close to that of the samples containing 3 wt% and 6 wt% ZnO.

At 77 K, it was considered that due to the decrease of the amount of the initial electrons emitted from the triple-junction, the amount of the secondary electrons decreased. Although the trap levels of ZnO/EP composites were deeper, the effect of the levels of traps was not as significant as that of the temperature. Therefore, with the increase of the content of ZnO, the surface flashover voltages tended to increase, but the increase was slight.

Moreover, according to Blaise G [11], the depolarization process, during which much energy would be released, can contribute to the surface flashover. The depolarization process can be triggered if trapping centers were hit by electrons. It was assumed that the electrons with a lower density were less likely to trigger the depolarization process compared to those with a higher density. Therefore, the depolarization process might initiate more easily at room temperature than at 77 K for samples containing the same weight percentage of ZnO.

4. Conclusion

Compared with pure EP, composites that doped with ZnO nanoparticles manifested higher surface flashover voltages at both room temperature and 77 K. Moreover, the increase of the surface flashover voltages was more observably at room temperature. The results indicated that ZnO nano-particles introduced more deep traps thus increased the surface flashover voltage. Meanwhile, for the composites with the same ZnO content, the surface flashover voltages at 77 K were higher than that at room temperature. It was due to the fact that with the decrease of the temperature, the the amount of the secondary electrons decreased, hence the surface flashover voltages increased.

5. Reference

[1] Baek S M, Kim H J, Cho J W and Ryoo H S 2018 *IEEE Trans. Appl. Superconduct.* **28** 1–4

[2] Tu Y, Zhou F, Jiang H, Bai F, Wang C, Lin J and Cheng Y 2018 *IEEE Trans. Dielect. Elect. Insulation.* **25** 308–1317

[3] Chen S, Wang S, Wang Y, Guo B, Li G, Chang Z and Zhang G 2017 *Appl. Surf. Sci.* **414** 107–113 .

[4] Kim S H, Baek S M, Kim Y S, Chung S Y and Joung J M 2002 *Cryogenics* **42** 415–418

[5] Chen Y, Cheng Y, Wu K, Nelson J K, Dissado L A and Li S 2009 *IEEE Trans. Plasma Sci.* **37** 195–203

[6] Li T F, Li S T, Yu S H and Zhu Y W 2015 *Proc. IEEE 11th Int. Conf. on the Properties and Applications of Dielectric Materials* (NewYork/IEEE) pp 52–55

[7] Donnelly K P and Varlow B R 2003 *IEEE Trans. Dielect. Elect. Insulation.* **10** 610–614

[8] Zhang Z, Zheng X, Jin Y, Wu J, Wu W and Lei W 2015 *Trans. Dielect. Elect. Insulation.* **22** 5951–57

[9] Li Y N, Xu W M and Zhang G Q 2015 *Mater. Sci. Eng.:IOP Conf. Series* **87** 012054

[10] Montanari G C et al 2005 *IEEE Guide for the Statistical Analysis of Electrical Insulation Breakdown Data* ed Turner M (NewYork: The Institute of Electrical and Electronics Engineers, Inc.) pp 5–7

[11] Blaise G and Gressus C L 1991 *J. Appl. Phys.* **69** 6334–39

[12] Singh S D, Mazumdar P S, Gartia R K and Deb N C 1998 *J. Phys. D: Appl. Phys.* **31** 231–234

Acknowledgments

This work was supported by the National Natural Science Foundation of China (Grant Nos.: 51577185, 51522705, and 51877209), the National Magnetic Confinement Fusion Science Program (Grant Nos.: 2015GB121001, and 2017YFE0301403), Strategic Priority Research Program of Chinese Academy of Sciences (Grant No. XDXDB25000000), Key Research Program of Frontier Sciences, CAS, (Grant No. QYZDB-SSW-JSC042) and National program for support of top-notch Young professional and Youth Innovation Promotion Association of Chinese Academy of Sciences (2013020).

IOP Conf. Series: Materials Science and Engineering **756** (2020) 012005 doi:10.1088/1757-899X/756/1/012005

Compression and thermal conductivity tests of Cryogel® Z for use in the ultra-transparent cryostats of FCC detector solenoids

V Ilardi[1, 2], L N Busch[1, 3], A Dudarev[1], T Koettig[1], P Borges de Sousa[1], J Liberadzka[1], H Silva[1] and H H J ten Kate[1]

[1] CERN, POB Geneva 23, 1211 Geneva, Switzerland
[2] University of Twente, Faculty of Science and Technology, POB 217, 7500AA Enschede, The Netherlands
[3] Karlsruhe Institute of Technology, Karlsruhe, Germany

E-mail corresponding author: Veronica.Ilardi@cern.ch

Abstract. The Future Circular Collider (FCC) study includes the design of the detector magnets for the FCC-ee$^+$ (electron-positron) collider, requiring a 2 T solenoid for particle spectrometry, and for the FCC-hh (proton-proton) collider, with a 4 T detector solenoid. For both solenoids and their cryostats, CERN is developing an innovative and challenging design in which the solenoids are positioned inside the calorimeters, directly surrounding the inner tracker. For this purpose, the cryostats must be optimized to have maximum radiation transparency. They are structured as a sandwich of thinnest possible metallic shells for achieving vacuum tightness, supported by layers of low density and highly radiation transparent insulation material, still providing sufficient mechanical resistance and low thermal conductivity. In this respect, thermal and mechanical analysis of innovative insulation materials are currently being carried out. The first material of interest, Cryogel® Z, is shaped as a flexible composite blanket, which combines silica aerogel with reinforcing fibers and a density of 160 kg/m^3. It allows a 4 m bore, 6 m long FCC-ee$^+$ detector solenoid cryostat with a total thickness of 250 mm. CERN has investigated the compression of Cryogel® Z under 1 bar equivalent mechanical load and its thermal conductivity between 10 K and room temperature, as well as the critical phenomena of thermal shrinkage and outgassing. We present the test results, as a first overview on the material.

1. Introduction

In the frame of the ongoing development of the next generation of particle colliders, CERN is designing multiple variants of Detector Magnets for the projected Physics Experiments at the Future Circular Collider (FCC), which includes electron-positron (ee$^+$), electron-hadron (eh) and hadron-hadron (hh) collision detectors [1]. A first step machine may be the high-luminosity, high-precision FCC-ee$^+$. A cutting-edge design shows the superconducting solenoid positioned inside the calorimeter, providing magnetic field for particle tracking in the inner detector. Its 3-D model, inspired by the so-called IDEA detector (International Detector Electron Accelerators), is shown in Figure 1.

The 4 m bore, 6 m long, 2 T and 170 MJ solenoid requires a cold mass and cryostat with extremely high radiation transparency and therefore lowest possible thickness and density, in order to minimize particle scattering [2, 3]. The classical cryostat sandwich comprising a relatively thick metallic vacuum vessel, radiation shield, MLI and solenoid cold mass, is in this concept replaced by a structure of very thin metallic vacuum vessel walls, supported by an insulation material with sufficient mechanical resistance, as pictured schematically in Figure 2.

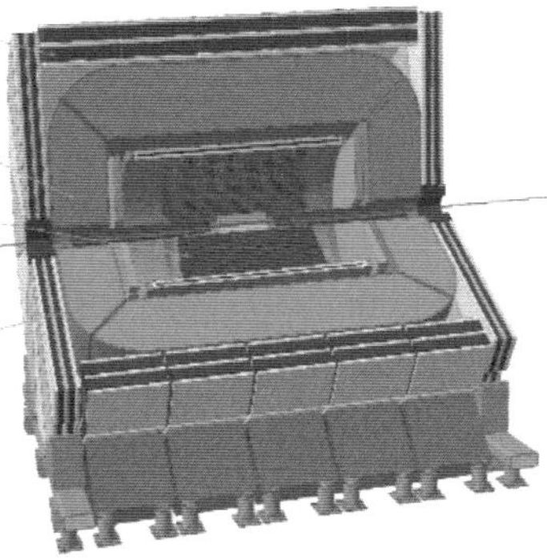

Figure 1. The FCC-ee+ IDEA detector with an ultra-thin solenoid inside the calorimeter.

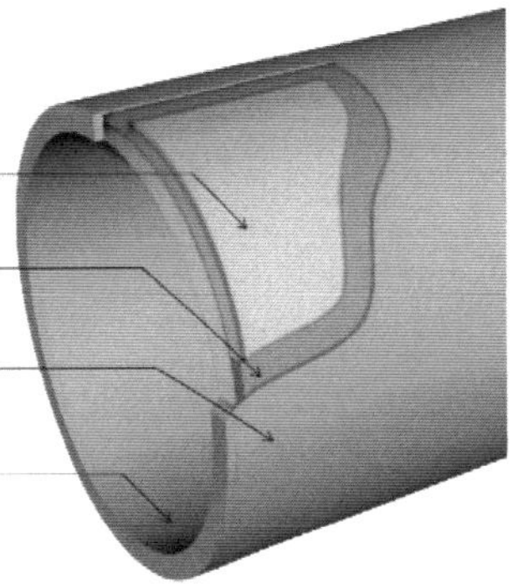

Figure 2. Design of the FCC-ee+ cryostat for a 2 T, 4 m bore, 6 m long solenoid.

The same concept can even be applied to the much more demanding detector for the FCC-hh, comprising an ultra-thin 4T/4m bore solenoid around the inner detector. The innovative design has been developed starting from the research on novel and compatible insulation materials, which has led, among others, to Cryogel® Z as manufactured by Aspen Aerogels Inc. It is shaped as a flexible aerogel composite blanket, designed for insulating low temperature environments ranging from cryogenic temperatures to ambient. It combines silica aerogel with reinforcing fibers and it has a density of 160 kg/m^3 [4, 5].

Here we present the results of the compression and thermal conductivity tests conducted on Cryogel® Z and the effects of possible thermal shrinkage and outgassing of the material.

2. Mechanical behaviour of Cryogel® Z under different conditions of temperature and pressure

The innovative superconducting solenoid features ultra-thin vessel walls, which are supposed to wrinkle under vacuum load while maintaining vacuum tightness. Dense layers of Cryogel® Z filling the space between the thin vacuum wall and solenoid cold mass are required to sustain the atmospheric pressure. Two tests, performed under different conditions and setups, are conducted. First, the compression of Cryogel® Z at room temperature and atmospheric pressure is estimated by applying a compressive mechanical load of 1 bar. During the second test, Cryogel® Z, under vacuum, is cooled down in a bath of liquid nitrogen, in order to check critical aspects as outgassing and shrinkage of the material, in addition to the percentage of overall compression reached.

2.1. Compression test in 10 cycles at room temperature

Cryogel® Z comes as a blanket of 10 mm thickness. Ten cycles of mechanical compression were conducted on a stack of material consisting of 10 blankets, each 10 mm thick and with an area of 100 mm x 100 mm. The results of the first cycle of compression are shown in Figure 3.

Under a mechanical load equivalent to 1 bar, the compression of Cryogel® Z is 29.0% of the initial thickness at the first cycle and it increases up to the 29.8% at the last cycle. Compression hysteresis is observed as, by removing the load, the percentage of material recovered is equal to the 20% of the initial thickness at the first cycle and it decreases to 17% at the last cycle. It can be concluded that Cryogel® Z shows roughly the same mechanical behavior after 10 cycles, giving evidence that it does not get seriously damaged under 1 bar compression cycles. However, the creep behavior of the material should be investigated.

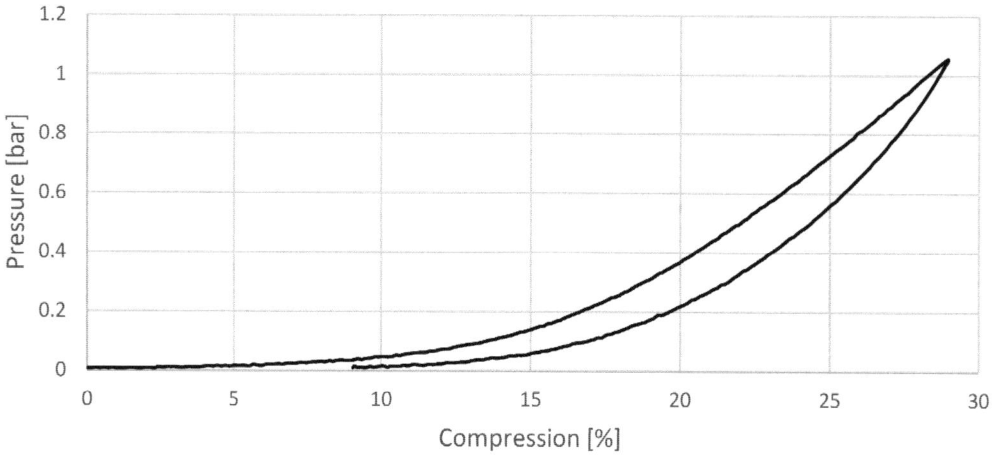

Figure 3. First compression cycle of Cryogel® Z at room temperature. Compressive pressure versus percentage of thickness reduction.

2.2. *Thermal shrinkage test of compressed Cryogel® Z*

In order to use Cryogel® Z as cryostat insulation, it is important to analyze the outgassing during pumping and the shrinkage of the material at low temperature. Figure 4 shows the setup that is vacuum pumped and cooled down in liquid nitrogen.

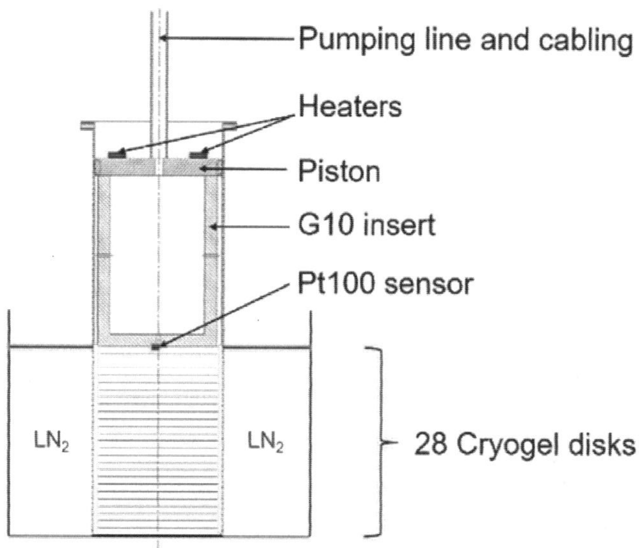

Figure 4. Test setup for contraction measurements of compressed Cryogel® Z.

It consists of a G10 cylinder of 484 mm height, filled with 28 discs of Cryogel® Z blankets of 155 mm diameter in close contact to the cylinder walls. A stainless steel piston with an O-ring seal ensures the leak tightness of the setup while the O-ring is kept at room temperature. To facilitate this, two electric heaters are placed on the piston and a 210 mm high G10 insert separates the O-ring from the stack of Cryogel® Z blankets. A Pt100 temperature sensor is placed on top of the Cryogel® Z discs.

The first pumping cycle compresses Cryogel® Z to 198.5 ± 0.5 mm, corresponding to 71% of the initial 280 mm height. Thus, the stack compressed of 29%, consistent with the test described in Section 2.1. Roughly 1 hour after releasing the load, the Cryogel® Z stack measures 265 mm, showing that the material recovered to 95% of the initial height (i.e. it kept the 5% height reduction). A second pumping cycle compresses the Cryogel® Z stack by 67.0 ± 0.5 mm, bringing the stack back to a 71% of the initial height.
The pressure decreasing at the first pumping cycle, see Figure 5, shows a clear holding level at 2.2 mbar, indicating outgassing of the material. By placing the setup inside the liquid N_2 bath, the minimum temperature measured by the Pt100 sensor is 180 K, so an average minimum temperature between 80 K and 180 K can be assumed for Cryogel® Z. No relevant displacement of the piston is observed, given the accuracy of ± 0.5 mm. Knowing the thermal expansion of G10 [6], the combined contraction of vessel walls and insert is 0.5 mm. Thus, Cryogel® Z may contract within our measurement accuracy, which is between 0 and 1 mm (0÷0.5% of the compressed height of Cryogel® Z stacks). Since the phenomena needs deeper study, new tests, with higher instrumentation accuracy, will be performed.

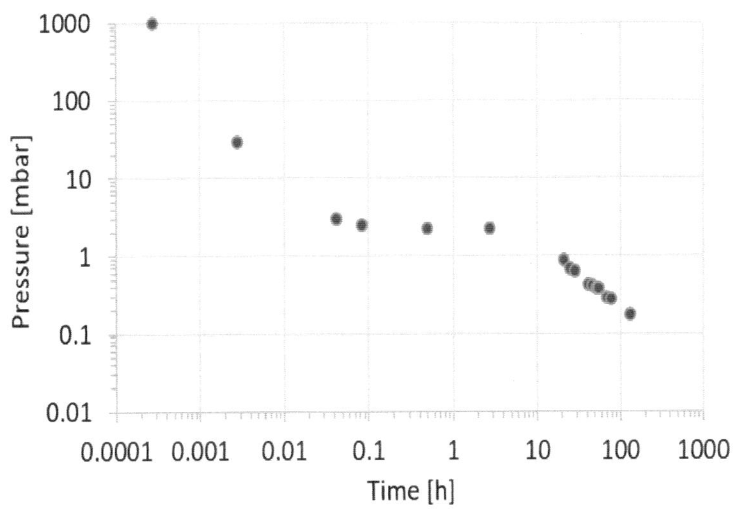

Figure 5. First pump down of a stack of Cryogel Z compressed layers at 293 K, pressure versus time.

3. Thermal conductivity measurements of Cryogel Z

At the CERN Central Cryogenic Laboratory, a thermal conductivity study was conducted through small-scale measurements. Cryogel® Z samples were tested between 10 K and 275 K, at atmospheric pressure, while compressed to 70% of their initial thickness. Two tests were conducted, based on different measurement methodologies: the integral and differential one [7]. The setup and results obtained from the integral method are described in our previous publication [8]. In the following paragraphs, we briefly outline the main features of the setup and compare the integral and differential approaches and their results.

3.1. Test setup and sample description

Two Cryogel® Z samples of 22 mm diameter and 10 mm thickness are mounted in between three copper plates of the sample holder. The discs are compressed in transverse direction, from 10 mm to an effective thermal length L of 7 mm, corresponding to 30% densification. The compression is provided by three screws between the top and bottom plates. To provide thermal contact, three copper braids are bolted to the external plates. Top and bottom plates are, therefore, kept at the same temperature, while the central plate is thermally decoupled from them, so that the heat load is only transferred through

Cryogel® Z. The sample holder is bolted to the second stage of a Sumitomo two-stage pulse tube refrigerator (PTR), which delivers 1 W of cooling power at 4.2 K on the second stage [9]. For both tests, a calibrated TVO temperature sensor is positioned on the bottom plate, while two Cernox sensors are placed on top and center plates, calibrated against the TVO. The heat load of 0 to 425 mW is provided by electric heaters.

In order to minimize radiation losses, the second stage has its own copper thermal shield wrapped in MLI. An overview of the setup is given in Figure 6.

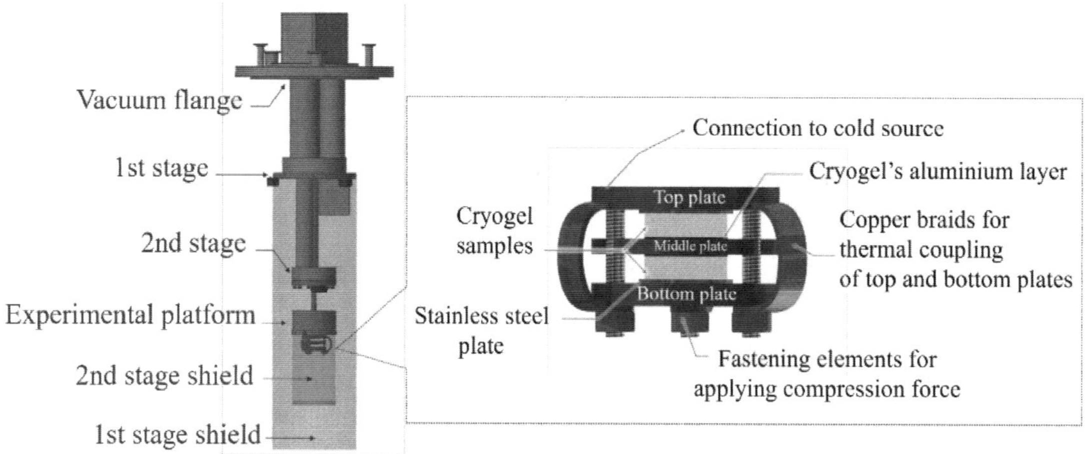

Figure 6. Test setup for Cryogel® Z thermal conductivity measurements [10].

3.2. *Integral and differential measurement methods*

A characteristic measurement procedure is to stabilise the experimental platform at its lowest possible temperature (about 2.6 K) and then incrementally increasing the heat load on the heated plates. The heated plates' temperatures are allowed to stabilise and a measurement point is taken, after which the heat load is increased for a new data point.

The thermal conductivity integral method, followed during the first test, consists of keeping the low temperature ends of the sample, the top and bottom copper plates, at a constant temperature, while progressively heating the other end, the center copper plate, to a higher temperature, and measuring the steady-state temperature gradient established between both ends.

The thermal conductivity integral is defined from the Fourier equation for steady state unidirectional conductive heat flux through a constant cross-section area as following:

$$I = Q\,\frac{L}{A} = \int_{T_c}^{T_h} \lambda(T)\,\mathrm{d}T, \tag{1}$$

where Q is the effective heat flux transferred through the sample and corrected for radiation and wiring losses, A is Cryogel's cross-sectional area, L is the distance between the cold and warm ends that are at temperature T_c and T_h, respectively. All variables Q, A, L, T_c and T_h are measured. As T_c is kept constant, the thermal conductivity λ at T_h is equal to the gradient of the integral of the thermal conductivity:

$$\nabla I(T_c, T_h) = \frac{\partial}{\partial T_c} I(T_c, T_h) + \lambda(T_h) - \frac{\partial}{\partial T_h} \int \lambda(T)\,dT\Big|_{T_c} = \lambda(T_h). \tag{2}$$

From equations 1 and 2, the thermal conductivity at T_h can be expressed as function of the measured change in effective heat flux. Thus, the thermal conductivity data points are plotted as a function of T_h.

For the second test, the thermal conductivity differential method is applied: once the experimental platform is stabilized at its lowest possible temperature of 2.6 K, it starts to be heated up, keeping a ΔT between the low temperature top and bottom copper plates and the high temperature center copper plate.

The ΔT is kept as low as possible, below 300 mK, so that, using equation (1), λ can be considered constant for every ΔT analyzed:

$$\lambda = Q\,\frac{L}{A}\frac{1}{(T_h - T_c)}. \quad (3)$$

3.3. Measurement results

The results are shown in Figure 7. Within the limit set by the accuracy of the instruments, the trends of the data points are fairly consistent. The data took into account some corrections of the effective heat load transferred across the samples. The conduction loss through both Cernox sensor and electric heaters was estimated at 3% of the applied heat load, while the heat loss by radiation of the middle plate to the inner thermal shield was 8% of the applied heating power for the highest plate temperature. The estimated maximum measuring error is 30%, taking into account the accuracy of voltmeters and current source for heaters and sensors, of the calibrated Vernier caliper for the sample cross section, active length and the thermal shrinkage of Cryogel® Z.

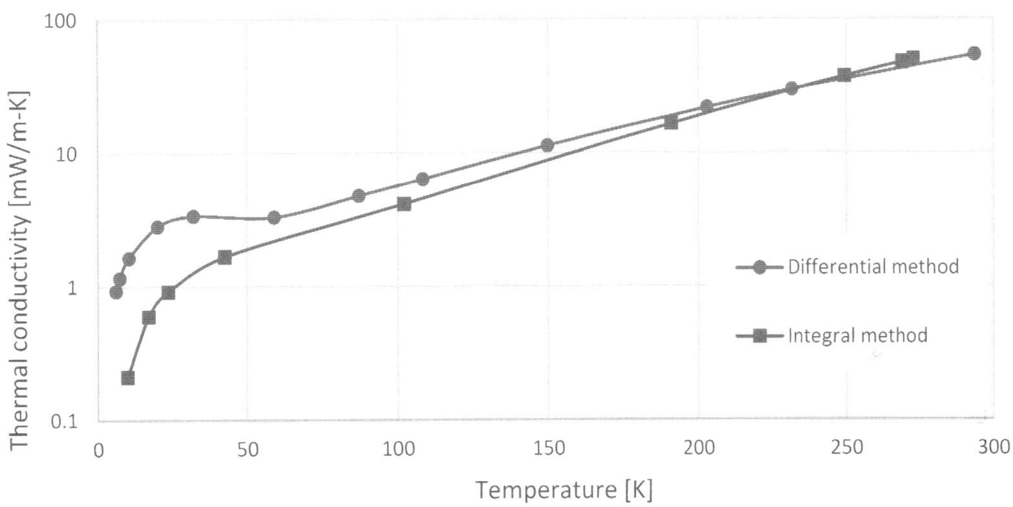

Figure 7. Comparison between the thermal conductivity measurements conducted on Cryogel® Z through the differential and integral methods.

4. Conclusion

From the compressive tests realized, Cryogel Z shows, under 1 bar mechanical pressure, a displacement varying from -29.0% to -29.8% of the initial height, over ten compressive cycles. The compression hysteresis is also consistent with a recovered material percentage between 17 and 20%. Therefore, the mechanical behavior of the material can be considered fairly stable and we can assume that Cryogel® Z does not get heavily damaged under 1 bar pressure.

The thermal conductivity of Cryogel® Z, measured on a small-scale setup, ranges from 0.2 mW/m-K at 10 K, to 50 mW/m-K at 273 K. This identifies Cryogel® Z as a promising insulation material for use in the ultra-thin cryostats of the FCC detector magnets. However, the thermal conductivity was measured for the bulk material. The cryostats of the detector solenoids of the FCC-ee+ and FCC-hh would be filled with several blankets of Cryogel® Z. The interfaces between those should, in principle, reduce the heat load transferred. In addition, because of the small dimension of the samples' cross section used for the tests, the measurements may be affected by the samples' boundary conditions. Therefore, a large-scale test setup has been designed and manufactured at CERN, comprising vacuum vessel walls, thermal shield and cold mass. A heat transfer analysis, using multiple blankets of Cryogel® Z as interposer between the parts, is currently being carried out.

IOP Conf. Series: Materials Science and Engineering **756** (2020) 012005 doi:10.1088/1757-899X/756/1/012005

References

[1] FCC Collaboration, Conceptual Design Report of the Future Circular Collider, 2019.
[2] Bielert E. *et al.* 2019 Superconducting detector magnets baseline designs for particle physics experiments at the Future Circular Collider *IEEE Trans. Appl. Supercond.* **29** no. 5.
[3] Kulenkampff T., Dudarev A., Ilardi V., Mentink M., Silva H., Bielert E. and Ten Kate H.H.J., Conceptual development of a novel ultra-thin and transparent 2 T superconducting detector solenoid for the Future Circular Collider, *IEEE Trans. Appl. Supercond.* **29** no. 5, 2019.
[4] Fesmire J. E., Rouanet S. and Ryu J., Aerogel-based cryogenic superinsulation, *Adv. in Cryog. Eng.* **44** 219-26, 1998.
[5] Aspen Aerogels, Cryogel Z technical specification sheet, 2017.
[6] Ekin J.W., *Experimental techniques for low-temperature measurements,* Book, OUP Oxford, 2006.
[7] Hust J.G. and Lankford A.B., Comments on the measurement of thermal conductivity and presentation of a thermal conductivity integral method, *Intern. Journ. Thermophysics* **3** 67-77, 1982.
[8] Ilardi V. *et al.*, Development of ligth and highly radiation transaprent cryostats for FCC detector magnets: first analyses of insulation materials, *IEEE Trans. Appl. Supercond.* **28** 2019.
[9] Sumitomo RP-082B2 pulse tube cryocooler technical specification sheet.
[10] Borges de Sousa P., Koettig T., Measurement report : Cryogel® Z thermal conductivity measurements between 3 K and 273 K, 2017.
[11] American Society for Testing and Materials. ASTM Standard C177, 2019.

Effect of Non-Uniformity of Electric Field on Breakdown Strength of Cryogenic Gaseous Insulation Media for Superconducting Power Applications

S. Telikapalli[a,b], M. Kvitkovicova[a,b], P. Cheetham[b], C. Park[c], C.H. Kim[b], L. Graber[c], S. Pamidi[a,b]

[a] FAMU-FSU College of Engineering, Department of Electrical & Computer Engineering, Tallahassee, FL, 32310, USA
[b] Center for Advanced Power Systems, Florida State University, Tallahassee, FL, 32310, USA
[c] Georgia Institute of Technology College of Engineering, Department of Electrical & Computer Engineering, Atlanta, GA, 30332, USA
st18b@my.fsu.edu

Abstract:

A new experimental set up for measuring the breakdown strength of gas media in weakly non-uniform electric field at cryogenic temperatures and high pressures is described. Measure of breakdown strength of helium gas at 77 K and 293 K in a non-uniform electric field with a field efficiency factor 62.5% are presented. The results suggest that the breakdown strength in weakly non-uniform electric field relate to that of uniform electric field by the field efficiency factor, η. This relationship holds good for both 293 K and 77 K data. This observation expands our previously reported systematics of dielectric strength in uniform electric field to the weakly non-uniform electric field conditions, which is important for designing HTS power applications. The established relationship eliminates the need for costly experiments for measuring the dielectric strength at a specific operating temperature and pressure and in the required non-uniformity of the electric field.

1. Introduction

High temperature superconductors offer a novel solution to meet the increasing demand for electrical power in the urban areas creating bottlenecks [1]. High temperature superconducting (HTS) devices offer significantly high power densities compared to conventional conductor technology. In HTS applications such as electric ships, electric aircraft, and high energy physics, where power density is important, the typical operating temperatures are in the range of 4 – 40 K. Hence, liquid nitrogen (LN2), the typical cryogen used in HTS applications cannot be used as it only allows operation between 63-80 K. Helium gas (GHe) is being considered due to its wide operating range (4-80 K). Helium gas is preferred over hydrogen gas which has a similar operating temperature (20-80 K) because of its lower condensation temperature, inert and non-flammable nature. The drawback of using GHe in terrestrial power grid applications is that the operating voltage is reduced significantly due to its lower dielectric strength compared to LN2. The problem of low dielectric strength of GHe can partly be overcome by adding a small mol% of H2 gas to GHe. It was demonstrated that the dielectric strength of a He-H2 gas mixture is proportional to the mole fraction of hydrogen [2]. The relationship between the composition and the dielectric strength of various gas mixtures was studied by modeling using Boltzmann analysis [3]–[7]. To further improve the prospect of helium gas cooled HTS power cables operating at medium to high voltage levels, a first of a kind superconducting gas insulated cable (SGIL) where the circulating cryogen acts as the sole dielectric was demonstrated for different gas mixtures [8]–[11]. However, the increase observed in the inherent breakdown strength of a gas mixture does not fully translate into the corresponding increase in the dielectric performance of the SGIL. One of the suspected factors causing this inconsistency is non-uniformity of electric field in SGIL. The reported enhancements in the intrinsic dielectric strength of gas mixtures were based on the measurements in uniform electric field emulated using Bruce profile electrodes with a gap distance ranging from 1-2 mm. However, in the SGIL the electric field profile is a function of the diameter of the HTS conductor and the inner diameter of the cryostat. This geometry will cause a weak non-uniformity in the electric field typically represented as the field efficiency factor (η) [12]. The relationship between η and breakdown strength is understood for room temperature applications [12], [13]. To test if this relation is valid for cryogenic temperatures and high pressures experimental investigations of the dielectric strength of gas media at cryogenic temperatures at various pressure levels are necessary. Additionally, the electric field profile of HTS power devices depend on the design and it is useful to have a fundamental understanding of the effect of weak non-uniformity on the dielectric strength and if it is dependent of the gas composition. Without the established relationship between the non-uniformity of electric field and the dielectric strength of a gas media, the relative performance of various designs of HTS power device such as SGIL cannot be modeled. Fabrication and experimental evaluation of various designs is costly.

The focus of this study is to establish an experimental setup to measure the dielectric strength of gases in a defined weakly non-uniform electric field and compare it with the results of uniform electric field at room temperature and cryogenic temperatures. A setup with η of 62.5 % was fabricated keeping the dimensions of the electrodes similar to those used in the uniform electric field experiments. Experiments were performed on helium gas at several pressures.

2. Methodology

2.1. Uniform electric field

For a uniform field, breakdown voltage, V_{bu}, can be sufficiently approximated as E/d where E is the dielectric strength of the insulating material uniformly spread between 2 electrodes separated by gap distance d

$$V_{bn} = E_{max} d\eta \tag{1}$$

Where,

$$\eta = \frac{E_{average}}{E_{max}} \tag{2}$$

η is 1 for uniform electric fields. In a uniform field, every point in the field is equally stressed and the probability of breakdown is the same at every point. Uniform field systems are very ideal and hard to realize in practice due to system geometries and inherent electric field gradient they produce. The difficulty of maintaining the uniformity of the field increases significantly for higher gap distances[13]. In our experiments, to emulate uniform electric field, Bruce profile electrodes with a radius of 12.5 mm were used as shown in Figure 1. The electric profile is maintained for small gap distances of up to 4 mm with this set of electrodes. It is difficult to maintain the uniformity of the field for larger gap distances as the required radius of electrodes rises rapidly and the electric field in the edges will curve thus creating a drop in η. High voltage source (HV) is connected to the top electrode through a bushing rated for up to 300 psi while the bottom electrode is connected to common ground of the supply via the cryostat [2]. Gauge blocks were used to set a desired gap distance between the electrodes. The setup is designed to keep the gap distance consistent for varying temperatures.

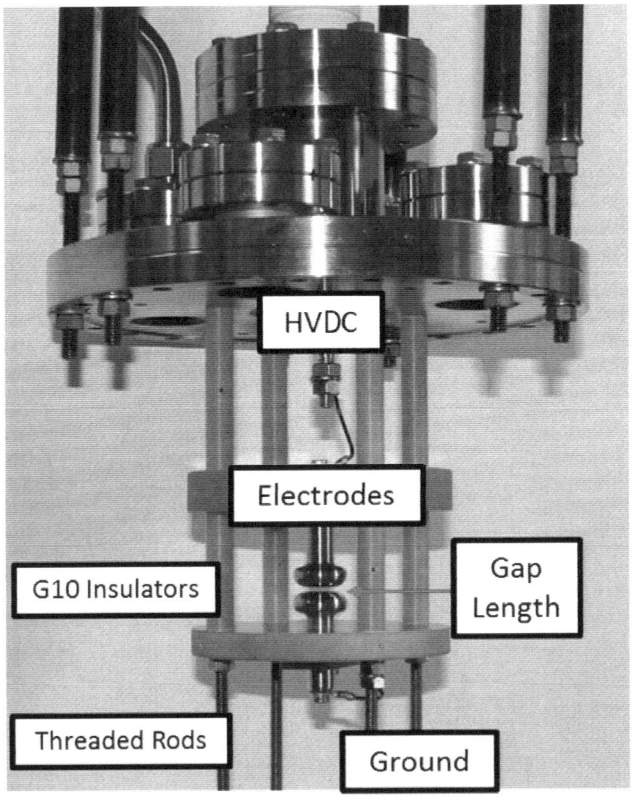

Figure 1: Experimental setup for breakdown measurements in uniform field

2.2. Weakly Non-uniform electric field

Equation (1) is also applicable for weakly non-uniform electric fields but, for weakly non-uniform fields η varies between 0 and less than 1 depending on the conductor geometry [14]. Table 1 shows the relation between η and radii of the conductors for different conductor configurations

Table 1: Field factors for different conductor geometries reproduced from [13]

Geometrical configuration	Maximum electric field	Field efficiency factor (η)
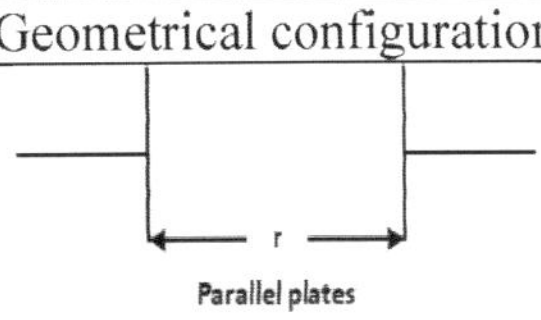 Parallel plates	$\frac{v}{r}$	1
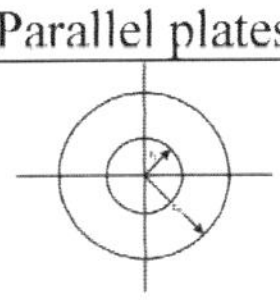 Concentric cylinders	$\frac{v}{r_i ln\left(\frac{r_o}{r_i}\right)}$	$\frac{r_o - r_i}{r_i ln(\frac{r_o}{r_i})}$
concentric spheres (Figure same as concentric cylinders)	$\frac{Vr_o}{r_i(r_o - r_i)}$	$\frac{R_o}{r_i}$

Table 1 shows that for increased dielectric strength in weakly non-uniform fields, increasing the gap distance is not always desirable. Figure 2 shows the breakdown strength for different inner radius configurations for a given outer radius using the relation for a coaxial system shown in Table 1. Maximum dielectric strength can be obtained when

$$ln\left(\frac{R_o}{R_i}\right) = 1 \quad (3)$$

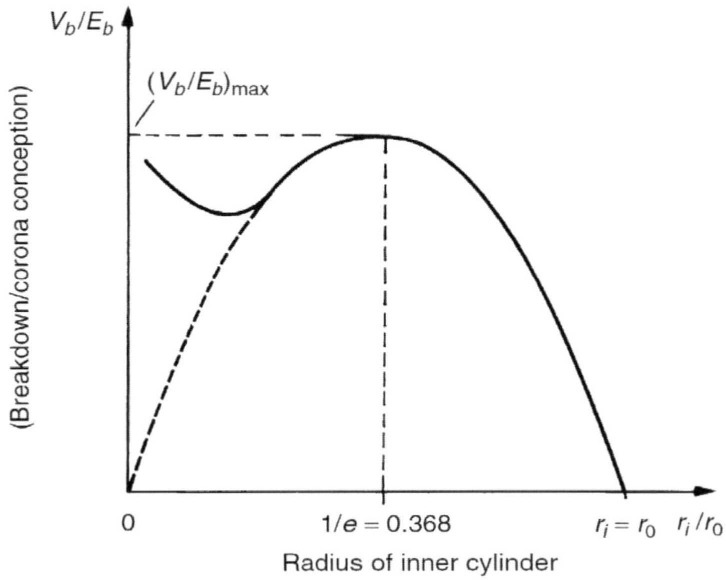

Figure 2: Conductor radii vs breakdown voltage relation for coaxial systems reproduced from [12]

and the maximum field efficiency obtainable is 58% [12]. To design a coaxial electric field experimental setup that is compatible with cryogenic temperatures without using any insulating spacers to obtain similar η is complicated and is non-reliable for temperature variations. Hence a hemispherical experimental setup was used to obtain the desired η. Keeping with dimensions of the electrodes previously used, a spherical electrode with a radius of 12.5 mm was used. From Table 1, to get a similar η, the radius of outer hemisphere must be 21.5 mm but an electrode with a 20 mm radius was used as shown in Figure 3 to get a η of 62.5%. This ensures that breakdown is not affected by corona discharges as shown in Figure 2.

Figure 4 shows the results of finite element analysis performed using 2d-axisymmetric COMSOL to ensure there is no field enhancement at the edges and to verify the electric profile is as needed. The spherical electrode is given a nominal test voltage of 1 kV and the hemispherical electrode is grounded. The setup is surrounded by helium gas with a relative permittivity of 1. The maximum field is observed on the surface of the conductor with a value of 0.2 kV/mm, this is similar to the situation in the SGIL cable studied previously, which has a maximum field of 0.2 kV/mm with insulating tubular spacers [8]. An enhancement was found near the edges and it was found that a fillet alone would not suffice to reduce its effects, thus an extra depth of 10 mm was given to give a sufficient fillet. A 10% enhancement was found which given the gap distance increase caused by the fillet was tolerable. The drawbacks of this setup are the slight difference in the η compared to that of an SGIL, the large gap distance necessary to provide the necessary field gradient to get desired non-uniformity and the optimistic breakdown strength predicted by this setup.

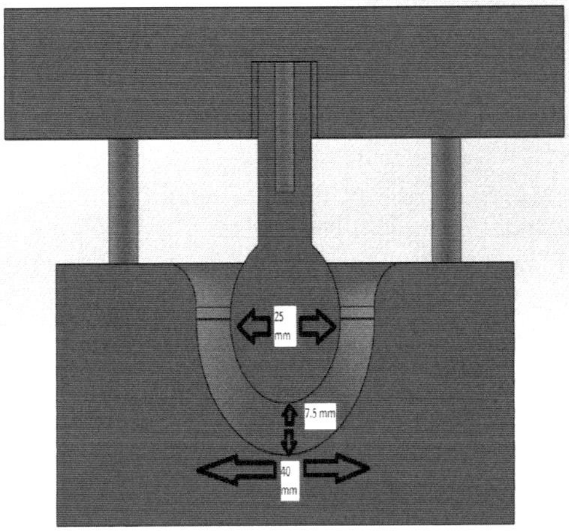

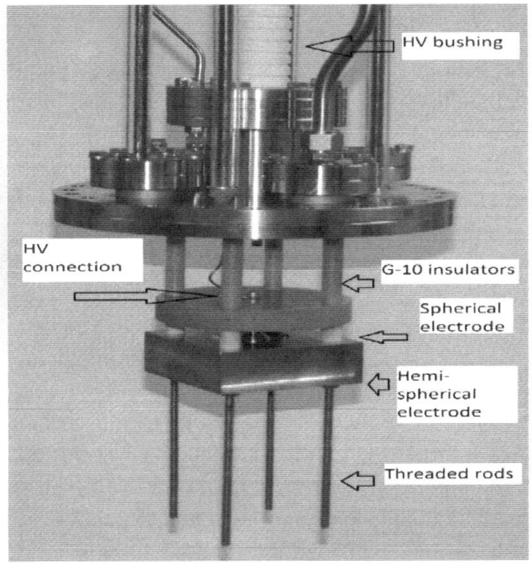

(a) (b)

Figure 3: A schematic (a) and a photograph (b) of the electrode arrangement for the breakdown experiments in weakly non-uniform electric field.

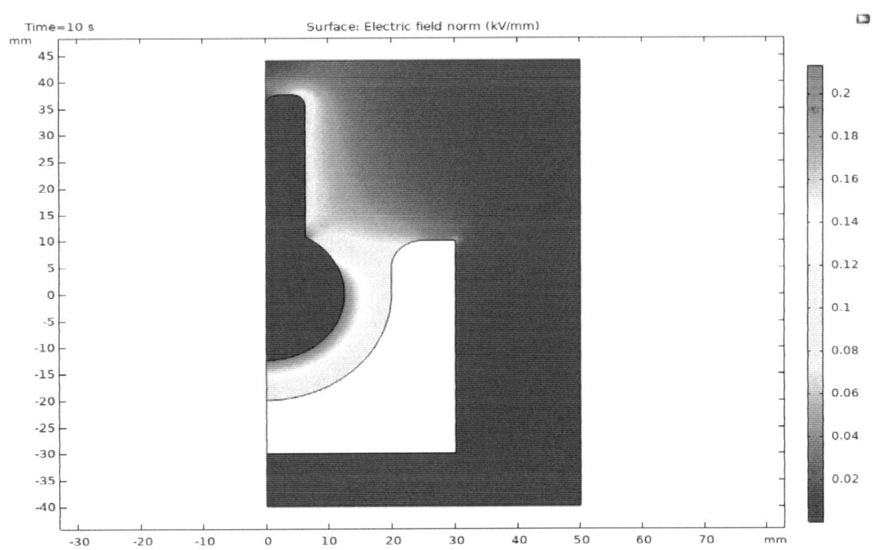

Figure 4: Finite element analysis of hemispherical setup's electric field

Figure 3b shows the assembled experimental setup mounted on the top plate of the pressure vessel used for the measurements. High voltage connection from the supply is connected to the top spherical electrode through a bushing rated for 300 psi and the setup was connected to a common ground by metal to metal connection (bottom hemispherical electrode to the threaded rod supports from the top-plate). To keep the spherical electrode at the proper height from the base of the hemisphere, 4 identical G-10 pieces were cut to precise length. The test setup was sealed inside a pressure vessel to test the dielectric strength for different gases at different pressure levels.

2.3. *Experimental procedure*

The pressure vessel is cleaned with alcohol and dry cloth to ensure there is no residual water vapor and dirt particles before sealing. To ensure the purity of gas in the pressure vessel, flush and evacuation cycles were conducted with nitrogen gas first to remove air and other impurities and later with industrial-grade helium (99.8% pure). Between the successive flushing cycles, the pressure vessel was evacuated to 2.2e-7 MPa in between these cycles using a dry-scroll pump. After the flushing cycles, the pressure vessel was filled with ultra-pure GHe (99.9999% pure) to 2 MPa at room temperature. The first ten measurements were considered as seasoning of the electrodes and were not used for the data. This is performed to ensure that the electrodes are free of any imperfections which may cause localized field enhancement.

Fifteen measurements were performed at each pressure level. A Glassman unit PS/SH2—R160j18 with a ramp rate of 0.3 kV/s was used as the high voltage source. DC voltage was ramped up until a breakdown. A two-minute wait time is included in the measurement protocol between successive measurements to ensure thermal equilibrium is attained inside the pressure vessel. The measurements were repeated for different pressure levels as required. Breakdown measurements were performed at room temperature, which was maintained at 293±1 K. For the measurements at 77 K, the pressure vessel was completely immersed in a bath of LN2.

3. **Experimental data**

The dielectric strength measurements of helium were performed at room temperature for 4 pressure levels in both uniform and weakly non-uniform electric fields. Figure 5 shows the experimental data of the dielectric strength of helium at room temperature in kV/mm in uniform and weakly non-uniform electric fields. An additional curve is included that is the dielectric strength in non-uniform electric field divided by the field efficiency factor, η, of 62.5% according to (4). The data obtained at room temperature is consistent with values predicted using(4).

$$v_{uniform} = \frac{v_{non-uniform}}{FEF} \tag{4}$$

Figure 6 shows the experimental data of the dielectric strength of helium at 77 K in kV/mm in uniform and weakly non-uniform electric fields. A comparison of the data in Figures 5 and 6 shows that the dielectric strength at 77 K is significantly higher than that at room temperature. This is as expected considering the higher gas density at 77 K [13]. Due to the larger gap distance used in the experiments in non-uniform field, the breakdown voltage value was high at 2 MPa and electrical breakdown occurred outside the pressure vessel and thus the data for 2 MPa could not be obtained.

It was previously demonstrated that the breakdown strength of gases scales with density and is independent of operating temperature [15]. Figure 7 demonstrates the dielectric strength of helium in kV/mm at 293 K and 77 K at equivalent densities, listed in Table 2.

Table 2: Equivalent density comparison

Pressure at RT MPa	Breakdown voltage kV/mm	Pressure at 77 K MPa	Breakdown voltage kV/mm
2	2.2	0.503	1.9
1.5	1.8	0.4	1.7
1	1.2	0.267	1.3
0.5	0.6	0.107	0.6

The scattering/pitting in the hemispherical electrode after performing the breakdown measurements was found to be uniformly distributed. This validates the proper centering of the electrodes and uniform gap distance throughout varying temperatures.

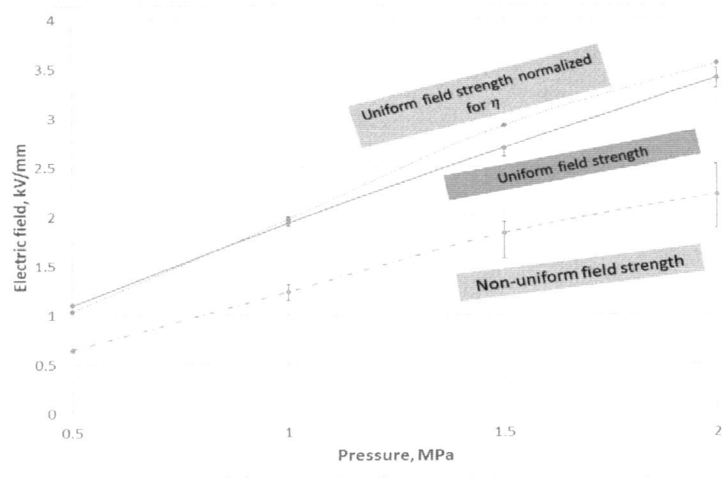

Figure 5: Dielectric strength of helium in kV/mm at room temperature for different pressure levels

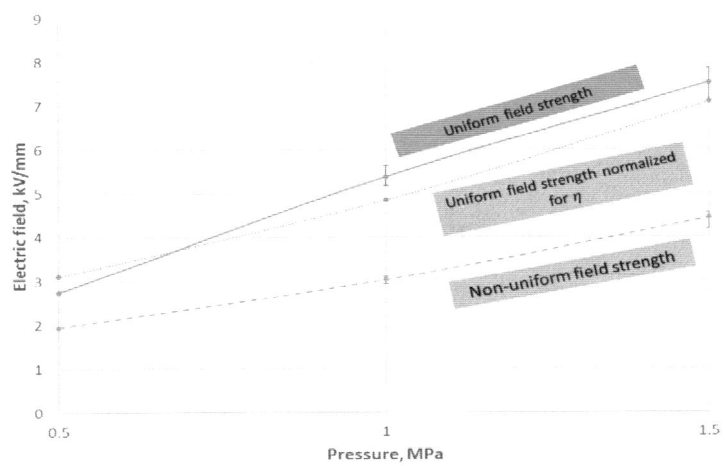

Figure 6: Dielectric strength of helium in kV/mm at 77 K for different pressure levels

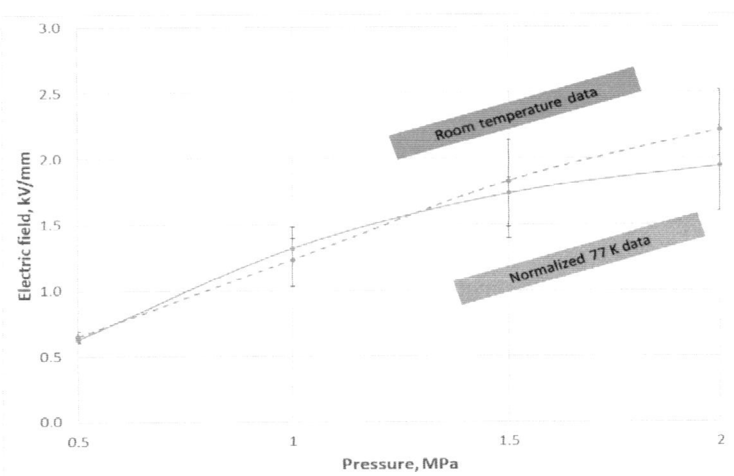

Figure 7: Comparison of breakdown voltages of helium gas at 293 K and 77 K at equivalent densities

4. Discussion:

The experimental set up established for conducting breakdown measurements in gas media in a weakly non-uniform electric field at cryogenic temperatures resulted in useful measurements. The dielectric strength in weakly non-uniform electric field is significantly lower than that in uniform electric field, as expected. The absence of any insulating spacers in the experimental setup prevented space charge accumulation and the associated uncertainties in the measurement data. The linearity in the dielectric strength data with pressure suggests that the gap distance was maintained consistent over varying temperatures and pressures.

The close match between the experimental data in uniform electric field and the data set in non-uniform electric field normalized with η shows that the field efficiency factor could be used to estimate the dielectric strength in non-uniform electric field conditions from uniform field data. This observation is useful in the design of HTS devices and estimated dielectric strength of gas media under non-uniform electric field and cryogenic temperatures without having to perform time-consuming measurements. The results also suggest that there is no fundamental difference in the dielectric behavior of gas media at room temperature and at cryogenic temperatures. The observations extend our previous understanding and assessment of the dielectric behavior of gas media that the temperature does not have a noticeable effect on the dielectric strength after accounting for the density of the gas [13].

5. Conclusions:

An experimental set up for measuring the breakdown strength of gas media at cryogenic temperatures and high pressures was established. The set up was used to measure the breakdown strength of helium gas at 77 K and 293 K in a weakly non-uniform electric field with a field efficiency factor 62.5%. The field efficiency factor is in the same range as encountered in a superconducting gas insulated line. The measured breakdown strength in weakly non-uniform electric field, when normalized by the field efficiency factor of 62.5% match with that in uniform electric field for both the 293 k and 77 K measurements. The identical relationship between the dielectric strength behavior in uniform and weakly non-uniform electric field at both 77 K and 293 K suggest that the temperature does not have any noticeable influence on the relationship. This expands our previously reported systematics of dielectric strength in uniform electric field at a broad range of temperatures and pressures which suggested that it scales with density and is independent of temperature and pressure of a gas medium to weakly non-uniform electric fields. This is an important conclusion that allows estimation of the dielectric strength of a gas medium for any application using the field efficiency factor, the operating temperature, and pressure from the data in uniform electric field at any temperature and pressure. This eliminated the need for costly experiments to measure the dielectric strength at the operating temperature and pressure and electrode geometry necessary to emulate the required non-uniformity of the electric field.

References:

[1] A. P. Malozemoff, "The power grid and the impact of high-temperature superconductor technology: An overview," *Superconductors in the Power Grid: Materials and Applications*, pp. 1–28, 2015.

[2] P. Cheetham, C. Park, C. H. Kim, L. Graber, and S. V. Pamidi, "Dielectric properties of cryogenic gas mixtures for superconducting power applications," *IOP Conf. Ser. Mater. Sci. Eng.*, vol. 278, no. 1, 2017.

[3] P. Cheetham, C. Park, C. H. Kim, L. Graber, and S. V Pamidi, "Dielectric properties of cryogenic gas mixtures for superconducting power applications," *IOP Conf. Ser. Mater. Sci. Eng.*, vol. 278, p. 012040, Dec. 2017.

[4] C. Park, L. Graber, P. Cheetham, J. G. Viquez, C. H. Kim, and S. Pamidi, "A versatile modeling technique for predicting dielectric strength improvements in gas mixtures for superconducting applications," *IEEE Transactions on Dielectrics and Electrical Insulation*,

vol. 24, no. 5. pp. 2755–2764, 2017.

[5] C. Park, S. Pamidi, and L. Graber, "Boltzmann Analysis of Cryogenic He – H_{2} Gas Mixtures as Dielectric Media for High-Temperature Superconducting Power Devices," *IEEE Trans. Appl. Supercond.*, vol. 27, no. 4, pp. 1–6, Jun. 2017.

[6] C. Park, L. Graber, and S. Pamidi, "The dielectric properties of gaseous cryogen mixtures of He, H 2 , Ne, and N 2 in a temperature range of 50–80 K at pressures up to 2.0 MPa," *J. Appl. Phys.*, vol. 121, no. 8, p. 083304, Feb. 2017.

[7] C. Park, J. Wei, P. Cheetham, C. H. Kim, S. Pamidi, and L. Graber, "The influence of temperature on the dielectric strength of gaseous cryogens," *Proc. IEEE Int. Conf. Prop. Appl. Dielectr. Mater.*, vol. 2018-May, pp. 164–167, 2018.

[8] P. Cheetham, J. Wagner, T. Stamm, C. H. Kim, L. Graber, and S. Pamidi, "Electric field analysis of insulating spacers for a superconducting gas-insulated transmission line," *Proc. IEEE Int. Conf. Prop. Appl. Dielectr. Mater.*, vol. 2018-May, pp. 185–189, 2018.

[9] P. Cheetham, J. Viquez, L. Graber, C. H. Kim, H. Rodrigo, and S. Pamidi, "Novel Design Concept and Demonstration of a Superconducting Gas-Insulated Transmission Line," *IEEE Trans. Appl. Supercond.*, vol. 27, no. 4, pp. 1–5, 2016.

[10] P. Cheetham, C. H. Kim, and L. Graber, "Optimization of a Superconducting Gas-Insulated Transmission Line."

[11] P. Cheetham, C. H. Kim, L. Graber, and S. Pamidi, "Practical Considerations for the Design of a Superconducting Gas-Insulated Transmission Line for Shipboard Applications," *ieee Electr. Sh. Tech. Symp.*, 2017.

[12] E. Kuffel, W. S. Zaengl, and J. Kuffel, "High Voltage Engineering - Fundamentals," *Electr. Eng.*, p. 539, 2000.

[13] Naidu, *High Volatage Engineering*, vol. 16, no. 6. 1996.

[14] R. Arora and W. Mosch, *High Voltage and Electrical Insulation Engineering*. 2011.

[15] A. Al-taie, C. Park, P. Cheetham, C. H. Kim, and L. Graber, "A New Representation of Paschen ' s Law Suitable for Variable Temperature Power Applications."

Thermal and Electrical Properties of Isocyanate Derived Organic Aerogels for Cryogenic Insulation Applications

G Churu[1], J A Demko[1], A Mole[1], R C Duckworth[2], H Lu[3], S Malakooti[3], and N Leventis[4]

[1]LeTourneau University, 2100 South Mobberly Avenue, Longview, TX 75607 USA
[2]Oak Ridge National Laboratory, One Bethel Valley Road, Oak Ridge, TN, 37831 USA
[3]The University of Texas at Dallas, Richardson, TX, 75080, USA
[4]Missouri University of Science and Technology, Rolla, MO 65409, USA

Abstract. The use and storage of cryogens such as liquefied nitrogen, helium, hydrogen among others requires reliable and efficient thermal insulation systems. Passive insulation from high performance materials that are well-known for their inherent low thermal conductivity would reduce the overall costs involved in design, manufacture and maintenance of such systems. One such class of materials are referred to as aerogels. These materials are known for their low density, high mesoporosity, high surface areas, low thermal conductivity and high acoustic impedance. Aerogels were invented by S.S. Kistler in 1931 and the most common type are those made of silica. However, the inherent fragility of silica aerogels makes them hard to mass produce, and therefore applications have been limited. A major breakthrough was introduced by our team almost 20 years ago with the invention of polymer crosslinked silica aerogels. Those materials shifted attention to all-polymer aerogels that have overcome all fragility issues associated with their inorganic counterparts. This study focuses on such polymeric aerogels that can be mass produced as large monoliths while maintaining the low thermal conductivity of traditional silica aerogels over a wide temperature range. Manufacturing flexibility of polymeric aerogels allows fabrication of blocks and sheets that can be applied in various configurations to insulate cryogenic and superconducting devices. The thermal conductivity with 80 K and room temperature boundary are reported as well as other properties (electrical), that need to be considered when designing devices for cryogenic applications.

1 Introduction

For effective control and to maintain the optimal cryogenic temperatures, high performance or low thermal conductivity insulation is needed to prevent heat transfer into the cryogenic piping and other reservoirs used to contain such fluids. Currently, multilayer and vacuum insulations [4] are widely used but the need to maintain vacuum in most cases adds another layer of complexity when designing and maintaining such systems. Aerogels [1-3] are being used in many space applications [5]. The current state of the art is to use silica aerogel embedded in lose fibers to form aerogel blankets [6]. This method is not very efficient since you still need a thick layer of insulation to increase Δx, the insulation thickness or, the distance the heat has to travel before it comes into contact with the cold surface. The use of such fibers also increases the effective thermal conductivity of the resulting materials, making them less effective than what would be expected if pure aerogels were used.

According to Grand View research Inc. [7] market research on aerogels, the aerogel market size is expected to grow to over $3.2 billion by the year 2025. This is mainly driven by the demand from the oil and gas industry. Companies that do subsea oil and gas exploration and in the polar regions especially need to insulate pipelines so that they can maintain the viscosity needed to keep the oil flowing when the surrounding temperatures are very low. Most oil companies such as British Petroleum (BP) and Exxon Mobil [8] use aerogel blankets consisting of fibers impregnated with silica aerogel powders. This is not ideal since the full potential of aerogels as an insulator is not utilized; the presence of other fibers increases thermal conductivity of the overall structure.

In most thermal insulation applications, silica aerogels are used either as powders entrapped inside blanket type materials as described above. Rarely are monoliths used because they are very hard to manufacture and to mass produce [9]. Silica aerogel monolithic materials are brittle and have very poor mechanical properties. In addition, they require supercritical drying to get the optimal properties. As such, the size that can be obtained is restricted to the size of the autoclave of the supercritical dryer, which further increases the cost of production. The fact that they also use chemicals that have high toxicity such as Tetraethylorthosilicate (TEOS) and tetramethyl-orthosilicate (TMOS) as the basic precursors makes mass production much more challenging [10].

2 Organic aerogel description

Two types of aerogels have been widely synthesized and studied. The first type consists of inorganic aerogels that are synthesized from transition metals and semimetal alkoxide. They include metal oxides aerogels such as TiO2, Al2O2 among others, as well as binary systems of metal oxides that form more complex systems like Al2O2/CuO aerogel [11]. Ternary aerogel systems that consist of more than two metal oxides such as ZrO2/CuO/ZnO etc have been synthesized [8] and used in catalytic converters in automobile exhaust systems to control emissions. Considering the large number of transition metals as well as many possible combinations that can form binary as well as ternary systems of oxides and semimetal alkoxides, the list of possible aerogels is large. This observation was first made by S. S. Kistler, concluding in his first paper on the subject that the “list can be extended indefinitely”, meaning the list of possible aerogels that can be fabricated is theoretically endless [3]. The process also gives us room to incorporate aerogels in areas such as in catalysis where nanoparticles can be assembled to maximize the contact surface area and improve catalytic efficiency compared to more compact arrangements.

The second type, organic aerogels [11], which are the focus of this work, are derivatives of polymeric systems and other natural organic products like cellulose, polyurethane, polyurea among others [12]. The work of R. Pekala initially involved synthesis and characterization of resorcinol-formaldehyde (RF) phenolic type organic aerogels. The aerogels of this study were made using a sol-gel process similar to one used by Leventis at al [12-14]. Isocynate was reacted with water in presence of a triethylamine catalyst. Various densities were made by varying the concentration of the isocynate in acetone that was used as the primary solvent. The density ranged from 0.075 g/cm^3 on the lower density end to 0.16 g/cm^3. The materials were allowed to gel and age in mold and later washed in acetone to remove the reaction products which mainly consisted of water and carbonic acid. The materials were then washed in pentane and dried in ambient conditions [14].

The organic aerogel has been made in various shapes such as thick (25.4 mm) and thin (6 mm) sheets and also in cylindrical shape as shown in Figures 1a, 1b, and 1c. The discs shown are the samples tested in this study and were cut from larger sheets. Shown in figure 1c is a solid cylinder and on top a hollow cylinder machined from a solid cylinder. This organic aerogel can be cast in different shapes provided a pattern for the mold is provided. Both the flat samples and the hollow cylinder of pipe were tested and results presented in this paper.

3 Flat sample thermal conductivity measurement

The thermal conductivity was measured for flat samples using boil-off calorimetry in a NASA cryostat-400. The procedure has been discussed in [13-16]. The samples are flat discs with a diameter of 203 mm and 25.4 mm thickness or stacks of four-layer samples that have the same total thickness. The pot shown in the figures is filled with liquid nitrogen to produce a temperature gradient across the sample. The boil-off is measured and is related to the heat transfer across the sample. The temperatures at the warm and cold boundaries are measured as shown in Figures 2a and 2b. For the four-layer tests, measurements of temperatures in between the layers was also performed for one test run.

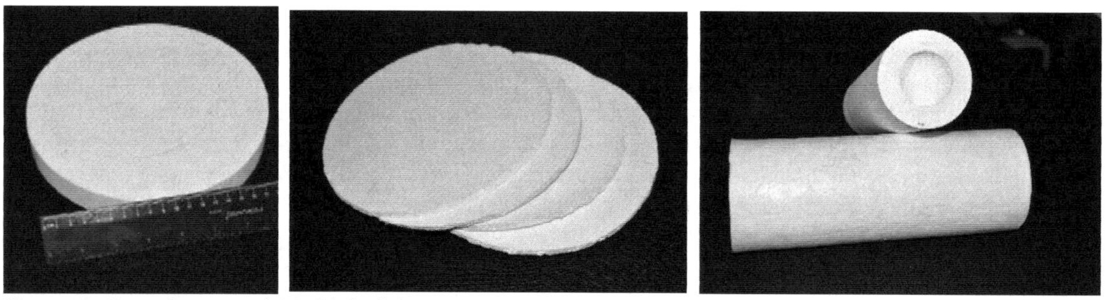

Figure 1. Organic aerogel (a) thick disk stack of thin disks, (b) series of thin disks, and (c) cylinder as cast and after machining into a hollow cylinder.

Knowing the size of the sample, the heat load, and temperature difference across the sample, the thermal conductivity can be calculated from the one-dimensional heat conduction equation as:

$$Q_{meas} = k_{meas} A_{eff} \frac{(T_{meas,warm} - T_{meas,cold})}{Thickness} \quad (1)$$

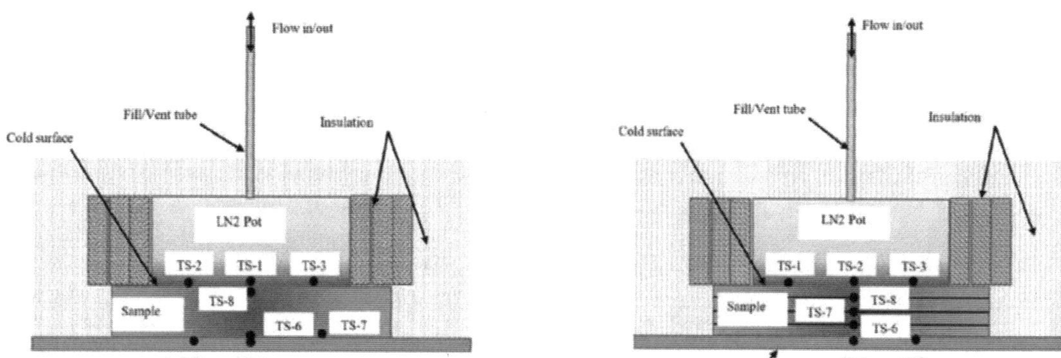

Figure 2. Test configuration for flat (a) solid and (b) the four-layer samples. The diameter of LN2 pot was 150mm, while the samples had a thickness of 25 mm, and a 200 mm diameter.

The Cryostat-400 is a comparative measurement and provides a reasonable estimate of the thermal conductivity. Most significant for this work is the relative comparison of the magnitude of the thermal conductivity of the materials measured. A summary of the measurements of two commercial insulations, Foamglass [18], and Aspen aerogel [19], along with the measurement of the organic aerogel are provided in Table 1. The thermal conductivity of the organic aerogel is seen to be slightly less than measurements of the Foamglass, but higher than the Aspen material. In general, the organic aerogel is comparable to these commercial materials. When making the organic aerogel, there is the possibility of controlling the density which tends to decrease the thermal conductivity as the density decreases.

Table 1. Summary of test results for the thermal conductivity, k, of organic aerogels compared to measurements on foam glass and three layers of an Aspen aerogel blanket. The sample boundaries were at a warm boundary temperature (WBT) and a cold boundary temperature (CBT).

Sample	WBT(K)	CBT(K)	k (mW/m-K)
Foamglass	293.0	82.0	64.7
Organic aerogel high density	280.1	82.0	53.9
Organic aerogel low density	292.2	81.9	49.0
Org-Solid-Sample-3	275.7	83.6	50.9
Organic aerogel 4 layer, (thermocouples between layers)	282.1	81.8	56.9
Organic aerogel 4 layer	274.3	81.3	55.8
Organic aerogel 4 layer	273.4	80.9	57.8
Three-layer Aspen aerogel blanket	278.7	82.3	48.0

4 Pipe shape insulation test results

A hollow cylinder or pipe shape sample was also tested in a device shown in Figure 3. The test apparatus is designed to do an absolute thermal conductivity measurement and was part of a two-year development as a Senior Design class project on the "Standardized Testing of Low Temperature Insulation" or STOLI [17]. The STOLI test section is shown in Figure 3b indicating the locations of temperature sensors. A test sample is placed over a heater as shown in the figure. The ends are insulated ad there are guard heaters at each end to maintain a constant temperature along the length of the sample. An electrical measurement of the heat applied to the test section and the warm and cold surface temperatures are measured along the sample. Knowing the length of test section, $L = 0.3048\ m$, the inner diameter of the sample corresponds to the outer diameter of a standard 2-inch pipe, $D_i = 0.06033\ m$, and the outer diameter corresponds to the inside diameter of $D_o = 0.110\ m$. The thermal conductivity is determined from the one-dimensional heat conduction equation in a cylinder given by:

$$Q = \frac{2\pi L k_{meas}(T_{Hot,meas} - T_{Cold,meas})}{\ln(D_o/D_i)} \tag{2}$$

In order to validate the results obtained in the STOLI apparatus, measurements were performed on commercially available materials including Foam-glass [18], and Aspen aerogel [19]. Table 2 shows the results of the known materials as well as the measurement of the organic aerogel material discussed in this paper.

5 Electrical Breakdown Characterization

Electrical breakdown characterization of Isocyanate Derived Organic Aerogels was performed at 77 K and room temperature as an initial screening to determine the functional limit of this insulation. Two aerogel insulation samples were provided, GC 7.75 and GC 11, and had average thickness of 0.187 mm and 0.197 mm respectively. For 77 K characterizations, 5-cm squares were placed between two rounded flat plate electrodes and immersed in liquid nitrogen (Fig. 4). AC voltage was applied across the electrodes using a

100 kV ac source that was ramped at a rate of 1 kV/min ramp rate. Two aerogel sample types were assessed, and their average breakdown strength were 15.1 kV/mm (GC 11) and 17.0 kV/mm (GC 7.75). Weibull distributions for each sample (Fig. 5) for each breakdown measured showed comparable uniformities with respect to the slopes of the failure probabilities with the performance of the GC 7.75 showing slightly higher performance.

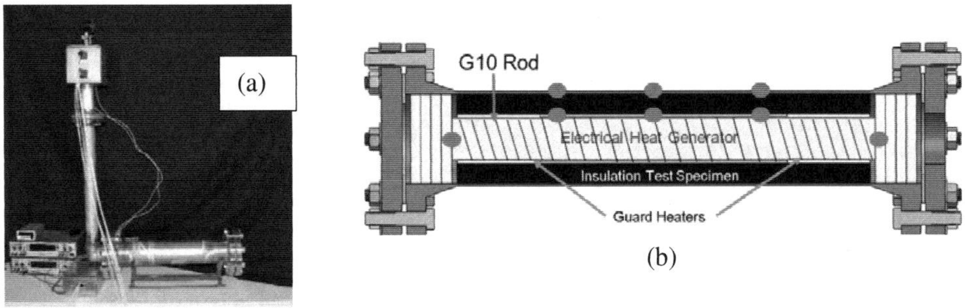

Figure 3. Pipe test apparatus (STOLI) (a) fully assembled and (b) details of test section. The insulation test specimen had an inner dimension of 60.3 mm, an outer diameter of 110 mm and length of 304 mm.

Table 2. Comparison of measured results and reported data for Foamglass, Aspen aerogel and organic aerogel.

Material	WBT (K)	CBT(K)	k (mW/m-k) measured	k (mW/m-K) Reported data
Foamglass [16]	335	309	42.1	40 – 42
Foamglass [16]	310	279	40.2	36 – 40
Aspen Aerogel [17]	335	275	17.9	12 – 18
Organic aerogel	324	298	51.8	N/A

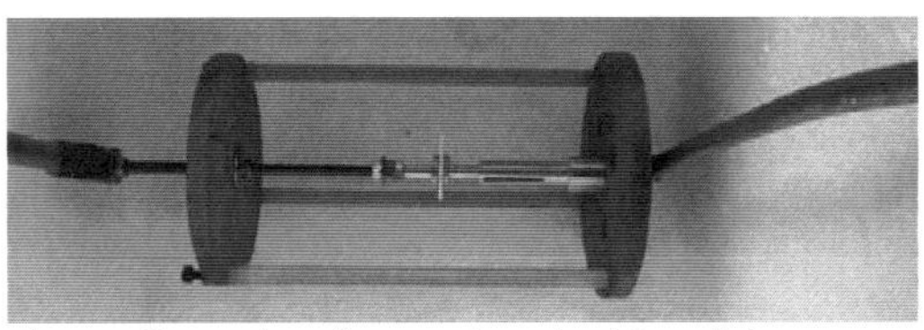

Fig. 4. Electrode and sample configuration for ac electrical breakdown measurement of aerogel insulations at 77 K.

For measurements at room temperature, the difference in the average breakdown strength and functional dependence of the breakdown strength grew between the two sample aerogel sample types. At room temperature, aerogel samples were immersed in Exxon Univolt transformer oil and placed in a Type TC/IEC test cell with adjustable, 36-mm diameter spherical dome electrodes to minimize possible flashover and to localize the electrical field across the sample (Fig. 6). AC voltage was applied to the electrodes using a 60 kV ac source with a 0.5 kV/min ramp rate. The average breakdown strengths for the two sample types, GC 11 and GC 7.75, were 11.9 kV/mm and 17.0 kV/mm respectively. Weibull distributions for each sample (Fig. 7) for each breakdown showed that the differences between the two samples was consistent of the samples measured and not from sample defects that could have impacted the failure distribution and average electrical breakdown strength. When compared to electrical breakdown strength for the Exxon Univolt measured with a fixed electrode separation comparable to the aerogel insulation thickness, a measurable improvement in electrical breakdown strength from the aerogel is demonstrated. However, this improvement lags behind when compared to other dielectrics like polyimides and cross-linked polyethylene that have dielectric breakdown strengths closer to 100 kV/mm. This difference is not surprising given the known porosity of the aerogels, but knowledge of the dielectric breakdown strength could be beneficial to applications where the functional use of the aerogel has to serve thermal and electrical isolation. Additional systematic investigation and refinement that cover topics ranging from the influence of the aerogel structure on interfaces that could result in partial discharge to dimensional measurement of electrical properties to screen for volumetric scaling and defects could be beneficial.

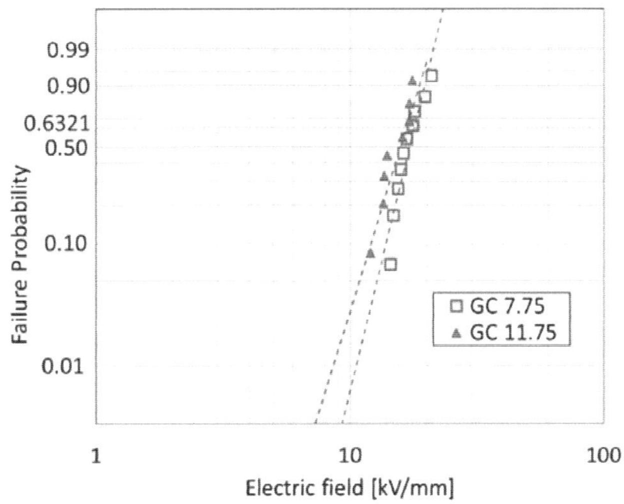

Fig. 5. Weibull distribution for ac electrical breakdown measurements at 77 K for two aerogel samples.

Fig. 6. Electrode and sample configuration for ac electrical breakdown measurement of aerogel insulations at room temperature.

6 Summary, Conclusions and Recommendations for Future Work

An organic aerogel has been developed and measurements of the dielectric strength and thermal conductivity have been measured for applications at liquid nitrogen temperatures. The date was obtained over a limited temperature range for the flat plate tests. The main conclusion from these tests is that the thermal performance for this material is comparable to other insulations. The material can be processed into different shapes such as plates, and cylinders depending on the mold used. Testing was performed on these two common shapes which showed that the thermal performance is not significantly affected by the manufactured shape.

There was a correlation between the thermal and dielectric constant with density, with lower density materials showing a lot of promise. More research on production of such materials is warranted and ongoing. Other recommendation is to conduct additional tests in the STOLI apparatus to obtain low temperature thermal conductivity data. Future plans are to also conduct mechanical property measurements of tensile and compressive strength under different static and cyclic loading conditions.

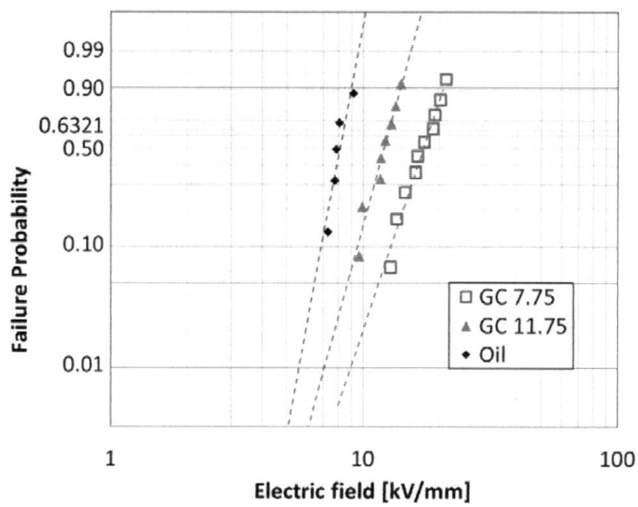

Fig. 7. Weibull distribution for ac electrical breakdown measurements at room temperature for two aerogel samples.

7 References

[1] S. S. Kistler, Coherent expanded aerogels and jellies, Nature 127: 741 (1931)

[2] S. S. Kistler, Coherent Expanded-Aerogels, J. Phys. Chem. 36, 52-64 (1932)

[3] S. S Kistler, A.G Caldwell, "Thermal conductivity of silica aerogels", Ind. Eng. Chem. 26: 658 (1934).

[4] Kropschot, R.H. et al., "Multiple-layer insulation," Advances in Cryogenic Engineering, 5:579-586, Vol. 5, Plenum, 1960, pp. 189-198.

[5] Fesmire J.E., "Aerogel insulation for space launch applications" Cryogenics 46 (2006) 111-117

[6] Nocentini K., Achard P., Biwole P., Stipetic M., "Hygro-thermal properties of silica aerogel blankets dried using microwave heating for building thermal insulation", Energy and buildings, 168, 2018; 165-166

[7] Grand View Research, "aerogels Market size worth $3.29 Billion by 2025", 2018.

[8] Hasan M. A., Sangashetty R., Ester A. C. M., Patil S. B., Sherikar B. N., Dey A, "Prospect of Thermal Insulation by Silica Aerogel: A Brief Review", J. Institute of Engineers (India), DOI 10.1007/s40033-017-0136-1, 2017.

[9] M. A. Aegerter, N. Leventis, M. M Koebel, "Aerogel Handbook", Springer Science + business media, pp. 3-45 (2011)

[10] Fricke and A. Emmerling, "Aerogels-Preparation, Properties, Application, Structure and Bonding ",77, Springer-Verlag Berlin Heidelberg (1992), pp. 37-87.

[11] Pekala R.W., "Organic aerogels from the polycondensation of resorcinol with formaldehyde", J. Of Materials Science, Vol. 24 iss. 9 pp 3221-3227, 1989.
[12] Leventis N, Sotiriou-Leventis C, Zhang G, Rawashdeh A-M M, (2002) Nanoengineering strong silica aerogels. Nano Lett 2: 957–960
[13] N. Leventis, C. Sotiriou-Leventis, N. Chandrasekaran, S. Mulik, Z.J. Larimore, H. Lu, G. Churu, M.T. Mang: Multifunctional polyurea aerogels from isocyanates and water. a structure– property case study, Chem. Mater. **22**(24) 6692-6710 (2010)
[14] S. Malakooti, H. G. Churu, A. Lee, T. Xu, H. Luo, N. Xiang, C. Sotiriou-Leventis, N. Leventis, H. Lu: Sound insulation properties in low-density, mechanically strong and ductile nanoporous polyurea aerogels, J. Non-Crystal. Solids **476** 36–45 (2017)
[15] Smith C. E., "Special Report: Aerogels advance in pipeline, tank storage use", Oil and gas Journal, 2006
[16] Fesmire, J.E., Johnson, W.L., Kelly, A.O., Meneghelli, B.J., and Swanger, A.S., "Flat Plate Boiloff Calorimeters for Testing of Thermal Insulation Systems," Advances in Cryogenic Engineering: Proceedings of the Cryogenic Engineering Conference (CEC) 2015, Vol. 101, Institute of Physics, 2016. https://iopscience.iop.org/article/10.1088/1757-899X/101/1/012057/pdf.
[17] J. E. Fesmire and S. D. Augustynowicz, "Thermal Insulation Testing Method and Apparatus," U.S. Patent 6,824,306, issued November 30, 2004.
[18] J. A. Demko, J. E. Fesmire, W. L. Johnson, and A. M. Swanger, "Cryogenic Insulation Standard Data and Methodologies," Advances in Cryogenic Engineering: Transactions of the Cryogenic Engineering Conference - CEC, American Institute of Physics, Vol. 59A, pp. 463-470, 2014.
[19] J A. Demko, J. E. Fesmire, J. Dookie, J. Bickley, and S. Kraft, "Comparison tests of cellular glass insulation for the development of cryogenic insulation standards," Advances in Cryogenic Engineering: Proceedings of the Cryogenic Engineering Conference (CEC) 2015, Vol. 101, Institute of Physics, 2016. http://iopscience.iop.org/article/10.1088/1757-899X/101/1/012016/pdf
[20] A. Kling, C. Wallace, A. Barrerras, I. Wiens, H. Mortensen, T. Davis, A. Hamid, D. Horn, J. Kiler, J. Demko, "Standardized Test of Low Temperature Insulation (STOLI)", Final Report for ENGR4823 Senior Design, LeTourneau University, Longview, Texas 75602, 2019.
[21] Foamglass data sheet, Pittsburg Corning Corporation, 800 Presque Isle Drive, Pittsburg, PA, USA 15239, www.foamglas.com, 2014.
[22] Aspen Aerogel data sheet, Aspen Aerogels, Inc., 30 Forbes Rd, Bldg B, Northborough, MA 01532 USA, 2019.
[23] ASTM C1774 - Standard Guide for Thermal Performance Testing of Cryogenic Insulation Systems. ASTM International, West Conshohocken, PA, USA (2013).

Acknowledgments

We would like to thank LeTourneau University School of Engineering and Engineering Technology for their generous support of this research effort. This manuscript has been authored in part by UT-Battelle, LLC, under contract DE-AC05-00OR22725 with the US Department of Energy (DOE). The US Government retains and the publisher, by accepting the article for publication, acknowledges that the US Government retains a nonexclusive, paid-up, irrevocable, worldwide license to publish or reproduce the published form of this manuscript, or allow others to do so, for US government purposes. DOE will provide public access to these results of federally sponsored research in accordance with the DOE Public Access Plan (http://energy.gov/downloads/doe-public-access-plan).

IOP Conf. Series: Materials Science and Engineering **756** (2020) 012008 doi:10.1088/1757-899X/756/1/012008

Macroflash boiloff calorimetry instrument for the measurement of heat transmission through materials

J E Fesmire[1], C D Bateman[2], and J L Thomas[3]

[1]NASA Kennedy Space Center, Cryogenics Test Laboratory, UB-R1, KSC, FL 32899 USA

[2]Cryotek LLC, Titusville, FL 32780 USA

[3]Solution One Industries, Cryogenics Test Laboratory (LASSO), KSC, FL 32899 USA

E-mail: james.e.fesmire@nasa.gov

Abstract. The Macroflash is a flat plate boiloff calorimeter that provides effective thermal conductivity (k_e) data for a wide range of materials from thermal insulation to structural composites to ceramics. The apparatus and method provide a practical, standardized way to measure heat transmission through materials under steady-state conditions at below-ambient temperatures and under different compressive loads. Another unique feature of this device is that it can provide test data at both large and small temperature differences. Using liquid nitrogen as a method to directly measure the heat flow rate, the device is applicable to testing under an ambient pressure environment at a wide range of temperatures, from 77 K to 403 K. Test specimens may be isotropic or non-isotropic; homogeneous or non-homogeneous. The Macroflash is currently calibrated in the range from approximately 10 mW/m-K to 1,000 mW/m-K using well-characterized materials. Reference data for hundreds of test specimens including foams, powders, aerogels, plastics, composites, carbon composites, wood, glass, ceramic, metal, and multi-layered composites have been compiled from Macroflash testing. The Macroflash apparatus is described and its operation, instrumentation, and control system discussed. The calibration approach is detailed as well as analysis of key data sets of standard materials.

1. Introduction

The advent of advanced polymers, composites, and novel material systems along with growing industrial needs in below-ambient temperature applications have brought about the Macroflash development. Accurate thermal performance information, including effective thermal conductivity data, are needed under relevant end-use conditions. The Macroflash is a practical tool for basic testing of common materials or research evaluation of advanced materials/systems.

The Macroflash is a flat plate boiloff calorimeter that provides thermal performance data for a wide range of materials from thermal insulation to structural composites. Boiloff calorimetry is the measurement principle for determining the effective thermal conductivity (k_e) and heat flux (q) of a test specimen under a wide range of conditions. The Macroflash is a comparative instrument that is calibrated in part from test data compiled using the Cryostat-500 apparatus, an absolute thermal performance instrument that is also for flat-plate specimens [2]. One of the greatest advantages of using liquid nitrogen (LN_2) boiloff calorimetry is its ultimate simplicity and provision of a direct energy measurement. The liquid provides a stable cold boundary temperature and serves as a sort of power meter. The approach also lends itself to testing under representative conditions (i.e., those that reflect the actual-use conditions) afforded by the very large temperature difference established by the liquid nitrogen.

The Macroflash apparatus and method were developed to provide a practical, standardized way to measure heat transmission through materials and systems under steady-state conditions at below-ambient temperatures. A unique feature of this device is that it can test at both small and large temperature differences. Using liquid nitrogen as a method to directly measure the heat flow rate, the device is applicable to testing under an ambient pressure environment at a wide range of temperatures, from 77 K to 403 K. Tests with the Macroflash follow the guidelines of standard ASTM C1774 (Annex A4), providing a cost-effective, field-representative methodology to test any material for below-ambient temperature applications to moderately elevated temperature conditions [1]. Materials include solids, foams, powders, composites, and multi-layered systems and may be isotropic or non-isotropic; homogeneous or non-homogeneous. From engineered materials development, to research testing, to quality control in manufacturing, the standardized device provides utility for the fields of energy, transportation, aerospace, construction, medical, and environment. The basic characteristics of two different flat-plate boiloff calorimeters are given in table 1.

Table 1. Thermal performance test cryostat instruments: flat-plate configuration.

Instrument	Type	Test Specimen Size	ASTM Test Standard	Environment	Heat Flux (W/m^2)
Cryostat-500	Absolute	203 mm diameter, up to 40 mm thick	C1774 Annex A3	Full range vacuum 77 K–403 K	0.4–400
Macroflash	Comparative	76 mm diameter, up to 10 mm thick	C1774 Annex A4	Ambient pressure 77 K–403 K	80–1000

2. Test apparatus design and setup

The Macroflash can be used to test solids, foams, or powders and the materials may be isotropic or non-isotropic; homogeneous or non-homogeneous. Test specimens are typically 75-mm in diameter and 6-mm in thickness. The cold boundary temperature (CBT) is maintained by LN_2 at 77 K while a heater assembly maintains a steady-state warm boundary temperature (WBT) from ambient up to 403 K. The steady boiloff of LN_2 provides a direct measure of the rate of heat being transmitted through the test specimen. Nitrogen or other gas is supplied to the instrument to establish a stable, moisture-free, ambient pressure environment. The CBT can be adjusted to any temperature between 77 K and approximately 300 K by the interposition of a known thermal resistance layer between the cold mass and the test specimen or by the use of intermediate temperature sensors.

The Macroflash is currently calibrated in the range from approximately 10 mW/m-K to 1000 mW/m-K using well-characterized materials. Reference data for hundreds of test specimens including foams, powders, aerogels, plastics, composites, carbon composites, wood, glass, ceramic, metal, and multi-layered composites have been compiled. The primary reference instrument for flat-plate boiloff calorimeter testing is Cryostat-500. The technology is built on the prior technology of flat-plate calorimeters developed by the Cryogenics Test Laboratory at NASA Kennedy Space Center (KSC) [3].

The Macroflash houses a cold mass test chamber centered directly over the test specimen, and wrapped by a custom designed multilayer aerogel blanket system to ensure the thermal isolation and stability necessary for accurate steady-state boiloff measurements. Test specimens with thicknesses from approximately 1 mm to 10 mm can be tested. For rigid-type materials, flatness is critical for ensuring good thermal contact on the cold side of the specimen. Compression loading is selected for 0 kPa, 14 kPa, or 34 kPa (0 psi, 2 psi, or 5 psi), with or without thermal grease, and the system can be readily adapted for other loads as required.

The Macroflash includes a heater controller connected to the heater plate assembly for control of the warm boundary temperature (WBT). All instruments are connected to a customized software interface using a National Instruments LabVIEW® data acquisition system for recording and monitoring. The mass flow rate of the boiloff can be measured by a weight scale or mass flow meter. A simplified schematic for the Macroflash instrument is given in figure 1.

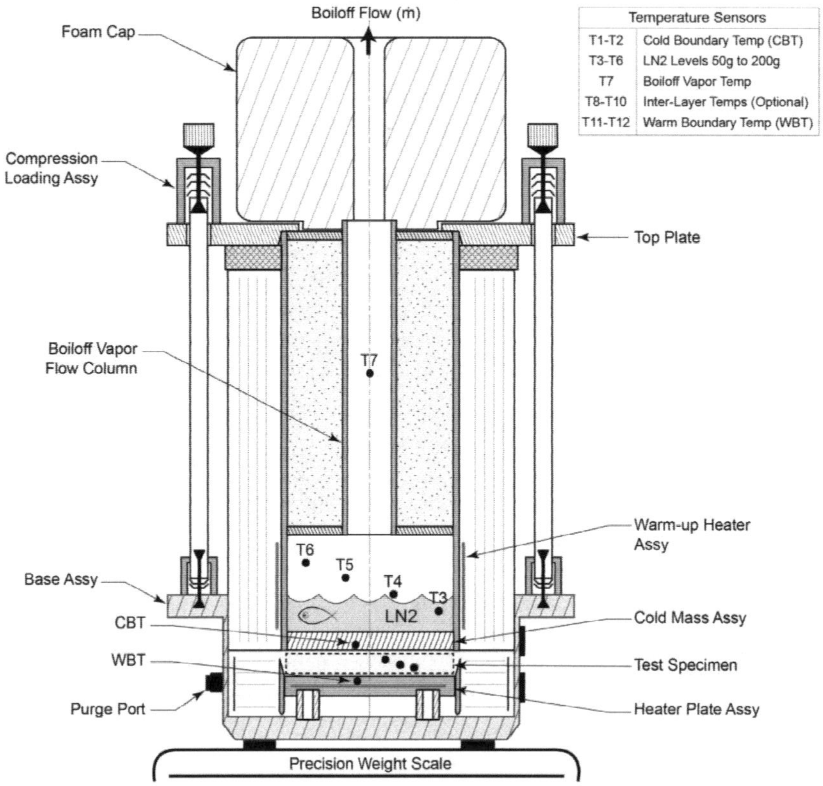

Figure 1. Macroflash: simplified schematic (not to scale). The cold mass assembly is 76-mm diameter by 250-mm tall. A test specimen is 76-mm diameter by 6-mm in thickness.

3. Testing methodology

The test measurement principle is LN_2 boiloff calorimetry where the mass flow rate of nitrogen gas is directly related to the rate of heat energy transmitted through the material [4]. The test conditions are representative of actual-use cryogenic applications; unless otherwise specified, standard is a WBT of 293 K and a CBT of 78 K.

All tests are performed at an ambient pressure gaseous nitrogen (GN_2) condition for consistency. Under steady-state flow conditions, the rate of heat flow through the test specimen is constant at all points through the thickness of the specimen. The boiloff mass flow rate, measured by a weight scale or mass flow meter, allows direct calculation of heat flux and effective thermal conductivity as given in equations 1-3.

$$Q = \dot{m} h_{fg} \tag{1}$$

$$q = \frac{Q}{A_e} \tag{2}$$

$$k_e = \frac{Qx}{(A_e \Delta T)} \tag{3}$$

The heat flow rate (Q) is the product of boiloff mass flow rate ($\dot{m}$) and enthalpy of vaporization of liquid nitrogen (h_{fg}). Knowing the boundary temperatures on either side of the specimen in a set test environment, the heat flux (q) can be calculated. Knowing the specimen's effective heat transfer area (A_e), as well as the thickness of the specimen (x) and the temperature difference (ΔT), the effective thermal conductivity (k_e) is easily calculated.

Based on using a mass flow meter with a full-scale accuracy of 1%, an uncertainty of approximately 5% has been calculated for the flat plate boiloff calorimeter apparatus [2]. For the basic weight scale used (Ohaus Explorer EX10201 with 10.2 kg capacity and 0.1 g readability) the accuracy is comparable. The condensed water vapor and ice is handled by design and procedure. The design of the unit with a series of insulated vent pathways including a top insulation cap allows the boiloff vapor to warm to near ambient temperature to minimize the condensate and frost. The standard procedure includes a topping off to 220 g or more of LN_2 and a cold soak phase from this level down to 100 g where the test measurement phase begins. The test phase is then repeated a minimum of three times to verify consistent results. The methodology provides a repeatability of <1%. The precision and bias for a given test are pending on future round-robin testing of identical test specimens at different laboratories.

The software interface includes test specimen preparations and test condition entries, and automated data analysis, calculations, and reporting. Primary inputs include thickness & diameter (mm), mass (g), WBT set point (K), and interlayer temperatures (if used). Test parameters include compression loading (psi), thermal greases (yes/no), GN_2 purge (yes/no), basis by weight scale or flow meter, and calibration range (LO, FULL, or HI). Primary outputs include k_e calibrated, $\dot{m}$ (g/s), Q (W), q (W/m^2), standard deviation (%), WBT (K), and CBT (K). Specimen outputs include the Structural-Thermal Figure-of-Merit (F_{ST}) and bulk density (kg/m^3).

Boiloff calorimetry provides the ability to test both simple uniform materials and complicated non-homogeneous, anisotropic, composite materials with equal ease [5]. The extent of the temperature difference is dependent on the cryogenic fluid and the heat source used, providing high sensitivity for the accurate measurement of highly thermally insulating materials or structural materials alike. In addition, optional intermediate temperature sensors can be used for calculating thermal conductivity (λ) as a function of mean temperature (T_m) over the wide range of temperatures provided in a single test [6].

Different materials and varied test objectives require an appropriate combination of apparatus and method. The type, thickness, density, flatness, compliance, and strength of the material are important considerations. The test specimens (examples shown in figure 2) are typically 76-mm diameter by 6.35-mm thickness and should be flat and smooth-faced or easily compressible to insure good thermal contact between the heater assembly and the cold mass.

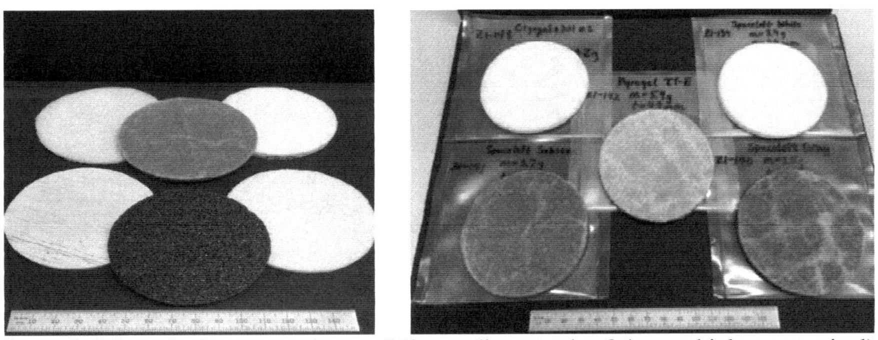

Figure 2. Macroflash test specimens (76-mm diameter by 6.4-mm thickness, typical).

4. Calibration and range of measurement

Initial measurement and calculation of comparative k_e values are calibrated to give the k_e values to be reported. The calibration is based on a linear fit of measured material thermal conductivities to standard reference data for commercial materials with known properties One challenge is the manufacturers' reported values for thermal conductivity are not usually for the boundary temperatures of 293 K and 78 K. Most typically, these data are taken at ambient temperature (around 293 K) with only a small temperature difference imposed by the commercial test instruments generally used. The standard reference data therefore have to be derived by careful research and certain selection from the available peer-reviewed technical literature of the world over many decades [9-11]. Some of these reference data are taken from the absolute cryostat test instruments, Cryostat-100 and Cryostat-500, of the Cryogenics Test Laboratory at NASA/KSC [12-15].

Standard methods of test specimen preparation, testing, thermal performance calculations, data recording and data reporting are all essential. The key parameters include density, compression, specimen thickness, diameter, and whether or not thermal contact grease is used. For standard reference work, many factors must match up including test specimen type, method, condition, etc., as listed below:

1. Test specimen: material, density, size, shape, surface finish
2. Test method: type of apparatus, principle of heat flow calculation, steady-state or transient
3. Test conditions: WBT, CBT, ΔT, calculated T_m, thermal grease, compression load, environment
4. Test results: type of thermal conductivity (λ, k_e, or other), large ΔT or small Δt, use of T_m, and method of calculation

More than 10 years of research and testing of hundreds of test specimens has provided a library of data and a foundation for calibration. Three ranges of calibration are listed: LO (k_e < 50 mW/m-K), FULL (k_e > 50 mW/m-K) and HI (k_e > 500 mW/m-K). The full range curve fit is the default and works well except for materials with a k_e < 30 mW/m-K. The high range curve is a current work in progress.

The reference materials for the calibration curves are found in the literature as discussed. As the Macroflash is a comparative instrument, testing of known materials with known thermal conductivity data under the same conditions is the process of calibration. The known thermal conductivity data under like conditions must come from an absolute instrument, making these data rare and unavailable but for a few materials. The materials used for the preliminary Macroflash calibration are listed in table 2. The compression loading and use of grease (or not) are selected to match, as closely as possible, the manner in which the materials were tested in the literature data. The low-end calibration curve is given in figure 3 while the full range calibration is presented in figure 4. The baseline test conditions for calibration remain as follows: 293 K and 78 K boundary temperatures; ambient pressure gaseous nitrogen environment.

Table 2. Reference materials used in Macroflash calibration.

Range		Compression† psi	Grease ?	Density* kg/m^3	Note
Low-end					
	Cryogel®	0	no	177	Aerogel composite blanket [13]
	SOFI BX-265	2	no	36	Polyisocyanurate spray-on foam insulation (SOFI) [14]
	Divinycell® H-45	2	no	50	Rigid foam insulation [15]
	Foamglas®	5	no	118	Cellular glass [11]
	Balsa wood	5	no	166	In-plane [9]
Full-range					
	Cryogel™	0	no	177	Aerogel composite blanket
	SOFI BX-265	2	no	36	Polyisocyanurate spray-on foam insulation (SOFI) [14]
	Pine wood	5	no	515	White pine (in plane) [10]
	PTFE	5	yes	2,120	Polytetrafluoroethylene [9]
	G10-CR	5	yes	1,939	Glass fiber reinforced composite (in plane) [9]

† Note: 1 psi = 6.895 kPa *As tested

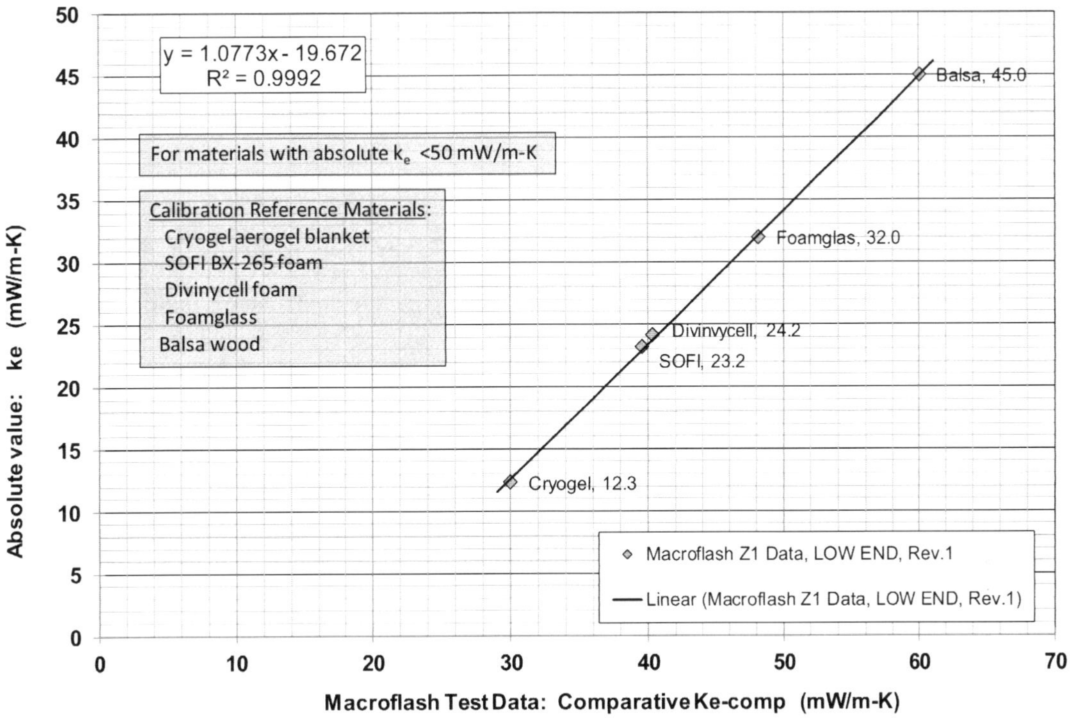

Figure 3. Macroflash calibration curve for effective thermal conductivity: low range (LO).

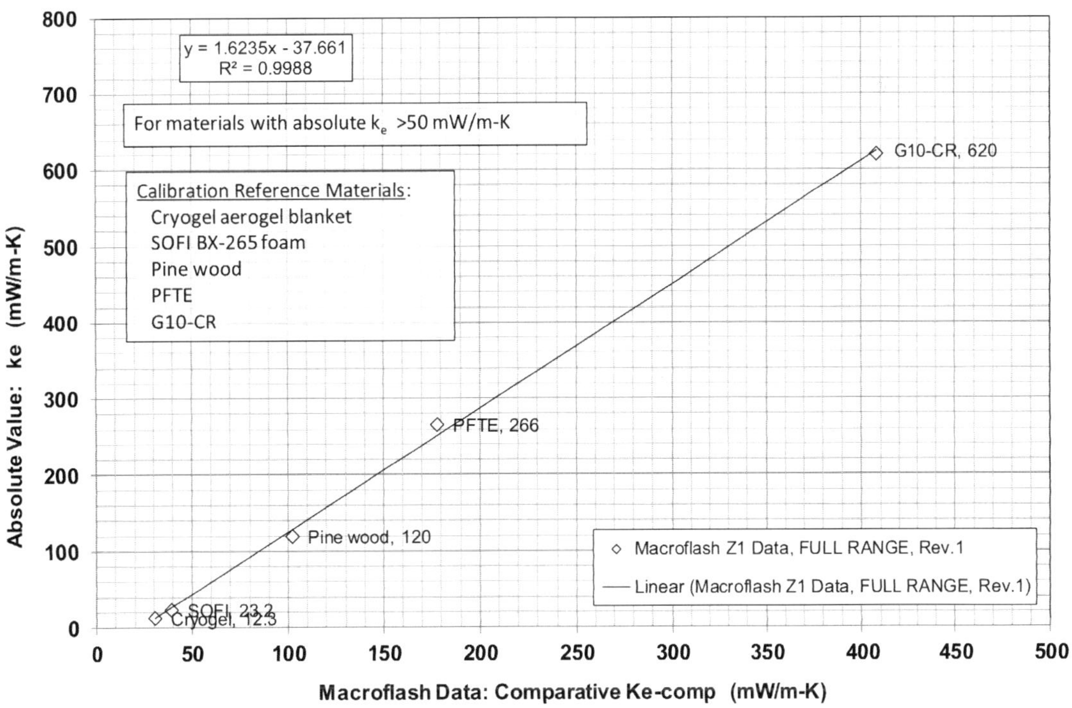

Figure 4. Macroflash calibration curve for effective thermal conductivity: full range (FULL).

5. Example test results and discussion

The test specimens cover a wide variety of materials from thermal insulators, composites, building materials, glasses, ceramics, and novel experimental materials such as syntactic micro-balloon nano-composites [16-17]. Combining values for thermal, mechanical, and density properties allows a rapidly obtained, quantitative screening parameter to identify potentially high performing structural thermal insulation candidates. Thermophysical data for a range of different structural-thermal materials used in cryogenic systems data are given in table 3. Also included are G10 composite, Teflon™, balsa wood, and polyisocyanurate spray foam for general reference and connection to the absolute data from literature [9, 11]. These k_e data were produced using a Macroflash, per ASTM C1774 Annex A4, for boundary temperatures of 78 K (CBT) and 293 K (WBT) and under a compressive load of 34 kPa. The structural-thermal figure-of-merit (F_{ST}) is calculated as in equation (4).

$$F_{ST} = \frac{\sigma}{\rho k_e} \text{ x } 10^6 \quad \left[\frac{\text{K} \cdot \text{m} \cdot \text{s}}{\text{g}}\right] \tag{4}$$

Where ρ is the bulk density in kg/m^3, k_e is the effective thermal conductivity in mW/m-K (with the prescribed CBT and WBT), and σ is the compressive strength in MPa (at ambient temperature).

Table 3. Thermophysical data for structural-thermal materials used in cryogenic systems

Material	†σ MPa	ρ kg/m^3	*k_e mW/m-K	F_{ST} K-m-s/g
G-10 (transverse direction)	448	1,939	467	495
Ultem® 2300 Glass Filled PEI	221	1500	212	695
Teflon™ PTFE	24.1	2,120	253	45
Rohacell® WF-300 PMI Foam (2 psi)	17.8	324	42.1	1,305
Balsa Wood (transverse direction)	7.0	166	45.9	919
AeroZero® polyimide aerogel	1.6	150	28.1	380
Foamglas® Cellular Glass Foam	0.8	118	32.3	210
Divinycell® H45 PVC Foam (2 psi)	0.6	50	23.8	504
Spray Foam Polyiso BX-265 (2 psi)	0.4	37	22.6	483

†At ambient temperature *Boundary temperatures 293 K / 78 K; compressive load 5 psi or as noted.

6. Conclusion

The Macroflash instrument, an easy-to-use flat-plate boiloff calorimeter, has been developed to provide effective thermal conductivity (k_e) data for a wide range of materials from thermal insulation to structural composites. Following the guidelines of standard ASTM C1774 (Annex A4), the Macroflash provides a cost-effective, field-representative methodology to test any material for below-ambient temperature applications to moderately elevated temperature conditions. Non-isotropic, layered, asymmetrical, or inhomogeneous types of materials are all acceptable; solids or powders can be tested; and foams, aerogels, wood, metals, ceramics, glass, or composites are readily tested. From engineered materials development, to research testing, to quality control in manufacturing, this standardized device provides utility for the fields of energy, transportation, aerospace, construction, medical, and environment.

7. References

[1] ASTM C1774 *Standard Guide for Thermal Performance Testing of Cryogenic Insulation Systems* 2013 (West Conshohocken, PA: ASTM International)

[2] Fesmire J, Johnson W, Swanger A, Kelly A, and Meneghelli B 2015 Flat plate boiloff calorimeters for testing of thermal insulation systems, IOP Conf. Series: Materials Science and Engineering 101 012057

[3] Fesmire J 2017 Boiloff calorimetry for measurement of very low heat flows, Part 4: Flat plate boiloff testing of insulation systems, Cold Facts, Cryogenic Society of America, Vol 33, No 1, pp 30-33

[4] Jacobs R B 1964 Theory of boil-off calorimetry *Rev. Sci. Instrum.* **35** No. 7 pp 828–32

[5] Fesmire J 2016 Boiloff calorimetry for measurement of very low heat flows, Part 2: Theory of boiloff calorimetry, Cold Facts, Cryogenic Society of America, **32**, No 5, pp 28-29, 34-35
[6] Demko J A, Fesmire J E, Johnson W L and Swanger A M 2014 Cryogenic insulation standard data and methodologies," *Adv. Cryog. Eng., AIP Conf. Proc.* 1573, pp 463–70
[7] Zarr R 2001 A history of testing heat insulators at the National Institute of Standards and Technology, ASHRAE Transactions, 107(2)
[8] Tye R 1990 Measurement of thermal insulation performance: The challenge of the next decade, Intl Journal of Thermophysics, 11(2)
[9] National Institute of Standards, NIST Cryogenic Materials Property Database Index, https://www.nist.gov/mml/acmd/nist-cryogenic-materials-property-database-index
[10] NIST Standard Reference Database 81, NIST Heat Transmission Properties of Insulating and Building Materials, Version 2.0, https://srdata.nist.gov/insulation/
[11] Flynn T 2005 Cryogenic Engineering, 2nd edition, Marcel Dekker, New York
[12] Fesmire J 2015 Standardization in cryogenic insulation systems testing and performance data, Physics Procedia, 67, pp 1089-1097
[13] Fesmire J, Ancipink J, Swanger A, White S and Yarbrough D 2017 Thermal conductivity of aerogel blanket insulation under cryogenic-vacuum conditions in different gas environments, Advances in Cryogenic Engineering, IOP 278 012198
[14] Fesmire J, Coffman B, Meneghelli B and Heckle K 2012 Spray-on foam insulations for launch vehicle cryogenic tanks, Cryogenics, 10.1016/j.cryogenics.2012.01.018
[15] Fesmire J, et al. 2002 Thermal performance testing of candidate tank insulation materials, Space Launch Initiative / 2nd Generation Launch Vehicles, Final Report to NASA-LaRC, Boeing, Northrup Grumman
[16] Dando K and Salem D 2017 The effect of nano-additive reinforcements on thermoplastic microballoon epoxy syntactic foam mechanical properties, Journal of Composite Materials
[17] Schmid E and Salem D 2015 Fabrication technique and thermal insulation properties of micro- and nano-channeled polymer composites, Acta Astronautica, 116:68-73

IOP Conf. Series: Materials Science and Engineering **756** (2020) 012009 doi:10.1088/1757-899X/756/1/012009

Cryogenic Power Electronics: Press-Pack IGBT Modules

C Park[1*], M J Mauger[1], T Damle[1], J Huh[1], S Steinhoff[2], and L Graber[1]

[1]School of Electrical and Computer Engineering, Georgia Institute of Technology, Atlanta, USA
[2]Littelfuse - IXYS UK, Chippenham, UK

*chanyeop.park@gatech.edu

Abstract. With the goal of enabling high-power-density cryogenic power converter technology and superconducting power applications for future aircraft and shipboard power systems, the dynamic and static performances of a press-pack IGBT module (T0160NB45A) at ambient and cryogenic conditions are reported. Compared to the wire-bond IGBT's, press-pack IGBT's are more suitable for cryogenic conditions as they do not have bonded connections and use fewer materials types, which reduces the risk of coefficient of thermal expansion (CTE) mismatch. The study has been conducted with a cryogenic testbed that provides a condensation-free condition during and after tests, which is essential for the preservation of the physical properties of IGBT's being tested. The dynamic performance characterization results show that the switching speeds of both turn-on and turn-off are improved with substantially reduced tail current and increased dv/dt at cryogenic conditions. Moreover, the static performance characterization results show a reduction in collector-emitter voltage drop, indicating higher conductivity of the IGBT at cryogenic conditions. Furthermore, the impact of clamping force and gate lead length on the press-pack IGBT's dynamic characteristics is reported. The findings of this study suggest that press-pack IGBT modules are suitable for cryogenic operation.

1. Introduction

One of the promising candidate technologies that is essential for the materialization of high-power-density power systems for future aircraft and shipboard applications is the superconductor technology. While superconducting cables [1,2], motors [3], generators [4,5], dielectric media and refrigeration systems have been researched extensively and are well understood [6–11], utilizing power electronics devices at cryogenic conditions is yet an existing challenge [12]. Power semiconductors, capacitors, and inductors are the key current-carrying components constituting any power electronics. The main challenge associated with operating these devices at cryogenic conditions originates from the mismatch of the coefficient of thermal expansion (CTE) among various materials comprising the devices [13]. Wire-bond IGBT modules, which are usually filled with gel, are prone to delamination at cryogenic conditions [14]. Electrolytic capacitors have no chance of surviving cryogenic conditions as the main charge carriers are ions, which will completely freeze. Inductors for cryogenic use will have to be designed to not have magnetic cores as eddy current loss will become substantial at cryogenic temperatures [15]. One of the ways of avoiding all of these challenges is by operating the power converter at ambient conditions while the rest of the system components are operated at cryogenic conditions. Since such solution is not ideal for maximizing the overall system efficiency, cryogenic compatible power electronics devices need to be developed for the materialization of a fully cryogenically operating power system that provides maximum thermal and system efficiency. The efficiency enhancement is achieved by eliminating the current flow between cryogenic and ambient sections by operating all components in the cryogenic system. That is, significant amount of thermal loss at the cryogenic to ambient joints will be eliminated by operating all devices in a single cryogenic cooling loop. In addition to the enhanced system efficiency gained by reducing the cooling cost, the performance of the IGBT's are expected to be enhanced at cryogenic conditions. Furthermore, the cryogenic operation provides the possibility of replacing the conventional conductors with thinner ones or with superconductors.

In this study, we focus on the cryogenic performance of press-pack IGBT modules. Although press-pack IGBT's are not originally designed for cryogenic use, press-pack modules have the most cryogenically compatible design among IGBT packaging technologies given the simplicity of their design and the lowest variety of materials comprising the devices. With a long term plan of modifying existing technologies and developing cryogenically compatible IGBT modules, we investigated the cryogenic performance of an off-the-shelf press-pack IGBT technology.

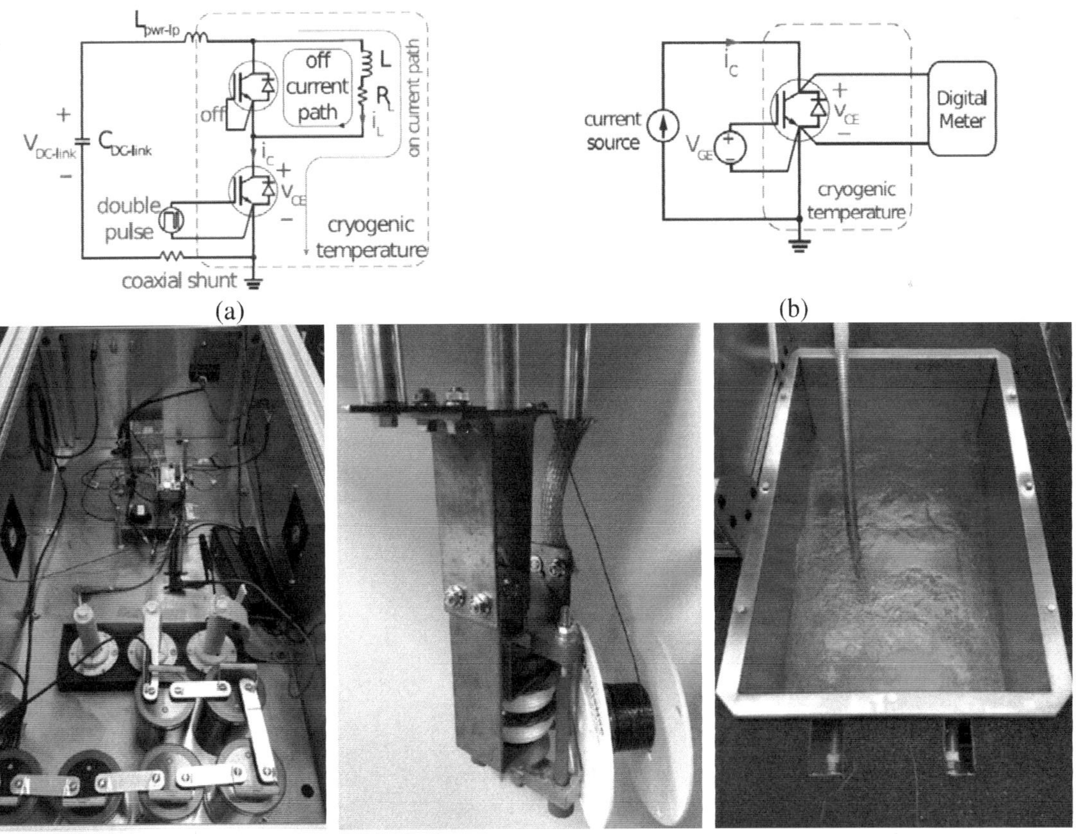

Figure 1. Cryogenic testbed used for characterizing the dynamic and static performance of the press-pack IGBT. (a) Double pulse test circuit [12], (b) Static test circuit [12], (c) Ambient section, (d) Press-pack IGBT and inductor, (e) Cryogenic section being filled with LN_2.

IOP Conf. Series: Materials Science and Engineering **756** (2020) 012009 doi:10.1088/1757-899X/756/1/012009

2. Experimental Setup

2.1. Cryogenic Testbed
For the dynamic and static characterization of the press-pack IGBT modules at cryogenic conditions, an experimental setup that consists of both cryogenic and ambient sections have been developed as shown in Figure 1. The cryogenic section is where the IGBT and the inductor are immersed in a bath of liquid nitrogen (LN_2). The ambient section is where the capacitor bank, microcontroller, and gate driver circuit are placed. For safety, we surrounded the ambient section with electrically grounded stainless-steel sheets, on which the control signal and the measurement terminals are mounted. The cryogenic and ambient sections are electrically connected by three conductor rods that are connected to the DC-link capacitors and the stack of IGBT modules. The three rods were fastened to a custom design nylon bushing mounted on the bottom plate of the ambient section via collar clamps, and the IGBT and the inductor were connected at the bottom of the rods. We mounted a large stainless-steel sink on a movable cart and thermally insulated the stainless steel container by covering it with foam boards. We used foam board adhesives to attach the boards on the stainless steel surface but also wrapped the foam boards with an elastic band to compensate for the possible loss of adhesion of the glue on cryogenic surfaces. Except for the stainless steel LN_2 container, all parts were mounted on the top ambient section, which also functioned as a lid.

2.2. Modification of the IGBT Module
For conventional applications, press-pack IGBTs are hermitically sealed and filled with a dielectric gas to deal with the high electric fields at the IGBT and diode die edge terminations. However, these gases would be incompatible with cryogenic conditions as they condense at low temperatures. Therefore, for the cryogenic experiments, the top plates of the IGBT devices were opened such that LN_2 could be filled as in Figure 2 (b). In this case, LN_2 serves as a dielectric medium, but also as a cooling medium for the chips. Replacing the dielectric gas with LN_2 does not pose significant impact in terms of dielectric strength nor in terms of cooling efficiency since LN_2 is both a good dielectric and cooling medium.

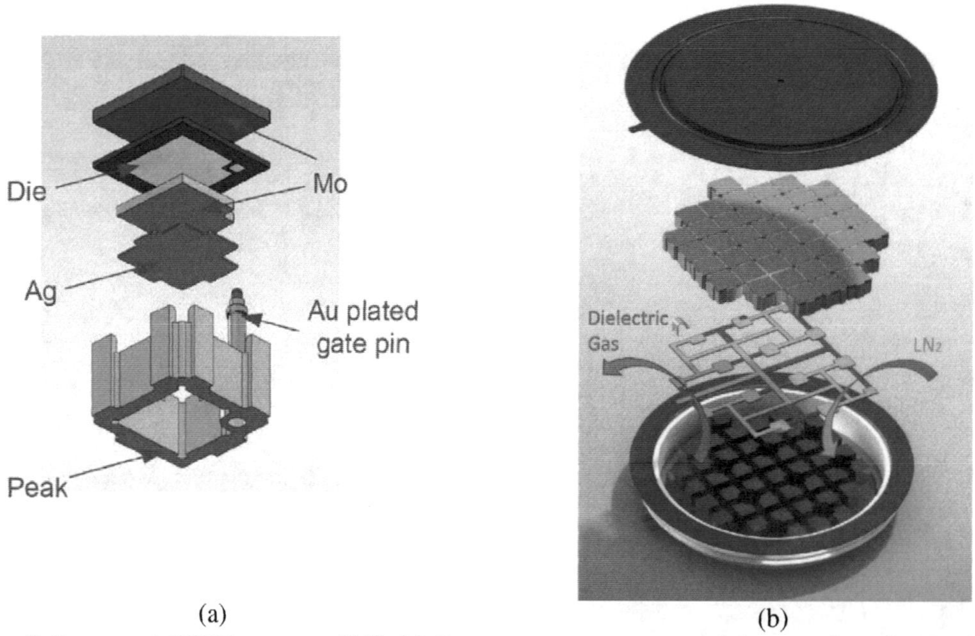

Figure 2. Press-pack IGBT structure [16]. (a) Cassette structure: many of these are installed in parallel in IGBT modules, (b) IGBT module: numerous cassettes are installed in parallel, the top plate was opened to displace the dielectric gas with liquid nitrogen.

2.3. Preventing Thermal Shock

The cooling process of a test subject should not be too fast. Rapid cooling such as dipping the test subject into a bath of LN_2 can lead to uneven thermal expansion/contraction within a device, which can generate substantial mechanical stress sufficient enough to permanently damage the device. Figure 3 shows a damaged IGBT die due to cold shock. The damage occurred when the press-pack IGBT module was immersed into a bath of prefilled LN_2. Initially, the device blocked the DC-link voltage of 300 V, but became short circuited when the gate signal of 15 V was applied. The phenomena suggest that the device was exposed to a thermal shock that cracked and weakened the semiconductor die, which were completely punctured during the first few turn on and off process. To prevent damage caused by thermal shock, the press-pack IGBT modules should be placed in the cryogenic container first, and then LN_2 should be filled gradually from the bottom of the container. Note that the level of precaution for the cooling process would vary depending on the test subject. In the envisioned applications where the power electronics would be operated in a cryostat, the devices would be cooled down progressively with no risk of cold shock.

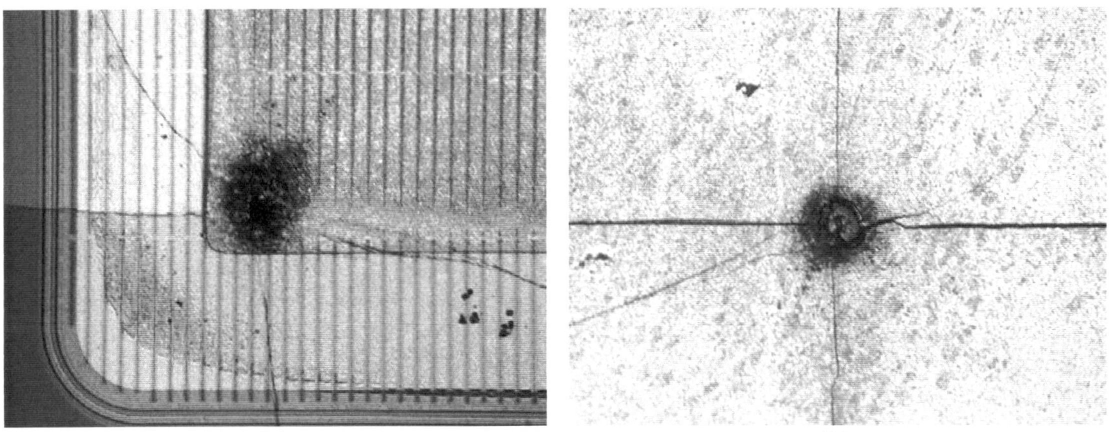

(a) (b)

Figure 3. Photo of a damaged IGBT die due to cold shock. The cracks occurred during a rapid cooling process, and the press-pack IGBT module failed short when 15 V gate signal was applied with the DC-link voltage of 300 V. (a) Emitter side damage, (b) Collector side damage.

2.4. Preventing Condensation

Preventing condensation is significant for maintaining the original physical properties of IGBT's over numerous iterations of cryogenic tests and experiments. In our earlier experiments, we have observed condensation of the IGBT modules deteriorating the performance of the IGBT's over many iterations of cryogenic tests. Figure 4 is a photo taken from the side of an IGBT cassette. A thick molybdenum platelet, which has a similar CTE value with silicon, is stacked on top of an IGBT die. As shown in Figure 4 (a), condensation generates whiskers on the molybdenum layer and causes damage at the interface of the IGBT die. To prevent condensation, we have equipped our new testbed with additional gas fittings that are used for filling the cryogenic section of the testbed with gaseous nitrogen while LN_2 is being drained. The cryogenic section is kept mildly pressurized above ambient pressure with nitrogen gas to prevent air from leaking into the system. The new measure has been proven to be effective for preventing condensation, and we have been able to conduct multiple iterations of cryogenic tests without noticeable performance degradations.

(a) (b)

Figure 4. Corroded IGBT cassettes due to condensation. This picture was taken from the side of the cassette after multiple iterations of cryogenic tests without properly executing the condensation prevention procedure. Whiskers formed on molybdenum plates. Silver layer detached from the IGBT die. (a) After corrosion. (b) Before corrosion.

3. Experimental Results

3.1. Dynamic Performance

The dynamic performance of a 4.5 kV, 160 A rated press-pack IGBT with anti-parallel diode (T0160NB45A) is assessed by measuring the switching voltage and current waveforms following a standard double pulse test procedure. Figure 1 (a) is the circuit diagram of a double pulse test experiment. The DC-link electrolytic capacitors, C_{dc}, providing the bulk of the energy to the pulse, are supplemented by a film capacitor, C_{bypass}, of lower parasitic inductance and resistance to offer a low impedance path to the switching currents from the stack. In this configuration, both the bottom IGBT module, and the top IGBT module's anti-parallel diode, are being tested although the focus of this paper will be on the behavior of the bottom IGBT. The gate of the top module IGBT is shorted to avoid spurious turn on of the switch while the gate of the bottom module IGBT is driven between -10 – 15 V through a commercial gate driver and microcontroller. The inductor L_{load} is first pre-charged to the target load current I_{load} by turning on the IGBT for a duration $t_{precharge}$, fixed by the DC-link voltage V_{dc} and the load inductance L_{load} as follows:

$$t_{precharge} = \frac{L_{laod} I_{target}}{V_{dc}}$$

At the end of the pre-charge, the IGBT is switched off and the load current freewheels through the anti-parallel diode of the top IGBT. The lower IGBT under test is subsequently turned on and off under the target load current and DC-link voltage and the corresponding switching waveforms are recorded to analyze the dynamic switching characteristics of the power device under test. The collector-emitter voltage of the IGBT under test is measured by a high-voltage differential probe connected to the bushing while the collector current is measured via a high precision shunt resistor inserted in the current path at room temperature.

Figures 5 and 6 are the voltage and current switching waveforms of the press-pack IGBT module. In both figures, the solid lines represent the measurements taken in LN_2 while the broken lines represent those taken at room temperature. Since the IGBT modules were opened to release the dielectric gas for the cryogenic experiments, we set the DC-link voltage for the double pulse test no higher than 1.5 kV to prevent unwanted dielectric failures. It should be noted that the voltage limitation of 1.5 kV was taken to ensure safety during the room temperature test, in which atmospheric pressure air is the dielectric media instead of the dielectric gas. In fact, we could have tested the device with higher DC-link voltage levels in LN_2, which is a good dielectric media, but stopped at 1.5 kV for a one-on-one comparison with the room temperature measurements. The standardized switching metrics [17] that are the current rise time at turn-on and fall time at turn-off, t_r and t_f, respectively, are averaged for the different tests at room temperature and in LN_2 and summarized in Table I. The average *dv/dt* at turn-on and turn-off are also provided, along with the average percentage change of the metrics at 77 K with the respect to the results at room temperature.

Table 1. Average switching metrics of the IGBT at Ambient and Cryogenic.

Condition	Average *dv/dt* Turn-on [V/μs]	Average *dv/dt* Turn-off [V/μs]	Average t_r Turn-on [μs]	Average t_f Turn-off [μs]
Ambient (293 K)	-558	1670	0.67	1.21
Cryogenic (77 K)	-783	2614	0.68	0.15
% Change at 77 K	40.5 %	56.6 %	1.5 %	-87.6 %

Figure 5 shows higher magnitudes of *dv/dt* at both turn on and turn off in LN_2. The turn-on voltage waveforms, Figure 5 (b) are of particular interest, with a sharp voltage drop, until ~40 % of the nominal DC-link voltage, followed by a lower *dv/dt* phase at lower voltages, identical to the switching at room temperature. These higher *dv/dt* are confirmed by Table 1 with an average *dv/dt* 40.5 % and 56.6 % higher at turn-on and turn-off, respectively, at 77 K. The *dv/dt* are lower at turn-on because of the configuration of the gate driver turn on channel following the standard practice to slow down the turn on of the IGBT to reduce the reverse recovery current from the opposite diode. The current waveforms at turn off shown in Figure 6 (c) demonstrates that the tail current is dramatically reduced at 77 K, owing to a higher recombination rate (*i.e.*, shorter carrier lifetime) of the carriers at cryogenic temperature [18–20]. This lead to a much faster fall time in LN_2 than at room temperature, with an average t_f reduced by as much as 87.6 % as in Table I. The impact on the current rise time t_r, however, seems to be very limited with very similar t_r in both cases. Further, the current waveforms at turn on, Figure 6 (b), show reduced reverse recovery currents from the antiparallel diodes at cryogenic temperature. This suggests that the Si diode also demonstrates better switching characteristics in LN_2, mostly because of reduced stored charges [18]. Finally, both voltage and current waveforms show that the turn-on process is delayed at 77 K, but the turn-off process occurs earlier than at room temperature. The phenomena are mainly caused by the increased threshold voltage of the press-pack IGBT at 77 K. This increase in the threshold voltage is mostly due to carrier freeze-out at cryogenic temperatures, as for any FET-based device [20,21]. This gives a better immunity to the common spurious turn-on in half-bridge configurations [22] and would potentially allow for a unipolar driving of the gate. The larger current magnitude at 77 K shown in Figure 6 (a) is mainly caused by the increased conductivity of the inductor which was also immersed in LN_2.

IOP Conf. Series: Materials Science and Engineering **756** (2020) 012009 doi:10.1088/1757-899X/756/1/012009

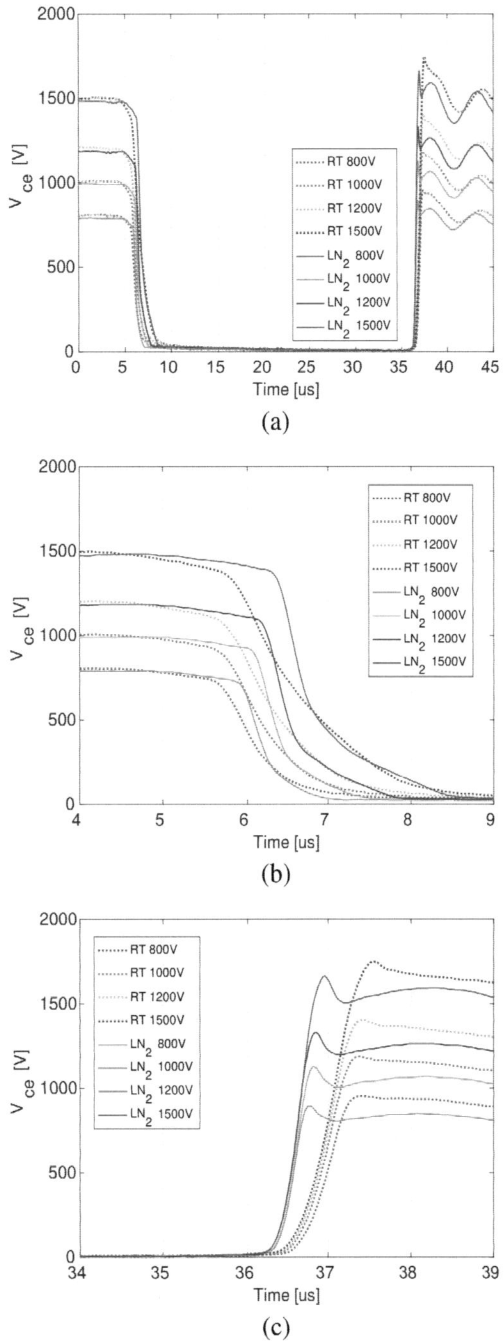

Figure 5. Double pulse test results: voltage waveforms of the press-pack IGBT module at cryogenic and ambient conditions. (a) Turn-on and off waveform, (b) Close up view of turn-on, (c) Close up view of turn-off

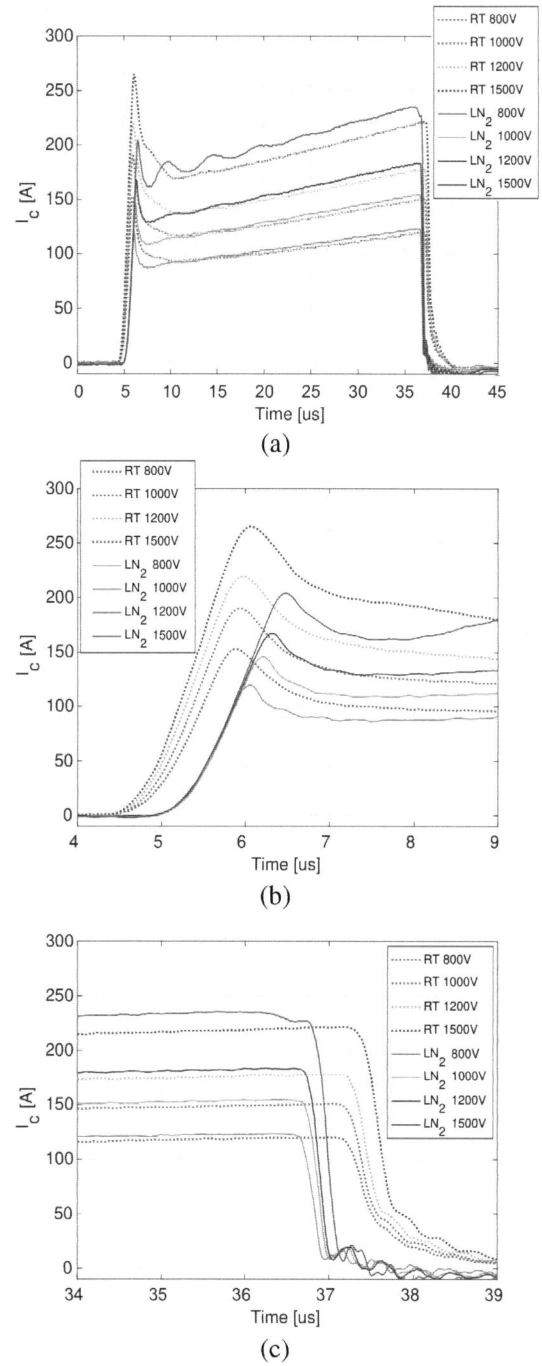

Figure 6. Double pulse test results: current waveforms of the press-pack IGBT module at cryogenic and ambient conditions. (a) Turn-on and off waveform, (b) Close up view of turn-on, (c) Close up view of turn-off

It is apparent that the IGBT not only operates normally at 77 K but also demonstrates faster switching dynamics than at room temperature, as shown from both the voltage and current waveforms in Figures 5 and 6, and the average switching metrics of Table I. The results suggest tremendous advantages for the cryogenic operation of power electronics as faster switching results in lower switching losses, and potentially allows for an increase in the switching frequency of the converter, and in turn a reduction in the size of the passive elements (*L* and *C*).

3.2. Static Performance

The static performances of the IGBT under test are assessed by measuring the collector-emitter saturation voltage $V_{ce(sat)}$ for different collector current I_c using the test circuit shown in Figure 1 (b). As shown in Figure 7 (a), $V_{ce(sat)}$ increases with higher collector current I_c both at room temperature and 77 K. While V_{ce} at room temperature is lower than that at 77 K when I_c is below 50 A, V_{ce} at room temperature continues to increase more rapidly than V_{ce} at 77 K when I_c is above 50 A. This is due to the IGBT's intrinsic structure and two conflicting phenomena observed at cryogenic conditions [19,21].

(i) Carrier freeze-out at cryogenic temperatures increases threshold voltage in the intrinsic PN junction region of the IGBT. This results in an increased knee-forward voltage in the I_c vs $V_{ce(sat)}$ characteristic of the device.

(ii) A large increase of the carrier mobility in the inversion layer of the IGBT at cryogenic conditions lead to much higher conductivity modulation in this region.

The total collector-emitter saturation voltage of the IGBT is the sum of the voltage drops across the two aforementioned regions. At 77 K and low collector currents (< 50 A), effect (i) is predominant and leads to the observed higher $V_{ce(sat)}$ than at room temperature. For higher collector currents (> 50 A), effect (ii) is dominant than (i) and lower $V_{ce(sat)}$ values are observed at 77 K than at RT [21]. As I_c keeps increasing, the reduction in $V_{ce(sat)}$ at 77 K becomes increasingly dominant. This is a significant finding showing that the current caliber of IGBT might be significantly increased at cryogenic temperature, as much as 50 % for the present device, for the same die area.
Figure 7 (b) shows decreasing V_{ce} values with increasing gate voltage V_{ge}. When the room temperature data are compared with those measured in LN_2, gate threshold voltage at 77 K is higher mainly due to carrier freeze-out.

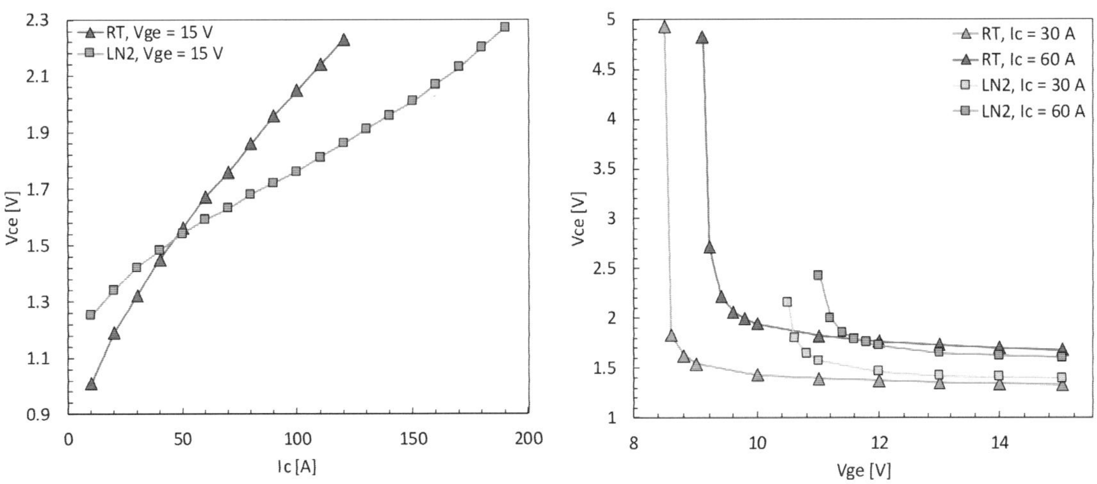

(a) (b)

Figure 7. Static performance test results. (a) V_{ce} as a function of I_c, (b) V_{ce} as a function of V_{ge}.

3.3. Impact of Clamping Force

The electrical connection of press-pack IGBT modules are formed by clamping mechanisms. Proper clamping force must be maintained throughout the lifetime of press-pack IGBT modules to ensure device performance. In this section, we discuss the impact of loose clamping force on press-pack IGBT modules as clamping force may be changed at cryogenic conditions due to thermal contraction. For the study, we manually adjusted the clamping force of the press-pack IGBT. Due to the lack of a clamping force measurement system, we only performed the double pulse test under tight and loose clamping force conditions. In this section, we decided to focus on the turn-off current waveform because it showed the most pronounced effect of the clamping force among other sections of the current and voltage waveform. As Figure 8 shows, the strength of clamping force has an impact on the switching performance of the IGBT. Low clamping force resulted in a substantial amount of current resonance during turn-off. As we tightened the clamp, the resonance reduced. Depending of the type of the clamping mechanism and the material that consists the clamping system, such reduction in the clamping force at cryogenic conditions may occur in various magnitudes, and pose a negative effect on the switching performance of IGBT's. It should be noted that the results shown in Figure 8 were obtained at room temperature to simply show the impact of clamping force on the switching performance of the press-pack IGBT. Hence, further study is necessary for the exact explanation on the clamping force variation and its impact on switching performance at cryogenic conditions.

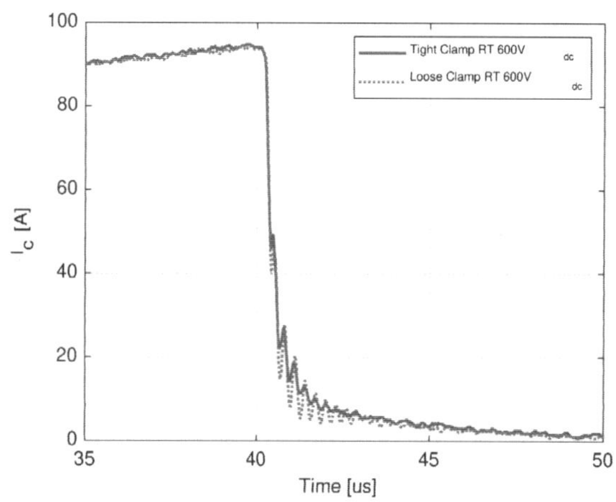

Figure 8. Impact of clamping force on the current turn-off characteristics. The measurements were taken at room temperature with a DC-link voltage of 600 V.

3.4. Short vs. Long Gate Lead

In an actual power application, long gate signal leads may be required if the control board and the gate drive board are located far away from the IGBT modules. The distance is likely to be longer in cryogenic power applications as there are ambient and cryogenic sections in the systems. Therefore, in this section, we discuss the impact of gate lead length on the gate signal reaching the IGBT side because the parasitic resistance and inductance of the gate lead could distort the gate signal when a long gate lead is used. Figure 9 shows the experimental results obtained from two twisted gate leads of different lengths. Figures 9 (a) and (b) show voltage at the gate driver output and that across the gate and emitter of the IGBT. As shown in the figures, the voltage across the gate and emitter of the IGBT is slightly lower in the case of the 101-inch-long gate lead. The difference is mainly attributed to the higher resistance in the longer lead. Although the impact of the gate lead length seems insignificant, even longer gate leads need to be tested to draw a clearer conclusion.

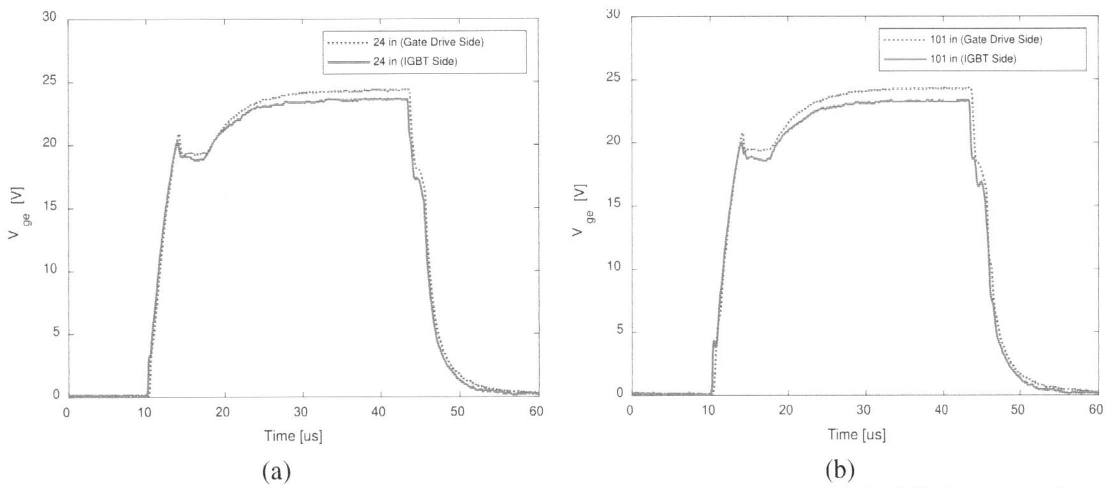

Figure 9. Gate voltage waveform comparison between the gate driver side and the IGBT side. (a) 24 inch (0.61 m) gate lead, (b) 101 inch (2.56 m) gate lead.

4. Conclusions

In this paper, we investigated the cryogenic compatibility and the electrical performance of a press-pack IGBT module. Dynamic and static performance, the impact of cold shock, corrosion caused by condensation, and the impact of clamping force and gate lead length on the electrical performance of the IGBT device were discussed. Both dynamic and static electrical performances were enhanced at cryogenic conditions, and the potential mechanical and thermal challenges were partially addressed. The findings of this study show promise for the successful development of cryogenic power electronics devices as well as the materialization of a fully-cryogenically operating power system.

References

[1] Honjo S, Mimura T, Kitoh Y, Noguchi Y, Masuda T, Yumura H, Watanabe M, Ikeuchi M, Yaguchi H and Hara T 2011 Status of superconducting cable demonstration project in Japan *IEEE Trans. Appl. Supercond.*

[2] Pamidi S, Kim C H and Graber L 2015 High-temperature superconducting (HTS) power cables cooled by helium gas in Superconductors in the Power Grid: Materials and Applications *Cambridge, UK Woodhead Publ. Ltd* **21** 225–60

[3] Eckels P W and Snitchler G 2005 5 MW High Temperature Superconductor Ship Propulsion Motor Design and Test Results *Nav. Eng. J.* **117** 31–6

[4] Snitchler G, Gamble B, King C and Winn P 2011 10 MW Class Superconductor Wind Turbine Generators *IEEE Trans. Appl. Supercond.* **21** 1089–92

[5] Gamble B, Snitchler G and MacDonald T 2011 Full Power Test of a 36.5 MW HTS Propulsion Motor *IEEE Trans. Appl. Supercond.* **21** 1083–8

[6] Park C, Cheetham P, Pamidi S, Kim C H, Viquez J G and Graber L 2017 A versatile modeling technique for predicting dielectric strength improvements in gas mixtures for superconducting applications *IEEE Trans. Dielectr. Electr. Insul.* **24** 2755–64

[7] Park C, Graber L and Pamidi S 2017 The dielectric properties of gaseous cryogen mixtures of He, H2, Ne, and N2 in a temperature range of 50–80 K at pressures up to 2.0 MPa *J. Appl. Phys.* **121** 83304

[8] Park C, Pamidi S and Graber L 2017 The critical electric field of gas mixtures over the extended range of cryogenic operating conditions *J. Appl. Phys.* **122** 153301

[9] Park C, Wei J, Cheetham P, Kim C H, Pamidi S and Graber L 2018 The influence of temperature on the dielectric strength of gaseous cryogens *Proc. IEEE Int. Conf. Prop. Appl. Dielectr. Mater.* **2018-May** 164–7
[10] Park C, Pamidi S and Graber L 2018 Evaluating the dielectric strength of helium-nitrogen gas mixtures by plasma parameter measurements *Phys. Plasmas* **25** 43520
[11] Cheetham P, Ravindra H, Stamm T, Park C, Kim C, Graber L, Steurer M and Pamidi S High Temperature Superconducting Power Cables for MVDC Power Systems of Navy Ships
[12] Graber L, Saeedifard M, Mauger M J, Yang Q, Park C, Niebur T, Pamidi S V and Steinhoff S 2017 Cryogenic Power Electronics at Megawatt-Scale Using a New Type of Press-Pack IGBT *IOP Conf. Ser. CEC-ICMC*
[13] Park C, Wei J, Singh S, Narra S, Imperatore M and Graber L 2018 The Characteristics of Film Capacitors at Room Temperature and in Liquid Nitrogen *AIAA/IEEE Electr. Aircr. Technol. Symp. AIAA Propuls. Energy Forum*
[14] Mainka K, Thoben M and Schilling O 2011 Lifetime calculation for power modules, application and theory of models and counting methods *Proc. 2011 14th Eur. Conf. Power Electron. Appl.* 1–8
[15] J. H. Claassen, "Inductor design for cryogenic power electronics," in IEEE Transactions on Applied Superconductivity, vol. 15, no. 2, pp. 2385-2388, June 2005
[16] F. Wakeman ; J. Pitman ; S. Steinhoff 2016 Long term short-circuit stability in Press-pack IGBTs *18th European Conference on Power Electronics and Applications* pp 1–10
[17] IEC 2007 IEC Standard 60747-9: Semiconductor devices – Discrete devices –Part 9: Insulated-gate bipolar transistors (IGBTs) *IEC*
[18] Baliga R S and B J 1998 *Cryogenic Operation of Silicon Power Devices* (Boston, MA: Springer US)
[19] A. Caiafa, A. Snezhko, J. L. Hudgins, E. Santi and R P 2004 IGBT operation at cryogenic temperatures: non-punch-through and punch-through comparison *in Power Electronics Specialists Conference* pp 2960–6 vol. 4
[20] A. J. Forsyth, S. y Yang, P. A. Mawby and P I 2006 Measurement and modelling of power electronic devices at cryogenic temperatures *IEE Proceedings - Circuits, Devices and Systems* pp 407–415 vol. 153, no. 5
[21] Baliga R S and B J 1992 Cryogenic operation of asymmetric N-channel IGBTs *Proceedings of the 4th International Symposium on Power Semiconductor Devices and ICs* pp 188–193
[22] S. Jahdi, O. Alatise, J. Ortiz-Gonzalez, P. Gammon, L. Ran and P. Mawby, "Investigation of parasitic turn-ON in silicon IGBT and Silicon Carbide MOSFET devices: A technology evaluation," 2015 17th European Conference on Power Electronics and Applications (EPE'15 ECCE-Europe), Geneva, 2015, pp. 1-8

IOP Conf. Series: Materials Science and Engineering **756** (2020) 012010 doi:10.1088/1757-899X/756/1/012010

Cryogenic Power Electronics: Capacitors and Inductors

C Park*, O Obadolagbonyi, and L Graber

School of Electrical and Computer Engineering, Georgia Institute of Technology, Atlanta, GA 30332, USA
*chanyeop.park@gatech.edu

Abstract. Capacitors and inductors that are suitable for cryogenic use are presented in this study. With the long-term goal of developing power electronic converters for cryogenic use, we studied various off-the-shelf metalized polypropylene film capacitors at cryogenic and ambient conditions. Capacitance and breakdown voltage of the film capacitors were the main parameters measured at room temperature and in liquid nitrogen. The results show that the material of dielectric film and the method of packaging play a role in the characteristics of breakdown voltage and capacitance in cryogenic conditions. In general, both capacitance and voltage rating of the capacitors were comparable if not better at cryogenic conditions. Moreover, with the long-term goal of developing inductors for cryogenic applications, we built and tested inductors with and without a magnetic core. The resistance, inductance, maximum current, and energy density were measured and compared. According to the results, the energy density of the cryogenic inductor without a magnetic core can be designed to be much higher than its room temperature counterpart mainly due to the superior cooling power of liquid nitrogen and the reduced resistivity of the windings at cryogenic temperatures.

1. Introduction

Superconducting technology is one of the promising candidate technologies that enables to build power systems of high power density, which is essential for future aircraft and shipboard applications. Superconducting cable, motor, and generator technologies have been demonstrated [1–4]. Also, the cryogenic dielectric and cooling media for these applications have been studied extensively [5–9]. However, developing power electronics devices suitable for cryogenic conditions is yet an existing challenge [10,11]. In particular, the experience with the development of cryogenic passive components is incomplete. The mismatch of the coefficient of thermal expansion (CTE) among various layers and parts of materials comprising passive components, and the increased eddy current loss of magnetic cores are the main challenges. Electrolytic capacitors fail to function at cryogenic conditions as the main charge carriers are ions, which are immobile due to the frozen electrolyte at such conditions. In addition, as eddy current loss in magnetic cores becomes substantial at cryogenic temperatures due to the enhanced conductivity of the material, cryogenic inductors will have to be designed either with warm cores or without cores at all. For the development of a fully cryogenically operating power system, which is expected to show substantially improved system efficiency and life expectancy, the successful development of cryogenically compatible passive components is essential.

In this study, we discuss the capacitor and inductor technologies. For the capacitors, we took several off-the-shelf metalized polypropylene film capacitors and tested them at both room temperature and at 77 K. We focused on observing the variation of capacitance and breakdown voltage as a function of temperature, dielectric film material, and packaging method. To understand the feasibility of developing cryogenic inductors without a magnetic core, we built and tested two prototype inductors; one with and one without a magnetic core.

2. Cryogenic Capacitor

We took several types of metallized film capacitors and tested in both ambient and cryogenic conditions. Electrolytic capacitors were not investigated as they do not work at low temperatures, at which ionic charge carriers freeze out. Capacitance and voltage withstand capabilities were measured a room temperature and in a bath of liquid nitrogen as shown in figure 1. Initially, small-scale metallized film capacitors were tested with the test setup [11]. Capacitance was measured with an LCR meter and the voltage withstand capability was measured based on the self-healing process of metallized film capacitors that is achieved when current is limited while ramping up the voltage of the power supply [11]. The results showed increased capacitance and voltage withstand capabilities in liquid nitrogen. The increased capacitance can be attributed to the variation in permittivity and the shrinkage of the dielectric layers. As the results were promising, we decided to continue the investigation with larger film capacitors. Figure 1 (a) shows the KEMET C44HLGR6600AASJ film capacitor that was tested, and Figure 1 (b) shows the experimental setup used for the study.

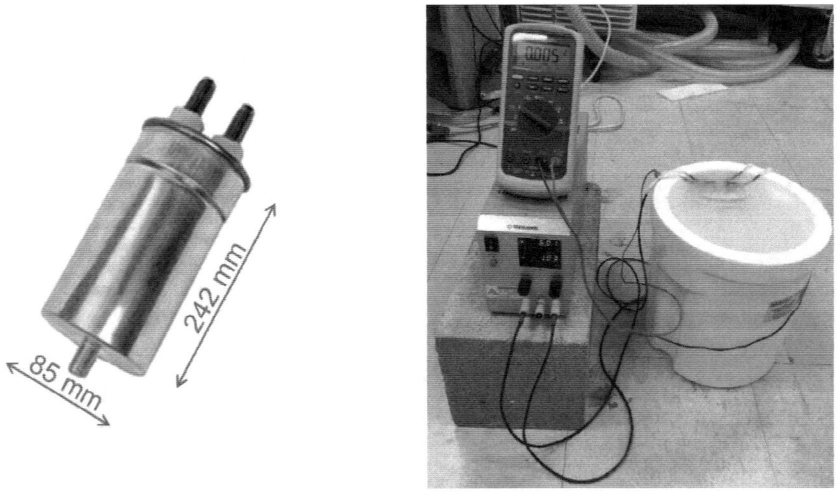

(a) (b)

Figure 1. Cryogenic testbed used for characterizing the capacitance and the withstand voltage of a polypropylene metallized film capacitor. (a) KEMET C44HLGR6600AASJ (600 μF, 700 V_{dc}), (b) Experimental setup.

The capacitance was measured with an LCR meter and the voltage was tested up to the voltage of 600 V. Safety concerns based on the large size of the capacitor prevented testing up to the breakdown voltage of the capacitor. The capacitor test results are shown in Table 1. Consistent with the small-scale capacitor test results of our previous study [11], the large film capacitor also showed increased capacitance and no sign of degradation in voltage withstand capability at cryogenic temperature.

Table 1. Cryogenic Capacitor Test Results Measured at 1 kHz.

Condition	Capacitance [μF]	Tested DC Voltage [V]
Ambient (RT)	597	600
Cryogenic (LN_2)	620	600

3. Cryogenic Inductor

The main challenge in developing a feasible cryogenic inductor is not being able to use magnetic cores, which generates a substantial amount of loss due to increased eddy currents. The only ways to avoid such increased loss are (a) by designing cryogenic inductors without magnetic cores, which in turn compromises the achievable inductance of the inductor, or (b) keep the core warm, which adds complexity and reduces the power density. Nevertheless, there are several advantages of operating inductors in cryogenic conditions, which are the enhanced conductivity of the windings, increased cooling power provided by liquid nitrogen, lower mass, and higher dielectric strength. The reduced resistivity of the windings allows higher current densities. The substantial amount of heat transfer provided by liquid nitrogen keeps the resistivity of the windings low at high currents. The reduction in mass achieved by not using a magnetic core is essential for applications that require light weight. The higher dielectric strength provided by liquid nitrogen, which has a dielectric strength typically higher than 35 kV/mm [12] (*i.e.*, more than an order of magnitude higher than air), allows cryogenic inductors to be operated at medium to high voltage power applications.

(a) (b)

Figure 2. Inductance comparison. (a) A total gap distance of 6.76 mm was obtained by inserting two 3.38 mm thick G10 plates between the upper and lower ferrite magnetic cores. The resulting inductance was 244 μH. (b) The inductance of the coreless inductor was 21.3 μH.

3.1. Inductance: RT vs. 77 K

To understand the performance of a coreless cryogenic inductor, we wound a copper wire 27 times around a bobbin such that it can perform both as an ambient inductor with a magnetic core as well as a coreless cryogenic inductor. The inductors used for this study are shown in Figure 2. We used two U-shaped ferrite cores, separated with two 3.38 mm thick G10 (high-pressure fiberglass laminate) plates on each side. The measured inductance value of the inductor with the core was 244 μH at 1 kHz. On the other hand, the inductor without the magnetic core showed the inductance of 21.3 μH. The difference in inductance is mainly due to the existence of the magnetic core.

3.2. Dissipated Power, Resistance, and Energy Density: RT vs. 77 K

With the ultimate goal of measuring the energy density of the inductor at room temperature and 77 K, we conducted a maximum current experiment as shown in Figure 3. We used a high-current power supply that can output up to 900 A in current control mode. A shunt resistor was installed in series to measure the current more accurately that what could be read on the display of the power supply. The tested inductor, which was either placed at ambient or in a bath of LN_2, was connected through a copper braid to prevent excessive voltage drop in the lead. The voltage across the inductor and the shunt resistor was measured by a National Instruments Data Acquisition system (NI DAQ).

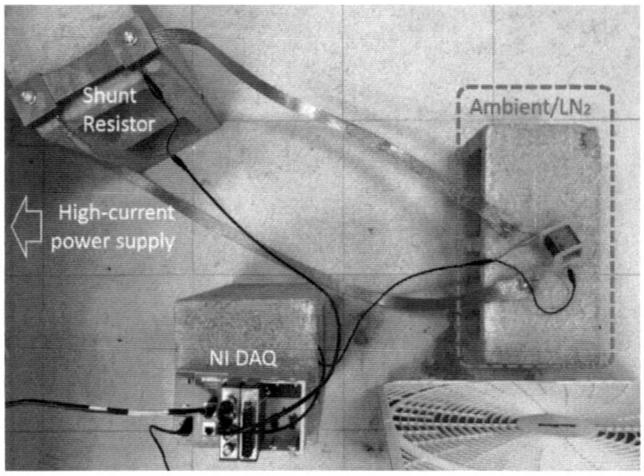

Figure 3. Testbed for measuring the maximum current limitation of the inductor.

The cryogenic test was conducted in an open bath of LN_2. The current was increased from 10 A with a 10 A increment as high as 240 A, at which point the inductor failed (contact between wire and current lead overheated and melted, shown in Figure 4(a)). The experiment at room temperature was conducted with the identical setup with the only difference of the inductor being placed at ambient. Current was increased from 10 A with a 5 A increment up to 55 A, at which the bobbin started to turn brown.

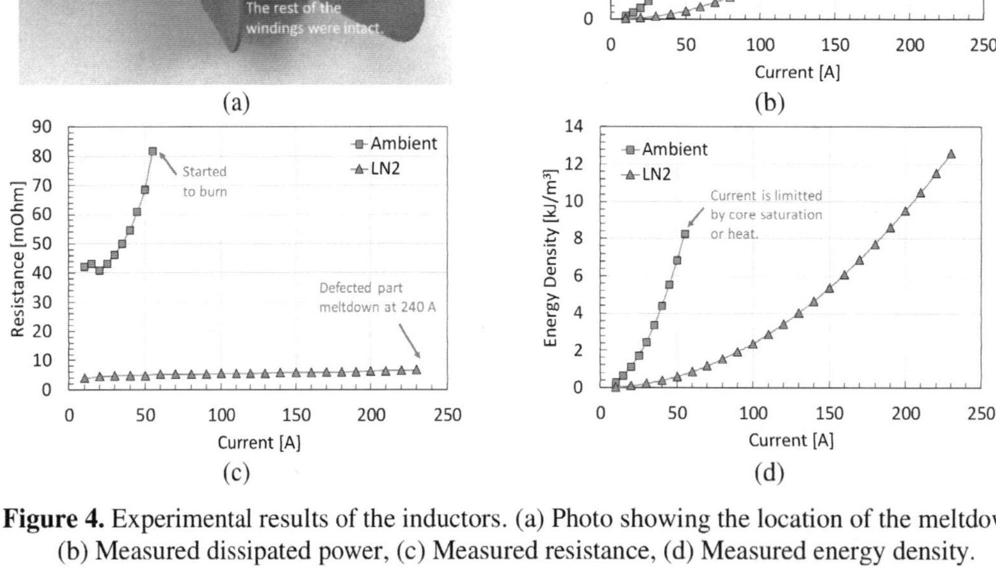

Figure 4. Experimental results of the inductors. (a) Photo showing the location of the meltdown, (b) Measured dissipated power, (c) Measured resistance, (d) Measured energy density.

Based on the measured voltage and current values, dissipated power, resistance, and the energy density were calculated as shown in Figure 4 (b) – (d). Compared to the measurements taken at 77 K, measurements taken at room temperature commonly show a steeper increase. The dissipated power (Figure 4 (b)) and the measured resistance (Figure 4 (c)) increased faster at room temperature due to the higher resistance of the copper windings. At room temperature, current was increased up to 55 A, which is when the bobbin started to burn. However, at cryogenic temperature, current was increased up to 240 A, which is when the copper wire melted and opened the circuit. Due to the difference in the maximum current conducting capability of the copper winding in a bath of liquid nitrogen and at room temperature, the achievable energy density of coreless inductors at cryogenic condition can be much higher than the ambient temperature inductor even though it cannot use magnetic cores. In other words, the same energy density can be achieved by using thinner wires with lower ampacity but with more number of turns to increase inductance. Since energy stored in an inductor is proportional to the square of the current and the inductance is proportional to the square of the number of turns, reducing the diameter of the conductor and winding more number of turns will provide similar results in terms of energy density as long as the cooling power is maintained. The results show a significant advantage in terms of volumetric power density and gravimetric power density for future cryogenic power applications.

4. Outlook

In the near future, we plan to investigate the capacitor and inductor performance in greater detail. Especially, when it comes to the inductor design, we plan to add more layers of windings. The associated cooling, current density, inductance, and energy density of the inductor with multiple windings will be investigated. To exploit the superior cooling power of liquid nitrogen, spacers will be introduced in the inductor design and the impact of orientation on the inductor cooing and energy density will be studied. Furthermore, various means of keeping the magnetic core in cryogenic conditions will be considered since having a magnetic core is a huge advantage for maximizing inductance.

5. Conclusions

The feasibility of developing cryogenic capacitors and inductors was studied. We reported a variety of metallized film capacitors tested at ambient and at cryogenic conditions, and the results showed an increase in capacitance and voltage rating in a bath of liquid nitrogen. In addition, we compared the energy density of an ambient inductor that has a ferrite magnetic core and a cryogenic inductor that does not have a core. The results showed that the energy density of a coreless inductor operated in a bath of liquid nitrogen could be higher than an inductor operated at room temperature with a core. Although inductance was lower due to the lack of magnetic cores, much larger current could be used in liquid nitrogen due to the substantially enhanced cooling that liquid nitrogen provides and the enhanced electrical conductivity of the wires. As a result, the energy density of coreless cryogenic inductors can be comparable or higher than that of ambient inductors with magnetic cores. In summary, the findings of this study show great promise in utilizing conventional technologies in cryogenic applications as well as in developing cryogenic passive components for cryogenic power electronics.

6. References

[1] Eckels P W and Snitchler G 2005 5 MW High Temperature Superconductor Ship Propulsion Motor Design and Test Results *Nav. Eng. J.* **117** 31–6
[2] Gamble B, Snitchler G and MacDonald T 2011 Full Power Test of a 36.5 MW HTS Propulsion Motor *IEEE Trans. Appl. Supercond.* **21** 1083–8
[3] Pamidi S, Kim C H, Kim J-H, Crook D and Dale S 2012 Cryogenic helium gas circulation system for advanced characterization of superconducting cables and other devices *Cryogenics (Guildf).* **52** 315–20

[4] Pamidi S, Kim C H and Graber L 2015 High-temperature superconducting (HTS) power cables cooled by helium gas in Superconductors in the Power Grid: Materials and Applications *Cambridge, UK Woodhead Publ. Ltd* **21** 225–60
[5] Park C, Cheetham P, Pamidi S, Kim C H, Viquez J G and Graber L 2017 A versatile modeling technique for predicting dielectric strength improvements in gas mixtures for superconducting applications *IEEE Trans. Dielectr. Electr. Insul.* **24** 2755–64
[6] Park C, Graber L and Pamidi S 2017 The dielectric properties of gaseous cryogen mixtures of He, H2, Ne, and N2 in a temperature range of 50–80 K at pressures up to 2.0 MPa *J. Appl. Phys.* **121** 83304
[7] Park C, Pamidi S and Graber L 2017 The critical electric field of gas mixtures over the extended range of cryogenic operating conditions *J. Appl. Phys.* **122** 153301
[8] Park C, Pamidi S and Graber L 2018 Evaluating the dielectric strength of helium-nitrogen gas mixtures by plasma parameter measurements *Phys. Plasmas* **25** 43520
[9] Cheetham P, Ravindra H, Stamm T, Park C, Kim C, Graber L, Steurer M and Pamidi S High Temperature Superconducting Power Cables for MVDC Power Systems of Navy Ships
[10] Graber L, Saeedifard M, Mauger M J, Yang Q, Park C, Niebur T, Pamidi S V and Steinhoff S 2017 Cryogenic Power Electronics at Megawatt-Scale Using a New Type of Press-Pack IGBT *IOP Conf. Ser. CEC-ICMC*
[11] Park C, Wei J, Singh S, Narra S, Imperatore M and Graber L 2018 The Characteristics of Film Capacitors at Room Temperature and in Liquid Nitrogen *AIAA/IEEE Electr. Aircr. Technol. Symp. AIAA Propuls. Energy Forum*
[12] Honjo S, Mimura T, Kitoh Y, Noguchi Y, Masuda T, Yumura H, Watanabe M, Ikeuchi M, Yaguchi H and Hara T 2011 Status of superconducting cable demonstration project in Japan *IEEE Trans. Appl. Supercond.*

IOP Conf. Series: Materials Science and Engineering **756** (2020) 012011 doi:10.1088/1757-899X/756/1/012011

Understanding Surface Flashover in Helium Gas Cooled High Temperature Superconducting Devices

A Al-Taie[1,2,3], P Cheetham[1,2], C H Kim[2], C Park[4], L Graber[5] and S V Pamidi[1,2]

[1] Florida A&M University - Florida State University College of Engineering, Tallahassee, FL 32310, USA
[2] Center for Advanced Power Systems, Florida State University, Tallahassee, FL 32310, USA
[3] University of Technology, Baghdad, 10066, Iraq
[4] Department of Electrical and Computer Engineering, Mississippi State University, Mississippi State, MS 39762
[5] Georgia Institute of Technology, Atlanta, GA 30332, USA

aha16b@my.fsu.edu

Abstract. Surface flashover voltage on solid insulators in gas cooled superconducting power devices is studied. The relationship between the surface flashover voltage and the dielectric strength of the gas media is established. Analysis of the data on surface flashover voltage measurements on cylindrical samples made of high-pressure fiberglass laminate (G10) in three different gas media with varying dielectric strength and pressure showed a positive correlation between the surface flashover voltage and the dielectric strength of the gas both at room temperature and 77 K. The positive relationship, however, is not linear. The results suggest that using gas media with higher dielectric strength, employing higher pressure, and lowering temperature in HTS power devices leads to higher surface flashover voltages and higher operating voltages because surface flashover is usually the limiting voltage in devices that use solid support structures and possess voltage gradients.

1. Introduction

High temperature superconducting (HTS) devices enable higher current densities and efficiencies compared to their conventional counterparts. For some applications, the required operating temperature is less than 50 K, and gaseous helium (GHe) is a potential cryogen. These applications include devices that use MgB_2 as the superconductor, and for HTS motors, generators, and magnets which operate with a higher internal magnetic field and require the lower operating temperature to achieve the necessary high current density [1]. To achieve high power ratings in the MW range, HTS devices need to operate at medium voltage between 5-20 kV. The cryogen is a part of the electrical insulation system of HTS devices, and the low dielectric strength of GHe limits the operating voltage of GHe cooled HTS devices [2], [3]. Our research over the last few years has focused on understanding the intrinsic dielectric properties of GHe and developing insulation materials and designs to increase voltage ratings of GHe cooled HTS devices [4]. We discovered that adding less than 7 mol% of hydrogen to GHe significantly increases its intrinsic dielectric strength without flammability concerns [5], [6]. It is now necessary to understand how this improvement translates to improved voltage ratings of GHe cooled HTS devices. The limiting voltage rating for an HTS device is not expected to be intrinsic breakdown in the gas medium, but instead through other mechanisms such as partial discharge and surface flashover, which occur at significantly lower voltages than the intrinsic breakdown.

Surface flashover is a phenomenon of electrical breakdown along the interface between the conductor and two electrical insulation materials with different electrical properties. The mechanisms which dictate surface flashover vary depending on whether an ac or dc voltage is applied. For ac voltages, surface flashover is typically dependent on the permittivity of the insulation materials and the electric field enhancement caused by the geometry. The permittivity of insulation materials plays a significant role in the field distribution under AC fields, and the temperature does not largely influence permittivity. However, under DC fields, the conductivity of insulation materials, which varies with temperature, governs the field distribution. Our previous work on gas mixtures shows that the intrinsic breakdown of a gas is a function of density and is independent of operating temperature [7]. To enable the potential of the dielectric design of HTS prototype devices to be verified at room temperature instead of at cryogenic temperature, it is necessary to confirm whether surface flashover is a function of gas density, but not a function of temperature. The ability to validate the dielectric integrity of HTS devices at room temperature instead of cryogenic temperature will lead to substantial cost savings, rapid prototype development, and collection of statistically significant data.

The purpose of the experiments conducted for this study is to understand the effect of the dielectric strength of the gas medium on the surface flashover voltage and to see if the temperature has any effect on the phenomenon. Three different gas media were investigated, which include research grade helium (GHe) with 99.9999% purity, 4 mol% hydrogen + 96 mol% GHe, and nitrogen gas (GN_2). These gases have varying dielectric strength, and comparisons can be made between the gases. The 4 mol% H_2 helium-based gas mixture and GN_2 are stronger than GHe in dielectric strength at room temperature. For the same particle density, GN_2 has 6.7 times the dielectric strength of GHe. That means under the same conditions of electric field, gap distance, pressure, and temperature, GN_2 should have a breakdown voltage 6.7 times to that of GHe [8], [9]. Additionally, 4 mol% H_2 helium-based gas mixture has 80% higher dielectric strength than GHe [5], which means the gas mixture is 1.8 times stronger than pure helium. Accordingly, GN_2 is 3.7 times stronger than 4 mol% H_2 helium-based gas mixture. It is important to understand if the ratios between the gases for intrinsic breakdown hold true for surface flashover. Surface flashover involves a different mechanism compared to intrinsic breakdown. As surface flashover is expected to be the limiting voltage rating for HTS power devices, it is necessary to understand the relationship between the dielectric strength of the gas insulating medium and the surface flashover voltage and to assess the practical benefits of incorporating helium-gas mixtures with enhanced dielectric strength.

Experiments were conducted at room temperature and at 77 K in the same gas medium and solid surface combination. Surface flashover voltages were measured on samples with both ac and dc voltages at room temperature and 77 K to understand the influence of the dielectric strength of the gas medium on surface flashover voltage.

2. Experimental Setup

There is no standardized surface flashover experimental setup. In this study, a needle-plane arrangement was selected. The needle plan arrangement allows for surface flashover to occur in multiple radial paths due to its symmetry. The experimental setup was designed to ensure the electrodes remained in contact with the insulating specimen at both room and cryogenic temperatures. Previous surface flashover experiment setups have had issues with specimens becoming loose between electrodes due to the mismatch in coefficient of thermal expansion between the metal electrodes and the polymer-based insulator. The current experiment setup ensures that experimental results between room and cryogenic temperature can be easily compared to one another. A 12.4 cm diameter copper disc was fabricated from a copper sheet. A 12.7 mm circular hole at the center and four 6.35 mm circular holes separated by 90 degrees by the circumference were made on the copper disc by a waterjet. The centre hole was made to fit the cylindrical G10 samples, which was inserted through, and the other four holes were made to fit in the 6.35 mm threaded rods in the cylindrical G10 set that provides the structure for the setup. Cylindrical G10 samples of 12.7 mm in diameter and 15 mm in height were used as solid insulator samples. G10 is a commercially available glass reinforced epoxy which is compatible with cryogenic temperatures. A needle electrode having diameter of 2 mm with its tip placed on the center of the G10 sample was connected to the high voltage source through a bushing while the copper disc was connected to ground. The experimental arrangement of the sample and the electrode is shown in Figure 1 (a).

The upper circular surface of G10 specimen sits at a higher level than the copper disc by 3.4 mm, as shown in Figure 1 (b). The minimum total surface flashover path length between the tip of the needle

(a) (b)

Figure 1. (a) Surface flashover experimental setup (b) closer view of needle-plane arrangement

electrode and the copper disc is the sum of the 6.35 mm radius and 3.4 mm vertical height of the cylindrical G10 insulator, which equals to 9.75 mm.

3. Measurements and Results for G10 in Different Gases

The three different gases were tested with G10 specimens to understand the relationship between the dielectric strength of the gas and the surface flashover voltage. Again, GN_2 was selected since it possesses higher dielectric strength compared to the other two gases, as discussed in the introduction.

At room temperature, the pressure levels used for GHe and 4 mol% H_2 helium-based gas mixture were 2.0, 1.5, 1.0 and 0.5 MPa. For GN_2 pressure levels of 2.0, 1.5, 1.0 and 0.5 MPa were also used, but in addition, the test was extended to lower pressure levels of 0.4, 0.3 and 0.2 MPa to have a wider pressure range, and accordingly wider gas mass density range. For the measurements at 77 K with GHe and 4 mol% H_2 helium-based gas mixture, the same pressure ranges were also used. For these two gases, besides the pressure levels of 2.0, 1.5 and 1.0 MPa at 77 K, lower pressure levels of 0.54, 0.4, 0.27 and 0.14 MPa were chosen to have the same gas densities at 2.0, 1.5, 1.0 and 0.5 MPa at room temperature, respectively. Pressure was monitored by CMM type of HEISE® dial pressure gauge of which accuracy is ±0.02 MPa.

AC measurements were performed using a Haefely high voltage test kit (100 kV/5 kVA AC, rms, 60 Hz). The DC measurements were performed using the Glassman high voltage source (100 kV/16 kW DC) since it has a stable DC voltage profile having 0.025% ripple factor and capability of limiting the output (discharge) current less than 3.2 mA (2% of its maximum). For both the AC and DC measurement, a ramp rate of approximately 300 V/s was applied until surface flashover occurred. Restricting the output current reduces the energy released during surface flashover, which enables more measurements to be taken before degradation of the sample occurs. Once degradation occurs on the sample by multiple experiments, measured surface flashover voltage was found more than 50% offset from the statistically meaningful average of the experimental value. With the limiting output current capability of the high voltage sources used in this study, degradation of the sample were not frequent or problematic.

Flashover voltage data obtained in the DC and AC measurements for room temperature and 77 K is shown in Figure 2. Each data point is an average of 5 measurements. The error bars in the figure show maximum and minimum flashover voltages obtained at each pressure level.

4. Discussion

The systematics of the results obtained are summarized below:

1. For all the cases studied, the DC surface flashover voltages are higher than the corresponding AC rms (60 Hz) surface flashover voltages for the same gas, temperature, and pressure combinations. Analysis of the surface flashover data obtained at 77 K to the previous obtained intrinsic dielectric strength of GHe and the 4 mol% H2 mixture demonstrates that the increase in DC surface flashover values is not solely based on representing the AC measurements in rms instead of peak voltage (Table 1).

   Table 1. Intrinsic dielectric strengths for the various gases

Pressure	2 MPa		1.5 MPa		1 MPa		0.5 MPa	
	AC	DC	AC	DC	AC	DC	AC	DC
GHe breakdown (kV) [5]	73.8	105.4	57.6	82.7	41.2	57.0	23.4	29.5
GHe Surface flashover (kV)	17.9	34.6	15.6	28.2	10.5	20.5	6.2	16.5
4 mol% H_2/ GHe Breakdown (kV) [5]	132.5	200.7	110.6	158.9	81.3	112.3	43.4	62.6
4 mol% H_2/ GHe Surface Flashover (kV)	19.4	39.9	18.1	37.0	15.5	28.6	10.5	20.7

   From Table 1 it can be seen that for the intrinsic dielectric strength comparison between AC and DC measurement the DC measurements are higher by $\sqrt{2}$. When the surface flashover data is compared between AC and DC voltages the DC measurements are between 1.8 – 2.7 times higher than the AC which is expected as AC and DC surface flashover are governed by different breakdown mechanisms. It should be noted that the DC voltage applied to the sample was done at a continuous ramp rate which may not have given sufficient time for space charge accumulation to influence the surface flashover voltage recorded. Table 1 also demonstrates the difference between intrinsic breakdown and surface flashover values. The experimental data on intrinsic breakdown taken from [5] was extrapolated to have the same gap distance (9.5 mm) as the surface flashover experiment. The significantly lower surface flashover values demonstrate why it is expected to be a limiting factor for HTS power devices.
2. Comparing the surface flashover voltage data of different gases studied, the 4 mol% hydrogen-helium mixture with higher dielectric strength resulted in higher surface flashover voltages compared to pure helium at the all the pressures measured at room temperature.
3. For both pure GHe and 4 mol% hydrogen-helium mixture, the 77 K measurements resulted in higher flashover voltages than the corresponding room temperature values. The dielectric strength of a gas for a given pressure at 77 K is 3.7 times stronger to that at room temperature due to the increase of gas density by the same number.
4. Surface flashover voltages in GN_2 are higher than those in pure helium or helium-hydrogen mixture. GN_2 is 6.7 times stronger dielectric medium than pure helium [8].
5. The needle plane experimental setup has a non-uniform electric field which means that different surface flashover voltages are expected if a negative DC voltage is applied to the needle compared to what has already been obtained for a positive DC voltage. As a continuation of this study, we intend to complete surface flashover measurements utilizing a modified experiment configuration with a negative DC voltage applied on the needle. Collection of these data will give a greater level of understanding of surface flashover in GHe and GHe mixtures for varying waveforms.

Analysis of all the results shows that increasing the dielectric strength of the gaseous medium, either by using a stronger gas or by increasing the pressure or decreasing the temperature, increases the surface flashover voltage.

To present the relationship between the dielectric strength of the gas and the surface flashover voltage, Table 2 summarizes all the data of surface flashover in the three different gas media and two

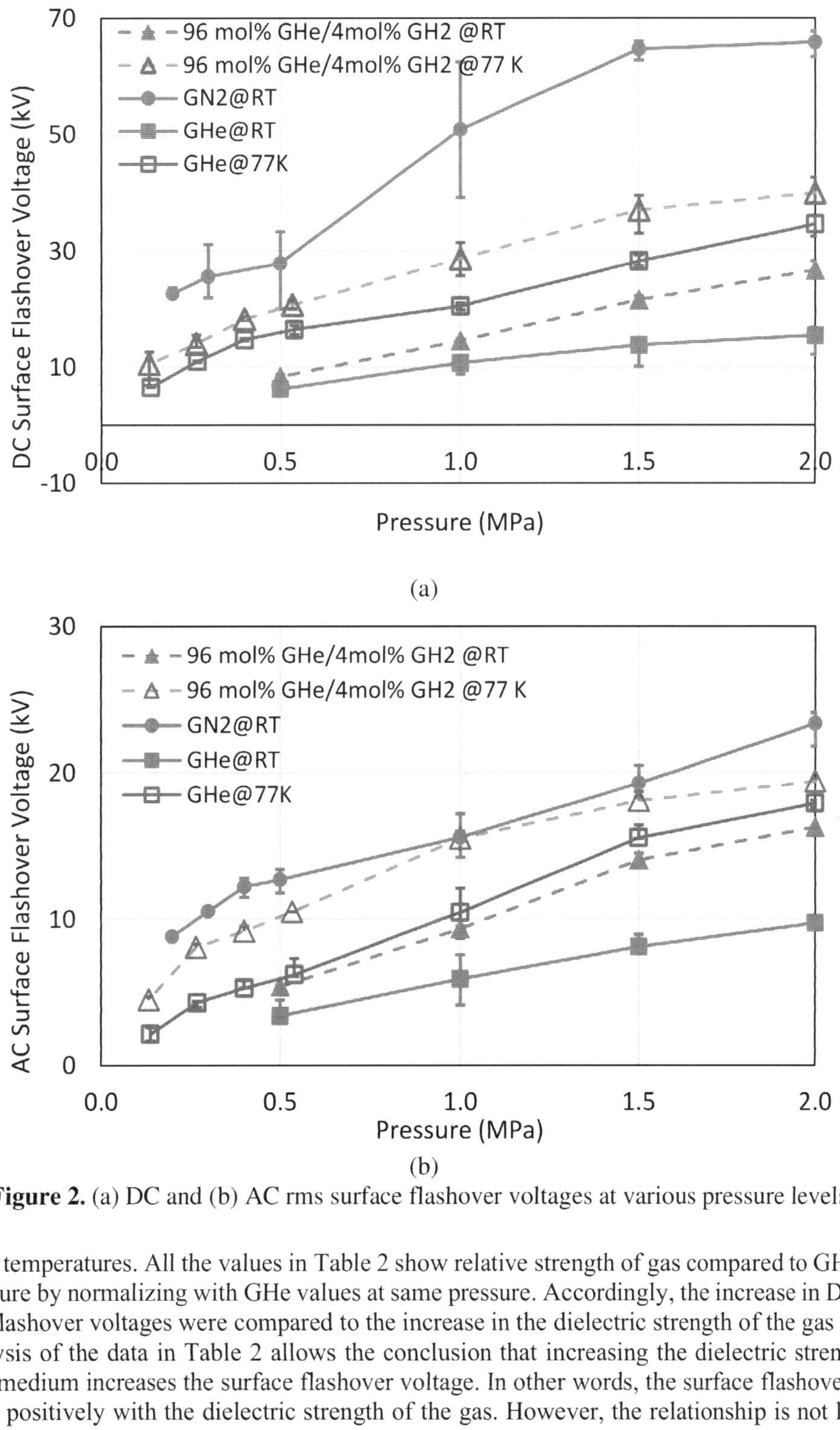

Figure 2. (a) DC and (b) AC rms surface flashover voltages at various pressure levels.

different temperatures. All the values in Table 2 show relative strength of gas compared to GHe at room temperature by normalizing with GHe values at same pressure. Accordingly, the increase in DC and AC surface flashover voltages were compared to the increase in the dielectric strength of the gas medium.

Analysis of the data in Table 2 allows the conclusion that increasing the dielectric strength of the gaseous medium increases the surface flashover voltage. In other words, the surface flashover voltages correlate positively with the dielectric strength of the gas. However, the relationship is not linear. For example, increasing the dielectric strength 3.7 times does not yield an increase of 3.7 times flashover voltage. Additionally, the data in table show that DC surface flashover voltages are generally higher than the corresponding AC surface flashover voltage.

Table 2. Dielectric strength of various gases, normalized to GHe at room temperature and the corresponding surface flashover voltages

	Both AC and DC Intrinsic Dielectric strength (ratio) [5], [8]	DC surface flashover voltage (ratio)	AC surface flashover voltage (ratio)
GHe at room temperature	1	1	1
H_2-He mixture at room temperature	1.8	1.3-1.7	1.6-1.7
GHe at 77 K	3.7	1.9-2.7	1.8-1.9
H_2-He mixture at 77 K	6.7	2.6-3.3	2.0-3.1
GN2 at room temperature	6.7	4.3-4.8	2.4-3.7

5. Conclusion

The surface flashover is the lower limit of the performance of an insulation system when solid surfaces and electric field gradients are present. Surface flashover measurements were performed at room temperature and at 77 K on G10 cylindrical samples in three different gas media at two different temperatures. The three different gas media, two different temperatures, and several pressures were used to change the dielectric strength of the gas media over a wide range. Increasing the dielectric strength by changing the gas media, increasing the pressure, or lowering the temperature leads to an increase in the surface flashover voltage, but not linearly. The non-linearity of the relationship between the dielectric strength of the gas medium and the flashover voltage need to be further investigated. The results of this study suggest that for gas-cooled HTS power applications, it is useful to employ a gas medium with higher dielectric strength to push the surface flashover voltages to higher values. Similarly, use of high pressures or lower temperatures are also useful in pushing the surface flashover voltages to higher values.

6. References

[1] Suzuki S and Fukuda K 2009 *Magnesium Diboride (MgB2) Superconductor Research* (Nova Science Publishers)

[2] Hazeyama M, Kobayashi T, Hayakawa N, Honjo S, Masuda T and Okubo H 2002 Partial discharge inception characteristics under butt gap condition in liquid nitrogen/PPLP® composite insulation system for high temperature superconducting cable *IEEE Trans. Dielectr. Electr. Insul.* **9** 939–44

[3] Cheetham P G, Kim W, Kim C H, Pamidi S V., Graber L and Rodrigo H 2017 Use of partial discharge inception voltage measurements to design a gaseous helium cooled high temperature superconducting power cable *IEEE Trans. Dielectr. Electr. Insul.* **24** 191–9

[4] Park C, Pamidi S and Graber L 2017 The critical electric field of gas mixtures over the extended range of cryogenic operating conditions *J. Appl. Phys.* **122** 153301

[5] Cheetham P, Kim W, Kim C H, Graber L, Rodrigo H and Pamidi S 2017 Enhancement of Dielectric Strength of Cryogenic Gaseous Helium by Addition of Small Mol% Hydrogen *IEEE Trans. Appl. Supercond.* **27**

[6] Park C, Graber L and Pamidi S 2017 The dielectric properties of gaseous cryogen mixtures of He, H2, Ne, and N2 in a temperature range of 50–80 K at pressures up to 2.0 MPa *J. Appl. Phys.* **121** 83304

[7] Park C, Wei J, Cheetham P, Kim C H, Pamidi S and Graber L 2018 The influence of temperature on the dielectric strength of gaseous cryogens *Proc. IEEE Int. Conf. Prop. Appl. Dielectr. Mater.* **2018-May** 164–7

[8] Vijh A K 1977 Electric Strength and Molecular Properties of Gaseous Dielectrics *IEEE Trans. Electr. Insul.* **EI-12** 313–5

[9] Vijh A K 1976 Intermolecular bonding and the electric strengths of dielectric gases *J. Mater. Sci.* **11** 1374–5

High throughput and variable temperature superconductor integrated circuit test and evaluation using ICE-T

A Sahu[1], B Chonigman[1], A Talalaevskii[1], V Dotsenko[1], S Ruotolo[1], J Tang[1] and D Gupta[1]

[1] HYPRES, Inc., Elmsford, NY 10523, USA

E-mail: asahu@hypres.com

Abstract. Superconductor digital integrated circuits (ICs) require rapid evaluation of multiple copies to obtain statistical operational data. These data are used for assessing model-to-hardware correlation and facilitate iterative IC design development. The Integrated Cryogenic Electronics Testbed (ICE-T) is a cryogen-free test platform, which can test multiple chips simultaneously with similar convenience to a liquid-helium immersion probe and with cool-down times of between 3.3 to 4.5 hours. We have developed a three-chip insert to increase the volume of chip testing and demonstrated simultaneous cooling of six chips with two such inserts. We report the test statistics collected from 27 chips across a single wafer. We have also used the ICE-T's convenient temperature control system to evaluate chips in the 3.5 - 6 K range. Such evaluation determines the robustness of circuit design and its tolerance to critical current fluctuations due to fabrication variation.

1. Introduction

As superconductor electronics mature from research prototype to productization, testing and characterizing a large volume of superconductor integrated circuits (ICs) over a short period of time (high throughput testing) becomes a must. One major barrier in testing, development, and use of superconductor electronics is the cost and availability of liquid helium to cool down these devices and ICs to the required temperature. Traditionally, liquid helium cryostats have allowed researchers to quickly cooldown and test these ICs using immersion probes [1]. However, refilling these cryostats and maintaining a liquid helium test facility contribute significantly to operational cost. Worldwide shortage of liquid helium, irregular supplies, and multifold price increases have been disrupting the operations of research labs in recent years [2]. Additionally, since most of these cryostats require frequent re-filling of liquid helium, they are difficult to use for long term continuous testing.

Cryocooled systems have been around for a while as an alternative to the liquid helium testing paradigm. However, they have suffered from their own drawbacks. These cryocooled systems have been designed and constructed to be rather experiment-specific. Since the configuration of the external interface varies between different devices and circuits, in terms of number of connections and their type (input vs output vs data), the user is required to spend significant time and effort to reconfigure the system which also requires an expertise in cryo-packaging. Even something as simple as switching out a chip and replacing it with a substitute of the exact same type requires the user to have extensive training in cryo-packaging and would require a day or two of the user's time [3].

To address these problems, we developed the Integrated Cryogenic Electronics Testbed (ICE-T) [4], a cryogen free testbed to do cost-effective high throughput testing and eliminate the need of cryo-

IOP Conf. Series: Materials Science and Engineering **756** (2020) 012012 doi:10.1088/1757-899X/756/1/012012

packaging for the end-user. ICE-T was designed with interchangeable top loading inserts, which house all the experiment specific cabling and cryo-packaging. This provides a user convenience similar to that of liquid helium immersion probes. At the same time, electricity and periodic maintenance are the only cost of running ICE-T, and these costs are rather low compared to the cost of liquid helium in a traditional setting [4]. The modularity and flexibility of ICE-T, as well as its fast cooldown and warmup time, make the ICE-T the preferred testing platform for superconducting ICs. This is especially true in an R&D environment where the ability to test chips of different designs quickly is a priority.

The ability to test multiple chips quickly and seamlessly is important at each phase of R&D work. In the screening phase this ability allows testers to quickly characterize the wafer and exclude poorly performing chips. In the iterative phase of IC design such an ability helps facilitate model-to-hardware correlation. And in the final phase of the design cycle the user can measure the robustness and tolerance of a design to variations in the fabrication process.

For a foundry, a testbed with the capacity for high throughput testing can be extremely valuable, as it enables the quick collection of full statistics on several fabricated wafers. This capacity allows for a robust analysis of fabrication parameters such as critical current density (J_C), critical current (I_C), missing area, wafer yield, etc.

Another significant advantage of ICE-T is its inbuilt temperature control system which allows a user to vary the testing temperature in a controlled manner; something which is extremely difficult to implement in liquid helium cryostats. This enables measurements of superconductor ICs, cryogenic semiconductor circuits, or a combination of both at different user-controlled temperatures. By varying the temperature during measurement one can thermally tune the critical current of superconductor circuits, which can be very useful for analyzing a circuit's behavior at different critical currents of the Josephson junctions (JJs) [4] [5].

2. Description of the Testbed

ICE-T is designed with seats for two interchangeable inserts which make thermal contact to the cold-head's two native temperature stages via a clamping mechanism. Each seat uses an ISO-100 vacuum port, which accommodates a cylindrical insert ~4" in outer diameter and extending ~19" inside the vacuum chamber. Both the inserts and the body of ICE-T have thermometers and heaters located at each stage. The placement of these in the system can be seen in figure 1. The thermometer and heater at the 2nd stage of the insert, together with the temperature controller, serve as a temperature regulator for variable temperature measurements. They also enable rapid (under 5 minutes) de-fluxing of superconductor ICs [6]. As ICE-T can accommodate two inserts for two independent experiments, and each insert has its own heaters, chip(s) in each insert can be de-fluxed independently.

Figure 1. CAD drawing of ICE-T with the inside of the vacuum chamber visible.

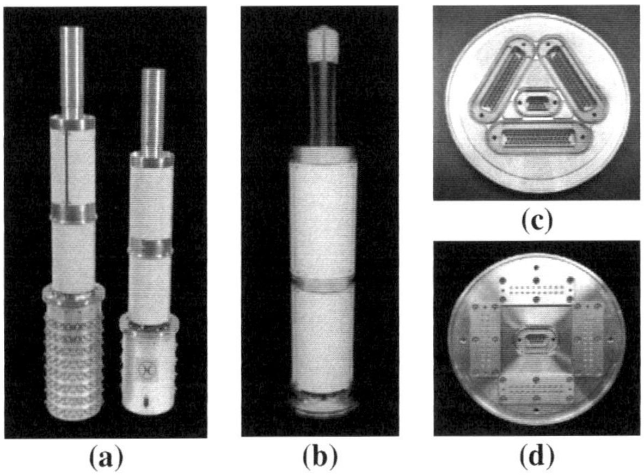

Figure 2. (a) 80 (left) and 40 (right) HF line universal inserts. (b) 3×40 LF line insert. (c) Vacuum feedthrough for the 3-chip LF insert. (d) Vacuum feedthrough for both 80 and 40 HF line universal insert.

2.1. Standard ICE-T Insert Design

ICE-T's modular design allows for many types of inserts customized to fit specific experimental requirements; each of which can be swapped out in minutes without having to open and reconfigure the entire testbed. For the testing presented in this paper we used some of the inserts that come as a standard option with the ICE-T. These standard inserts are designed to take either a 5x5 mm^2 or 10x10 mm^2 chip using a standardized pad layout; this accommodates the testing of a wide variety of different superconducting IC designs. The superconductor chips are pressure mounted in the sample holder of standard inserts the same way as in traditional helium-immersion probes [1].

One type of standard insert is what we call a "universal insert", as all the lines running from the room temperature connector panel to the chip are identical and can handle any frequency from DC to 26.5 GHz with < 2 dB loss, allowing the chip designer/tester to use any line for either input, output or bias [7]. The universal insert comes with the option of either 40 or 80 high frequency (HF) lines. The universal inserts and their vacuum flanges can be seen in figure 2(a) and figure 2(d) respectively. The benefit of these two inserts is the ability to test any chip matching the standardized pad layout, independent of the number of high or low speed lines the chip requires. This provides the greatest flexibility in designing and testing all chips with a 40-pad or 80-pad layout.

The other standard insert option is our 3×40 insert designed for low frequency (LF) basic functionality level testing of superconductor ICs. The insert, as shown in figure 2(b) with its vacuum flange shown in figure 2(c), is capable of supporting three chips simultaneously with 40 low speed lines running to each chip. The benefit of the 3-chip insert is a much higher throughput of testing. When loading two inserts the design allows a user to cool down a total of 6 chips simultaneously.

2.2. ICE-T Cooldown and Warmup Time

For a cryocooler based testbed, the testing cycle includes the cooldown time to the required temperature, the chip test time, and the warmup time to room temperature. For high throughput testing of superconductor ICs it is very important that the cooldown time and warmup time are acceptably low. For reference, Table 1 shows the cooldown and warmup times for ICE-T's standard inserts. As ICE-T can accommodate up to two inserts we show how long it takes to cool down one or two inserts at the same time as separate entries. The time was measured in HYPRES's ICE-T development unit. The number of chips per insert which can be cooled down simultaneously are shown too. For example, cooling down 6 chips in two 3×40 LF line inserts, will take 3.7 hours. Similarly, it would take 4.5 hours to cool down 2 chips in two 80 HF line universal inserts. The warmup time is only 2.5 hours for any combination of inserts when HYPRES's proprietary heating method for accelerated warmup is used.

Table 1. Cooldown and warmup times for standard ICE-T inserts.

Insert Type	# of Inserts	# of ICs	Cooldown[a] Time to 4.2 K (Hrs)	Warmup[b] Time to 293 K (Hrs)
3 x 40 LF Line (High Throughput) Insert	1	3	3.3	2.5
	2	6	3.7	
40 HF Line Universal Insert	1	1	3.7	
	2	2	4.0	
80 HF Line Universal Insert	1	1	4.0	
	2	2	4.5	

[a] As measured in HYPRES's test lab unit
[b] Using HYPRES's propriety heating system

3. Novel Testing Methods Using ICE-T

In this section we present the results of several experimental studies of superconductor circuits that are made possible only by the versatility, customizability, and modularity of the HYPRES ICE-T.

3.1. High Throughput Testing

The flexibility and ease of swapping a chip in an immersion probe and the speed of cooldown and warmup are often cited by researchers as reasons to use traditional liquid helium test setups. This is particularly true in an R&D environment, despite the increasing cost of liquid helium and the cost to build and maintain the liquid helium infrastructure. When compared to traditional cryocooled systems, which are highly specific to a narrow set of chips and require cryopackaging expertise and extensive labor to dismantle much of the system to access or swap a chip, the argument for traditional liquid helium testing in an R&D environment seems reasonable. ICE-T, however, has become a game changer in the liquid helium vs. cryocooled system debate. The ease of loading multiple chips into ICE-T inserts and speed of cooldown and warmup as mentioned in Table 1, eliminated the main disadvantages of cryocooled systems vs liquid helium thermostats and enabled ICE-T to test chips at higher throughput without requiring cryopackaging expertise on the user's side.

High throughput testing is an important step in design certification as it measures the tolerance of a design to several fabrication parameters and wafer runs. In this experiment we tested 27 chips from various locations of the same wafer at 4 K. The chips were tested in the ICE-T with two 3×40 LF line inserts, as seen in figure 2(b). This setup allowed us to cool down 6 chips simultaneously, test them and warm up in one extended 10-12 hour workday; providing sufficient time to swap the chips and cooldown overnight for testing the following day. At this pace we were able to test all 27 chips available over 5 extended workdays, in a total of 60 man-hours.

Each chip, as shown in figure 3, contains a set of 3-to-2 and 7-to-3 parallel counters [8] whose operating circuit bias margins were measured for functional level testing at low frequency (< 1 kHz) using the Octopux automated test system [9]. Test patterns were randomly generated at each bias point to measure the margins. The margin is the range of DC bias currents at which the circuit operates as designed and is characterized by the minimum and maximum bias current values as well as the center point.

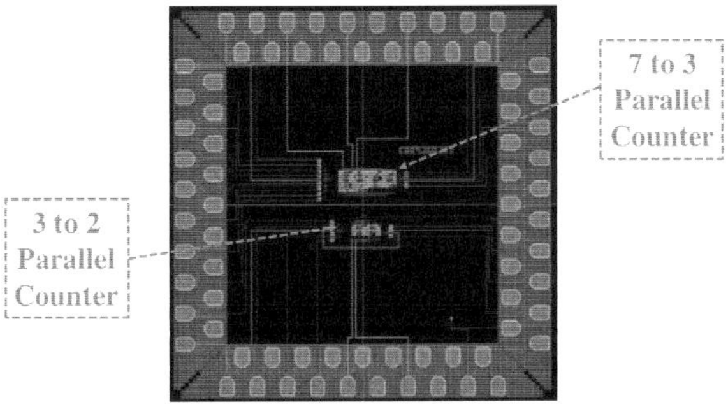

Figure 3. Layout of chip with 7-to-3 parallel counter and 3-to-2 parallel counter circuits.

Wider operating margins indicate potential higher frequency operation and better performance, thus signifying a better IC design and/or batter fabrication quality. When some chips of a particular design have wider operating margins than other chips of the same design, it indicates a higher fabrication quality of the wafer or of an area of the wafer.

Table 2 shows the average bias center in millivolts (mV) and bias margins which were calculated from measuring 27 chips, all from the same SFQ584-17-1-W3 wafer fabricated at MIT-LL [10] [11] [12]. The margins and values of both circuits' key biases were measured and analyzed. Critical bias margin is the lowest bias margin of all the biases. As seen from Table 2, logic bias is the critical bias for both circuits.

The spread and variation of critical margin of both circuits are shown in figure 4. The wafer location map is shown as provided by the MIT-LL foundry. From testing twenty-four 3-to-2 parallel counter we found that the average margin of all the biases was greater than ±20%. Having comparatively wide and uniform operating margins indicates that this design is very robust and tolerant to fabrication variation across the wafer. However, when testing the twenty-seven 7-to-3 parallel counters we found that the logic bias had a suppressed average margin of less than ±10% as compared to the others with greater than ±20%. Pre-layout simulations predicated more than a ±20% for all biases. A consistent suppression of the logic bias across the entire wafer indicates a mismatch between the simulation and post-fabrication parameters.

Table 2. Cooldown and warmup times for standard ICE-T inserts.

Circuit Name	# of Circuits Tested	Bias Name	Average Bias Center (mV)	Standard Deviation of Bias Center (mV)	Average ±% Margin
3-to-2 Parallel Counter	24	Data	2.40	0.08	36%
		Clock	2.02	0.08	33%
		Logic	2.29	0.07	25%
7-to-3 Parallel Counter	27	Data	2.63	0.10	24%
		Clock	2.26	0.09	30%
		Logic	2.03	0.08	9%

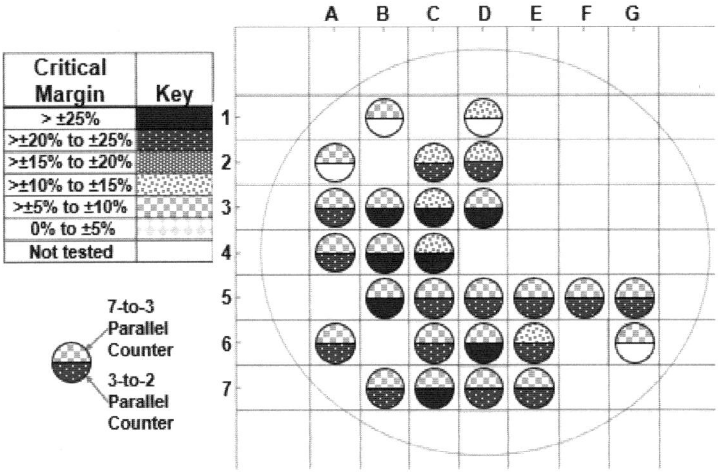

Figure 4. Performance of circuit as a function of critical margin across the wafer. The upper half of each circle represents the critical margin for the 7-to-3 parallel counter on the chip at that location; while lower half represents the critical margin for the 3-to-2 parallel counter circuit on the chip at that location. Different patterns for different critical margins are shown in table on left.

3.2. Variable Temperature Testing

Liquid helium (4He) environments restrict a chip's operating temperature to 4.2 K. Even though measuring superconductor IC above 4.2 K is possible in liquid helium immersion probes, such measurements are difficult and imprecise because controlling the temperature of the chip in helium vapor is difficult. ICE-T, being a cryocooled system with integrated native heaters at two temperature stages, provides complete control over the temperature environment. By varying the temperature, the critical currents of all the JJs on a chip can be adjusted for optimal circuit performance.

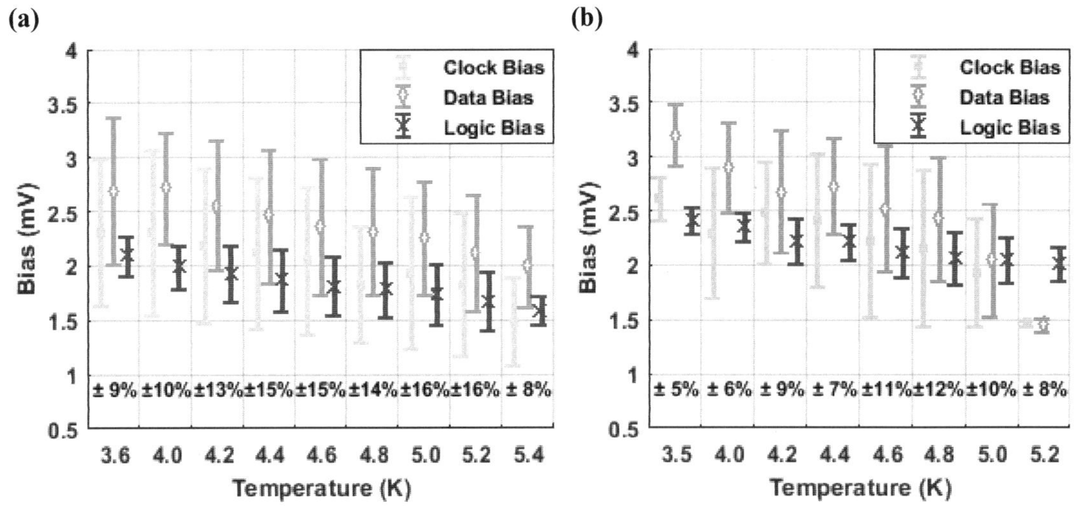

Figure 5. Operating bias margins of a 7-to-3 parallel counter as a function of temperature from a (a) 10 kA/cm^2 wafer and (b) 20 kA/cm^2 wafer fabricated at MIT Lincoln Laboratory.

Figure 5(a) and 5(b) show the result of varying the temperature of the superconductor chips from two different wafers. The chip design is same as shown in figure 3 with each chip containing a set of 3-to-2 and 7-to-3 parallel counters [8]. The operating bias margins were measured using the Octopux automated test system in the same way as described in the previous experiment. These margins were measured for all circuit biases at different temperatures, ranging from 3.5 K to 5.4 K. Each bias is plotted with its center point and the bias range. Each bin in figure 5(a) and 5(b) is the operating temperature of the chip and the critical bias margin is mentioned below the bias margins at that temperature.

The designs of the chips are normally optimized for the best IC performance at 4.2 K. However, some chips have better performance at a different temperature which can be due to either imperfect simulations at the design stage or fabrication process variations. As the temperature is raised from 3.5 K to 5.2 K, the critical margin increases from ±9 % at 3.6 K to ±16 % at 5.2 K and ±5% at 3.5 K to ±12% at 4.8 K for the chip in figure 5(a) and figure 5(b) respectively. The margins then start decreasing above 5.2 K and both chips stop working above 5.4 K. By increasing the temperature we are lowering the critical current of the all the JJs. This indicates a mismatch in the designed critical current and the fabricated critical current.

Similarly, a temperature vs bias margin plot is shown in figure 6 for a 256-bit long shift register with six different biases. The chip was tested in ICE-T from 3.3 – 6.0 K. The bias margin was measured in Octopux automated test setup [9]. The critical bias margin is above 15% in the 3.3 – 5.5 K temperature range and the chip works all the way to 6.0 K. This shows that the design is very robust and has a high tolerance to fluctuation to critical current. In fact, it works the best at temperatures below the nominally designed optimal of 4.2 K. This is the opposite trend of the previous chip design. It also points an anomaly in one of the biases identified as BK32. The bias center is lower compared to the other biases, which will result in lower overlapping margin.

This technique allows testers to find the optimal temperature for a particular chip as well as to characterize how closely the design and fabrication meet the spec parameters.

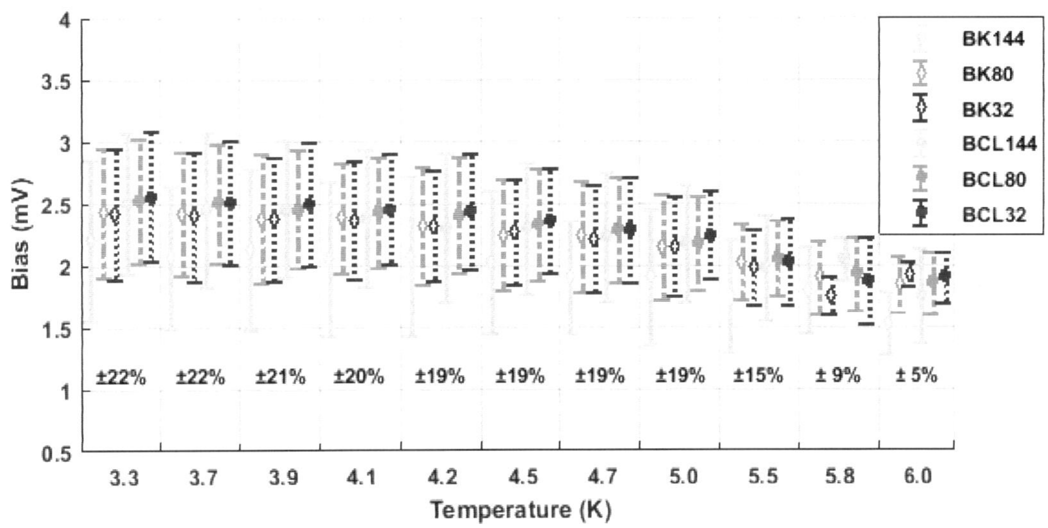

Figure 6. Bias margin as a function of temperature for 256-Bit long shift register. The width of biases is the bias margin. Each bin contains 6 bias margins at specific temperature.

4. Conclusions

High throughput and variable temperature testing are a relative new testing paradigm in the field of superconductor electronics. The ability to vary temperature precisely while testing the 3-to-2 and 7-to-3 parallel counters and the 256-bit shift register design allowed us to determine the critical current of the JJs that gave the most robust performance. When the optimal operating temperature is different than the nominal designed 4.2 K, how much and in what direction it differs provides important feedback to the circuit designer. Using this information, they can then run post-layout simulations varying the parasitics to match those found in the experimental results. From these empirically derived parasitic values the IC designs can then be optimized, and the model-to-hardware correlations can be iteratively improved. These provide better clues to circuits designers than testing at a single temperature point.

High throughput testing of the 7-to-3 parallel counter circuits uncovered wafer-wide reduced margins and low center points for the logic bias, while affirming the robustness of the 3-to-2 parallel counter. If only one or two chips were tested, some of the 3-to-2 chips showed suppressed operating margin while some of the 7-to-3 chips showed higher operating margins. This would have created a false impression of circuit performance and would lead to poor decisions in future designs. When enough statistics are collected the average circuit parameters provide far more reliable indicators for the designer.

References

[1] Gaidarenko D and Robertazzi R 1999 High performance packaging system for superconducting electronics *IEEE Trans. Appl. Supercond.* **9** 3668-3671

[2] Kramer D 2017 Erratic helium prices create research havoc *Physics Today* **70** 26-29

[3] Mukhanov O, Kirichenko D, Vernik I, Filippov T, Kirichenko A, Webber R, Dotsenko V, Talalaevskii A, Tang J, Sahu A, Shevchenko P, Miller R, Kaplan S, Sarwana S and Gupta D 2008 Superconductor digital-RF receiver systems *IEICE Trans. Electron.* **E91-C** 306-317

[4] Dotsenko V, Chonigman B, Sahu A, Tang J, Lehmann A, Sarwana S and Gupta D 2017 Superconductor Integrated Circuit (IC) Testing with the integrated cryogenic electronics testbed (ICE-T) *IEEE Trans. Appl. Supercond.* **27** 1-7

[5] Van Duzer T and Turner C 1999 *Principles of superconductive devices and circuits* (Upper Saddle River, NJ: Prentice Hall PTR)

[6] Narayana S, Polyakov Y and Semenov V 2009 Evaluation of flux trapping in superconducting circuits *IEEE Trans. Appl. Supercond.* 19 640-643

[7] Dotsenko V, Sahu A, Chonigman B, Tang J, Lehmann A, Gupta V, Talalevskii A, Ruotolo S, Sarwana S, Webber R and Gupta D 2017 Integrated cryogenic electronics testbed (ICE-T) for evaluation of superconductor and cryo-semiconductor integrated circuits *IOP Conf. Ser.: Mater. Sci. Eng.* 171 012145

[8] Filippov T, Sahu A, Kirichenko D, Celik M, Sarwana S, Lehmann A and Gupta D 2018 Parallel Counters for low-pass phase modulation–demodulation ADCs *IEEE Trans. Appl. Supercond.* **28** 1-5

[9] Zinoviev D and Polyakov Y 1997 Octopux: an advanced automated setup for testing superconductor circuits *IEEE Trans. Appl. Supercond.* **7** 3240-3243

[10] Tolpygo S, Bolkhovsky V, Weir T, Johnson L, Gouker M and Oliver W 2015 Fabrication process and properties of fully-planarized deep-submicron Nb/Al–AlO/Nb Josephson junctions for VLSI circuits *IEEE Trans. Appl. Supercond.* **25** 1-12

[11] Tolpygo S, Bolkhovsky V, Weir T, Wynn A, Oates D, Johnson L and Gouker M 2016 Advanced fabrication processes for superconducting very large scale integrated circuits *IEEE Trans. Appl. Supercond.* **26** 1-1

[12] Tolpygo S, Bolkhovsky V, Zarr S, Weir T, Wynn A, Day A, Johnson L and Gouker M 2017 Properties of unshunted and resistively shunted Nb/AlOx-Al/Nb Josephson junctions with critical current densities from 0.1 to 1 mA/μm^2 *IEEE Trans. Appl. Supercond.* **27** 1-15

Acknowledgments

The authors would like to thank A. Erik Lehmann and Saad Sarwana for their help in conceiving the product, Mustapha Habib for his help with data recording, and Timur Filippov for his editorial contributions.

Neutron irradiation effect on critical current of Nb_3Sn wire under 8 T to 15.5 T

A Nishimura[1,2]

[1] Technical Institute of Physics and Chemistry, Chinese Academy of Sciences, Beijing 100190, P. R. China
[2] National Institute for Fusion Science, Toki 5095292, Japan

e-mail: nishimura.arata@toki-fs.jp

Abstract. Superconducting magnets for fusion application will be irradiated by fast neutrons produced by fusion reaction, and it has been reported that the superconducting properties of the superconducting wires vary drastically by the irradiation. A bronze route Nb_3Sn wire and an internal tin process wire were irradiated in a fission reactor up to 1.7 x 10^{23} n/m^2 (> 0.1 MeV neutron), and the change in the critical current was measured by 15.5 T superconducting magnet and a variable temperature insert. The 4.9 x 10^{22} n/m^2 irradiation for the bronze route wire increased the critical current by 1.75 times, but the 7.9 x 10^{22} n/m^2 irradiation showed 1.03 times increase for the bronze route wire and 0.72 times decrease for the internal tin process wire. The 1.7 x 10^{23} n/m^2 irradiation for the internal tin process wire decreased the critical current severely and the critical magnetic field down to around 16 T. This was the first result that the high fluence neutron irradiation caused the degradation of both the critical current and the critical magnetic field in this type of a Nb_3Sn wire. Some irradiation defects will become the magnetic flux pining sites and strengthen the pining force. But the strong irradiation will damage the superconducting phase and degrade the properties.

1. Introduction

Large-scale plasma experimental devices have been designed and constructed in the world. One of the main targets is realization of fusion reaction, namely, deuterium - tritium reaction, which will generate 14 MeV neutrons. The ITER project is an international collaboration project (http://www.iter.org/) and the construction is in progress at Cadarache in France. The ITER is a pulse machine and will generate 14 MeV neutron periodically. To support the ITER project, JT-60SA project is running in Japan (https://www.qst.go.jp/site/jt60-english/). JT-60SA will not generate the fast neutrons, but the operation technology and the plasma research will give a lot of practical suggestions to the ITER project. The design of a helical system fusion reactor has been carried out and the three-dimensional neutron irradiation analysis was reported [1]. Regarding the neutronics analysis for ITER project, the pioneering research has been done and the local fluence distribution was reported [2]. In addition to the fusion application, high energy particle physics and the related accelerator engineering require the experimental data on the neutron irradiation effect of magnet materials when they design the devices. The energy level of the neutron coming out of the collision point is GeV class, and there is a big difference from the fusion. However, the systematic experimental facilities are anticipated in the world.

In 2000s, the investigation on the neutron irradiation effect on the superconducting magnet materials started at National Institute for Fusion Science under the collaboration research framework with Universities in Japan [3]. After the installation of 14 MeV neutron irradiation facility in Tokai campus of Japan Atomic Energy Agency (JAEA), the neutron irradiation research system with the fission reactor has been developed and 15.5 T superconducting magnet and a variable temperature insert (VTI) were installed in a radiation control area at Oarai center in Tohoku University [4,5]. The control system, the data acquisition system and others are still being improved little by little.

In this study, the neutron irradiation of 4.9 x 10^{22} n/m^2, 7.9 x 10^{22} n/m^2, 1.7 x 10^{23} n/m^2 were carried out and the critical current after irradiation was measured in the range of 8 T to 15.5 T. The neutron irradiation effect on the critical current of Nb_3Sn wires made by a bronze route and an internal tin process will be presented, and the mechanism of the change in the critical current will be discussed.

2. Test Facilities, Test Procedures and Sample Wire

The test facility for the superconducting properties measurement system consists of 15.5 T superconducting magnet, the VTI and the control and data acquisition systems. The sample current capacity is 500 A, and the sample is cooled down by thermal conduction with high purity aluminum rod (over 99.999%) connecting to two sets of G-M refrigerator. The samples were capsuled with helium gas and sent to Belgium Reactor II (BR2) in Belgium. After the irradiation in the fission reactor, the samples were sent back to the Oarai Center in Tohoku University, and the post irradiation experiments were carried out in the radiation control area. The recent status of the test facilities has been reported in [6,7].

There are three CERNOX sensors on the sample holder as shown in Figure 1. CU type sensors were attached on both positive and negative electrodes and AA type sensor was equipped on the positive electrode. One test result is shown in Figure 2. The temperatures were measured at the rate of 1 Hz or 2 Hz and the sample current was recorded at the rate of 1 or 2 Hz and 10 Hz, independently. It is noticed that the temperatures of the both electrodes rise by joule heating when the current flow started. There will be several reasons for the temperature rise. Here, the joule heating was considered as the strongest factor. The behavior of the temperature rise depended on each test. The sample temperature was not always constant because the sample was cooled down by thermal conductivity. But the test temperature was below 5 K. The base temperature will be shown at each data set.

Since the sample temperature rises during the current flow due to the joule heating, the relation

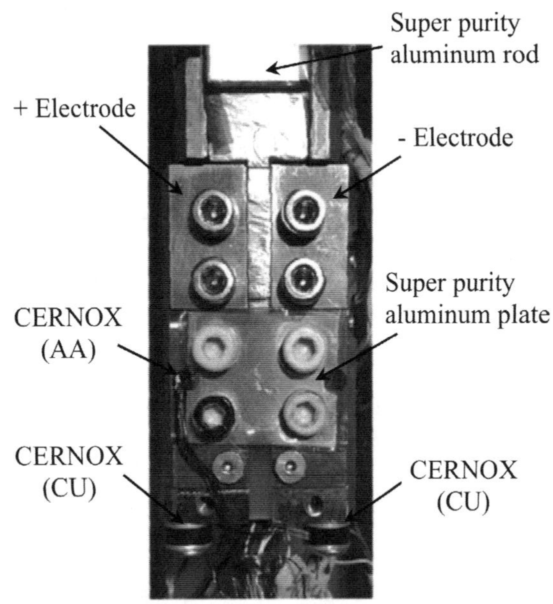

Figure 1. Sample holder. The space between bus bar and sample holder was filled with super purity aluminum. Sample was set on back side.

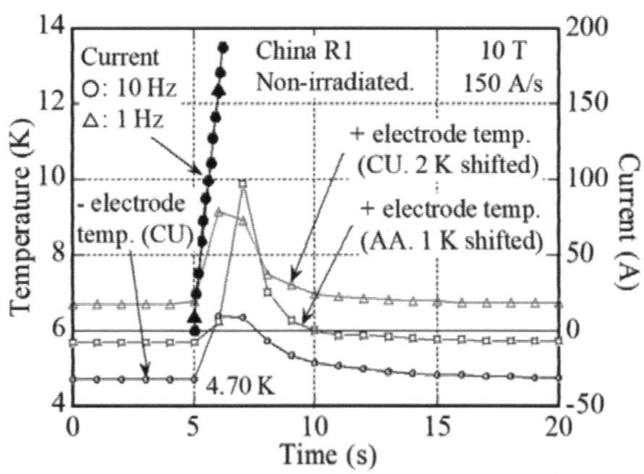

Figure 2. One example of the temperature rise of sample holder during the current flow.

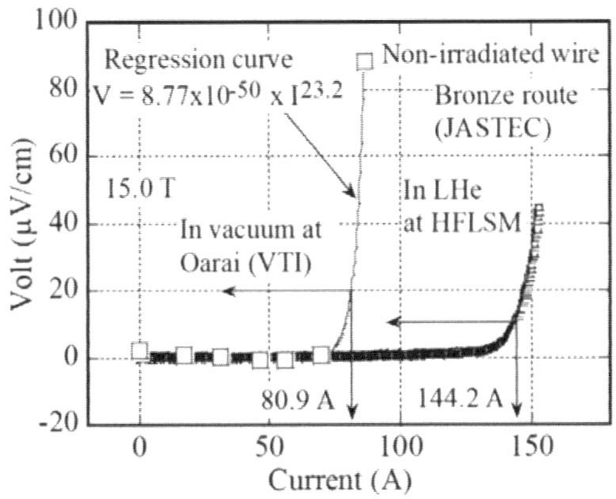

Figure 3. Comparison of V-I curves measured in vacuum with VTI at Oarai center and in LHe at HFLSM.

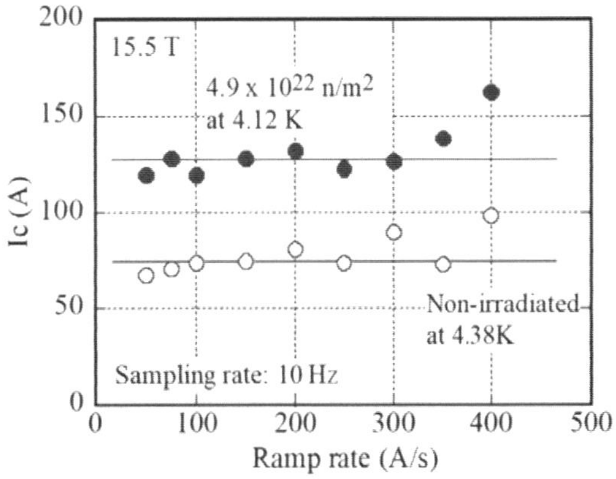

Figure 4. Effect of ramp rate on of irradiated and non-irradiated wires at 10 T.

between the voltage (the tap distance was about 5 mm) and the current (V-I curve) becomes different from that in LHe test. The example is shown in Figure 3 [7]. The wire was the bronze route Nb_3Sn. The V-I curve obtained with VTI was approximated by a power function of $V = C \times I^n$, where C and n are constants, and the current at 20 μV/cm was defined as the critical current of the wire under the given condition. In case of the critical current in LHe, the 10 μV/cm was adopted as show in Figure 3.

As the temperature rise by the joule heating depends on the ramp rate of the current, the effect on the ramp rate on the critical current was investigated. The one results are shown in Figure 4. It is noticed that the fast ramp rate shows rather large critical current because the thermal conduction needs a time to transfer the heat to the sample wire which is generated in the sample holder by the joule heating. In this study, the ramp rate of 150 A/s was adopted as a measurement reference condition. As the sampling rate of the V-I curve was 10 Hz in the VTI system, the lower ramp rate was adopted when a certain number of data sets was required.

Two types of Nb_3Sn samples were prepared. One is a bronze route Nb_3Sn wire and the other is an internal tin process Nb_3Sn wire. The cross section of both wires is shown in Figure 5. The bronze route wire provided JASTEC in Japan has very fine filaments and the internal tin wire provided by WST in China has 20 tin core filaments. Both wires were 32 mm long and 0.8 mm diameter, and they were heat treated at 650 degree in C for 240 hours. The total content of tin of both wires is not clear but the internal tin process wires looks contain more tin than the bronze process wire has.

The three neutron irradiation tests were carried out. In the first test, the neutron fluence (> 0.1 MeV) of 4.9×10^{22} n/m^2 was achieved, and 7.9×10^{22} n/m^2 and 1.7×10^{23} n/m^2 of the fluence were irradiated in the second and the third irradiation tests, respectively. It must be written about the

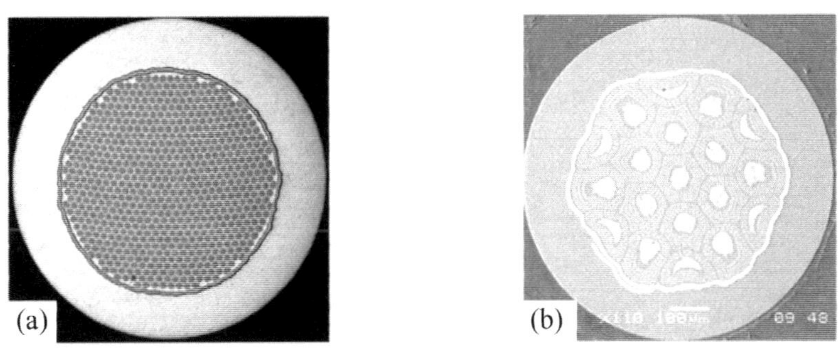

Figure 5. Cross section of tested Nb_3Sn wires. (a) JASTEC in Japan and (b) WST in China.

IOP Conf. Series: Materials Science and Engineering **756** (2020) 012013 doi:10.1088/1757-899X/756/1/012013

residual gamma ray dose after the irradiation. The same size wires were inserted in the aluminum capsule and irradiated in BR2. After the irradiation, the wires were sent back to Oarai center and the gamma ray on the surface was measured. So, the neutron fluence and the decay time were the same. The measured values were 30.00 mSv/h for the internal tin process wire and 0.03 mSv/h for the bronze route process wire. Although the detailed investigation is required, it suggests that the isotope of Sn would emit strong gamma ray after irradiation.

3. Results and Discussion

The change in the critical current of the bronze route wire irradiated to 4.9 x 10^{22} n/m^2 is shown in Figure 6 together with the results of I_C/I_{C0}. The I_C in LHe is also plotted in the figure. The I_C with VTI is smaller than that in LHe because of the temperature rise by the joule heating as explained before. When the wire was irradiated to 4.9 x 10^{22} n/m^2, the I_C increased in the range of 8 T to 15.5 T, and the ratio of the I_C of the 4.9 x 10^{22} n/m^2 irradiation to that of non-irradiated wire (I_{C0}) becomes about 1.75 and there is no clear tendency to increase or decrease of I_C/I_{C0} against the magnetic field. The results of the bronze route wire irradiated to 7.9 x 10^{22} n/m^2 is shown in Figure 7. The critical current after the irradiation deceased in the range of 8 T to 15.5 T, and the ratio of I_C/I_{C0} is about 0.72. It is noticed that the almost same scatter of I_C/I_{C0} could be seen in both diagrams. Since the I_C/I_{C0} keeps the same value in the region of 8 T to 15.5 T, it is considered that the B_{C2} would not change significantly under such irradiations.

The reason of the change in the critical current has been recognized as follows: The neutron irradiation would cause a lot of small damage in the matrix, which is an atom-level irregular order like a cascading or knocking, and these defects would become flux pinning sites. When a certain amount of such damage is installed in the matrix, the pinning force would be increased so that the critical current would improve. However, as total amount of defects exceeds, the matrix becomes dirty and the superconducting phase is deteriorated resulting in the decrease of the critical current.

The experimental result in the case of further irradiation are shown in Figure 8 and 9. The samples are the internal tin process wires and irradiated to 7.9 x 10^{22} n/m^2 (Figure 8) and 1.7 x 10^{23} n/m^2 (Figure 9). The data were obtained under the current ramp rate of 150 A/s. In Figure 8, the critical current measured in LHe are plotted together. As in the results shown in Figure 6 and 7, the I_C/I_{C0} becomes almost constant of 1.03 in the range from 8 T to 15.5 T. There is a difference of I_C/I_{C0} between the bronze route Nb_3Sn and the internal tin Nb_3Sn when the 7.9 x 10^{22} n/m^2 irradiation was carried out. The reason is not clear at present and the further investigation is expected to make it clear. In case of 1.7 x 10^{23} n/m^2 irradiation (Figure 9), the I_C/I_{C0} decreases as the magnetic field increases, and the result does not show the constant I_C/I_{C0}. When the critical current is below 50 A, there are only three or four sets of the data, and rather large uncertainties are considered. Therefore, the parentheses

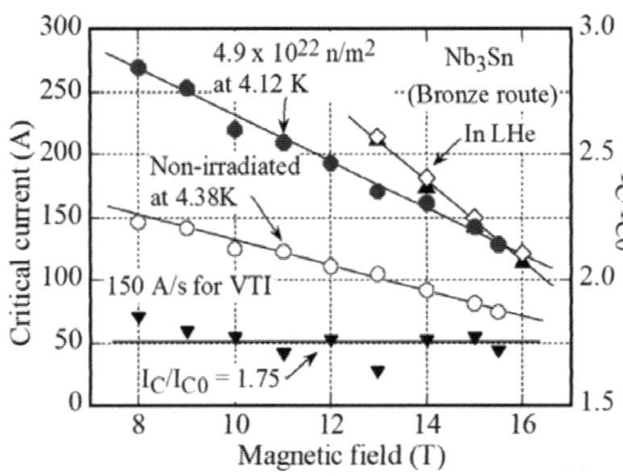

Figure 6. Change in critical current and I_C/I_{C0} against magnetic field. Bronze route Nb_3Sn wire. Neutron irradiated to 4.9 x 10^{22} n/m^2.

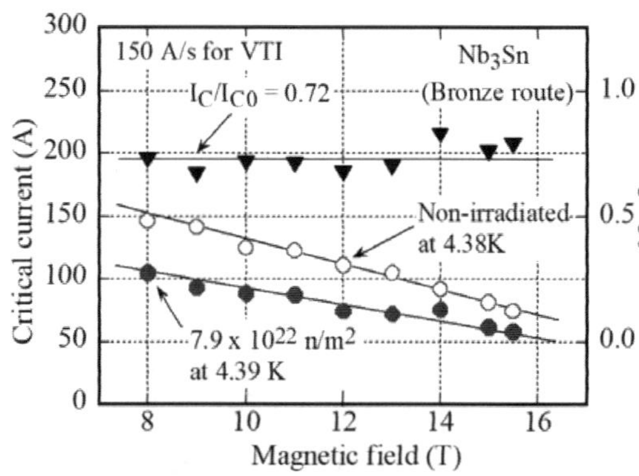

Figure 7. Change in critical current and I_C/I_{C0} against magnetic field. Bronze route Nb_3Sn wire. Neutron irradiated to 7.9 x 10^{22} n/m^2.

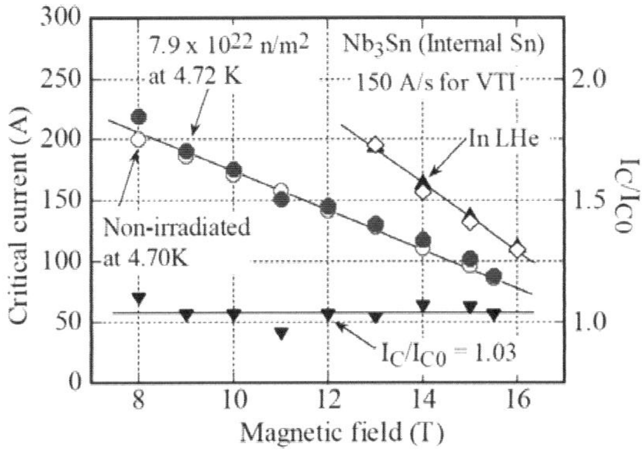

Figure 8. Change in critical current and I_C/I_{C0} against magnetic field. Internal tin Nb_3Sn wire. Neutron irradiated to 7.9 x 10^{22} n/m^2.

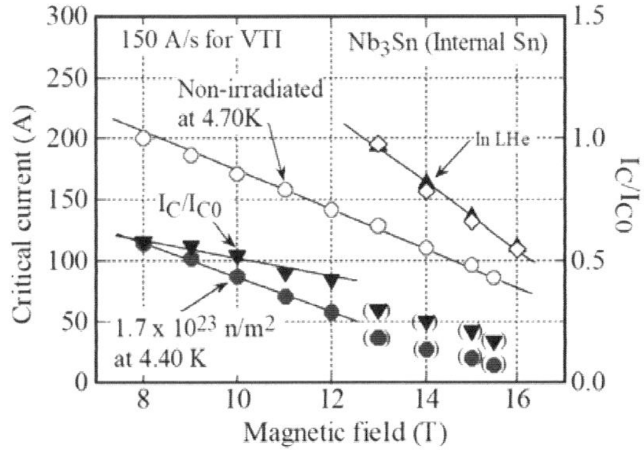

Figure 9. Change in critical current and I_C/I_{C0} against magnetic field. Internal tin Nb_3Sn wire. Neutron irradiated to 1.7 x 10^{23} n/m^2.

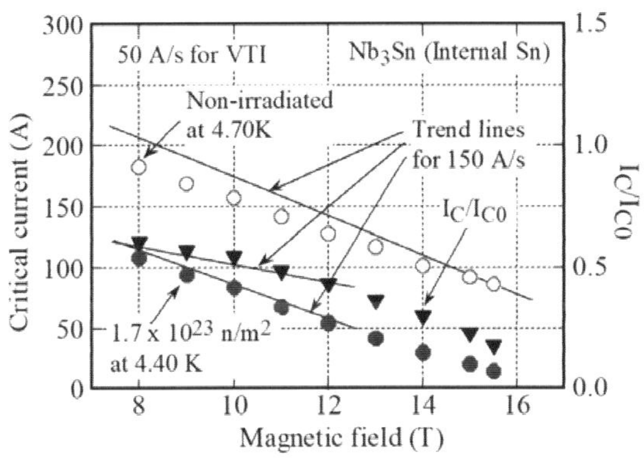

Figure 10. Change in critical current and I_C/I_{C0} against magnetic field. Internal tin Nb_3Sn wire. Neutron irradiated to 1.7 x 10^{23} n/m^2. Current ramp rate was 50 A/s.

are attached on these data. To make it sure, the sample ramp rate of 50 A/s was applied, and the larger number of experimental data points were obtained. The results are shown in Figure 10. The data of non-irradiated wire were also measured at the rate of 50 A/s. The trend lines in the Figure 10 are the same as those in Figure 9. The trend lines are still fitting the experimental data, so it is recognized that there is not significant difference between 150 A/s and 50 A/s data sets, and that the critical current decreases almost straightly to zero, though the critical current of non-irradiated wire decreases gradually. When these experimental plots are extrapolated to zero, the critical magnetic field is expected to be around 16 T. It means that the further neutron irradiation degrades the critical magnetic field and the wire is getting to lose the superconductivity, and finally it is considered that it will become the material which does not show the superconductivity anymore.

This experimental data on the deterioration of the critical magnetic field is the world first finding fact. If one can measure the critical current at 8 T or 10 T after the 1.7 x 10^{23} n/m^2 irradiation, it will be reported that there is still a certain critical current exist and a certain value of the I_C/I_{C0}. The important issue is that the superconducting wire is damaged and becomes dirty, and it is losing the superconductivity already, especially in the higher magnetic field. So, at this stage, a transition behavior from the superconducting to the normal conducting is being observed, which means no constant I_C/I_{C0} in the range of 8 T to 15.5 T. When the stronger magnetic field is applied to the sample irradiated to 1.7 x 10^{23} n/m^2, the flux pinning is easily coming off.

The obtained data on I_C/I_{C0} in this study were plotted against the neutron fluence as shown in Figure 11. In the figure, several experimental data sets of Nb_3Sn wires were collected from the

IOP Conf. Series: Materials Science and Engineering **756** (2020) 012013 doi:10.1088/1757-899X/756/1/012013

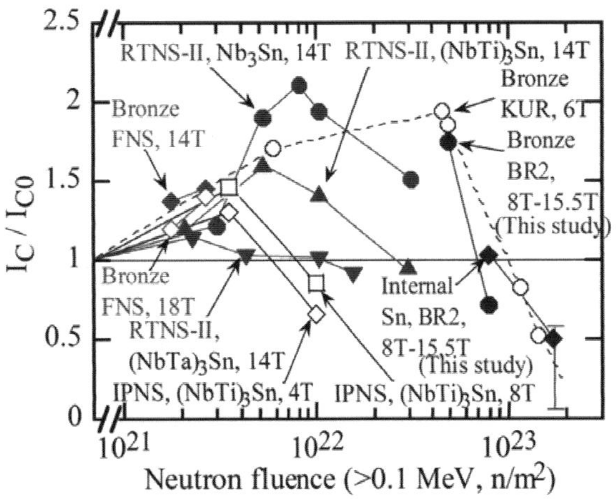

Figure 11. I_C/I_{C0} against neutron fluence. Published data were plotted together with the data obtained in this study.

published papers and plotted. (FNS; Fusion Neutron Sources, RTNS; Rotating Target Neutron Source, IPNS; Intense Pulsed Neutron Source, KUR; Kyoto University Reactor). The open round marks show the results of the bronze route Nb_3Sn wire irradiated at KUR and the I_C measurement was carried out at 6 T. The test results in this study are plotted on the almost same line as the data in KUR except for the data of 1.7 x 10^{23} n/m^2 irradiation. Therefore, it is noticed that there is no effect of the magnetic field under 8 T to 15.5 T when the I_C/I_{C0} is over about 1.0. It means that the induced irradiation defects have the enough ability to keep the flux pinning even under the 15.5 T, and that there is a potential that the critical current of the non-irradiated wire would be improved more. However, when the wire is irradiated much more, the I_C/I_{C0} drops below 1.0 and the dependence on the magnetic field comes up remarkable. Generally, the Nb_3Sn wire has the critical magnetic field of over 24 T. But, as the neutron fluence increases, the I_C-B_{C2} relationship would shift to downward, the critical magnetic field decreased down to around 16 T in case of 1.7 x 10^{23} n/m^2 irradiation. This is unexpected big drop of B_{C2}.

The further irradiation will install much irradiation defects and the matrix will become dirty. The dirty matrix will not sustain the flux pinning anymore resulting in the decrease of the critical magnetic field. Due to the degradation of the critical magnetic field, the dependence of the critical current on the magnetic field becomes significant and the critical current drops near to zero under the higher magnetic field.

To confirm the mechanisms of the change in the I_C/I_{C0} mentioned above, the further investigations are planned and would be carried out in near future.

4. Summary

The neutron irradiation tests on Nb_3Sn wire were carried out at BR2 in Belgium under the condition of the cooling water temperature of 50 degree in Celsius. The fluence of over 0.1 MeV neutron was 4.9 x 10^{22} n/m^2, 7.9 x 10^{22} n/m^2 and 1.7 x 10^{23} n/m^2.

The 4.9 x 10^{22} n/m^2 irradiation increases the I_C and the 7.9 x 10^{22} n/m^2 irradiation showed the almost same I_C as that of non-irradiated sample. The 1.7 x 10^{23} n/m^2 irradiation decreased the I_C and caused the drastic drop of B_{C2}.

The neutron irradiation would make a lot of irradiation defects in the matrix and these defects would become the flux pining sits and reinforce the flux pining resulting in the increase of I_C. However, further irradiation would damage the superconducting phase and decrease the I_C and B_{C2}. The big degradation of the B_{C2} was shown in this paper for the first time in the world.

The irradiated wires were prepared by the bronze route process and the internal Tin process. It was hard to make notes about the difference of the irradiation properties on both wires in this study. The further investigation must be carried out.

References

[1] Tanaka T, Sagara A, Youssef M Z 2006 *Fusion Eng. Design* **81** 2761
[2] Sato S, Iida H and Nishitani T 2002 *J. Nucle. Sci. Technol.* **39** 1237
[3] Nishimura A, Hishinuma Y, Tanaka T, Muroga T, Nishijima S, Shindo S, Takeuchi T, Ochiai K, Nishitani T and Okuno K 2005 *Fusion Eng. Design* **75-79** 173
[4] Nishimura A, Takeuchi T, Nishijima S, Ochiai K, Nishijima G, Watanabe K, Narui M, Kurishita H and Shikama T 2011 *J. Nucle. Mat.* **417** 842
[5] Nishimura A, Takeuchi T, Nishijima G, Watanabe K, Ochiai K, Li D and Shikama T 2013 *Fusion Eng. Design* **88** 1551
[6] Nishimura A, Miyata H, Yoshida M, Iio M, Suzuki K, Nakamoto T, Yamazaki M and Toyama T 2017 *Mat. Sci. Eng.* **279** 012029 doi:10.1088/1757-899X/279/1/012029
[7] Nishimura A, Nakamoto T, Yoshida M, Iio M, Yamazaki M, Toyama T, Hishinuma Y, Oguro H and Awaji S 2019 *Supercon. Sci. Technol.* **32** 024004 doi: 10.1088/1361-6668/aaf903

Acknowledgments

The author would like to express sincere thanks to Prof. Tatsuo Shikama at Hachinohe Institute of Technology, Prof. Yoshimitsu Hishinuma at National Institute for Fusion Science, Mr. Masanori Yamazaki at Tohoku University and Prof. Laifeng Li at Technical Institute of Physics and Chemistry for their encouragement and contributions.

This work was supported by the NIFS working budget (NIFS11UFFF036-2) and the Collaborative Research Program (NIFS17KECT001 and NIFS19KECF029). Also, it was supported by Grant-in-Aid for Scientific Research (19K03804) and Chinese Academy of Sciences President's International Fellowship Initiative, Grant No. 2019VEA0017.

This work was performed at the International Research Center for Nuclear Materials Science in the Institute for Materials Research, Tohoku University.

IOP Conf. Series: Materials Science and Engineering **756** (2020) 012014 doi:10.1088/1757-899X/756/1/012014

Effect of layer thickness on structural, morphological and superconducting properties of Nb_3Sn films fabricated by multilayer sequential sputtering

M N Sayeed[1,2], U Pudasaini[3], C E Reece[4], G V Eremeev[4], H E Elsayed-Ali[1,2]

[1]Department of Electrical and Computer Engineering, Old Dominion University, Norfolk, VA 23529, USA

[2]Applied Research Center, 12050 Jefferson Avenue, Newport News, VA 23606, USA

[3]Applied Science Department, The College of William and Mary, Williamsburg, VA 23185, USA

[4]Thomas Jefferson National Accelerator Facility, Newport News, VA 23606, USA

E-mail: msaye004@odu.edu

Abstract. Superconducting Nb_3Sn films can be synthesized by controlling the atomic concentration of Sn. Multilayer sequential sputtering of Nb and Sn thin films followed by high temperature annealing is considered as a method to fabricate Nb_3Sn films, where the Sn composition of the deposited films can be controlled by the thickness of alternating Nb and Sn layers. We report on the structural, morphological and superconducting properties of Nb_3Sn films fabricated by multilayer sequential sputtering of Nb and Sn films on sapphire substrates followed by annealing at 950 °C for 3 h. We have investigated the effect of Nb and Sn layer thickness and Nb:Sn ratio on the properties of the Nb_3Sn films. The crystal structure, surface morphology, surface topography, and film composition were characterized by X-ray diffraction (XRD), scanning electron microscopy (SEM), atomic force microscopy (AFM), and energy dispersive X-ray spectroscopy (EDS). The results showed Sn loss from the surface due to evaporation during annealing. Superconducting Nb_3Sn films of critical temperature up to 17.93 K were fabricated.

1. Introduction

Nb_3Sn is one of the type II superconductors that has wide applications from DC high-field magnets to radiofrequency cavities [1–5]. Magnetron sputtering is considered as one of the fabrication methods used to synthesize Nb_3Sn films. Nb_3Sn can be fabricated either from a stoichiometric Nb_3Sn single target [6, 7], or from sputtering of Nb and Sn followed by annealing [8]. Deposited films should maintain a controlled Sn composition range with atomic composition of 17–26% to obtain superconducting Nb_3Sn films [3]. For multilayer growth, the atomic Sn concentration of the films can be controlled by varying the thickness of the Nb and Sn layers.

We report on the structural, morphological and electrical properties of Nb_3Sn films fabricated by multilayer sequential sputtering of Nb and Sn films on sapphire substrates followed by annealing at 950 °C for 3 h. The temperature and time were selected based on previously reported work on Nb_3Sn fabrication [8]. The film properties were characterized for different layer thicknesses of Nb and Sn multilayers. The role of layer thickness was studied by varying the thickness of the Nb layers while keeping the Sn layer thickness constant, which adjusts the stoichiometry and by varying the thickness of both Nb and Sn layers while keeping the Nb:Sn thickness ratio constant.

2. Experimental details

2.1. Fabrication

Nb_3Sn films were grown by annealing sequentially-sputtered multilayers of Nb and Sn. Some details on the deposition chamber, substrate, and target materials were described elsewhere [8]. Multiple layers of Nb and Sn with different thicknesses were deposited sequentially with the first layer Sn and the final layer Nb. The final Nb layer minimizes Sn loss during post-deposition annealing. Films were deposited in a high vacuum chamber at 4×10^{-3} mbar Ar sputtering gas (99.999% purity) with a flow rate of 20 SCCM. The substrate holder was rotated at 30 rpm throughout the sputtering process to maintain a homogenous deposition. Information about the film thickness is shown in Table 1. In the study of film stoichiometry, we chose a constant thickness of 10 nm of Sn and four thicknesses (10, 20, 30 and 40 nm) of Nb layers. These four conditions are referred to as conditions 1–4 in our results. The total thickness for all four conditions was kept constant (~900 nm) by varying the number of layers. In another set of films, we kept the Nb-to-Sn thickness ratio 2:1. For all films on these conditions, the total thickness was ~1.2 μm. These four conditions will be referred as conditions 5–8 in our results. All films were annealed in a separate vacuum furnace at 950 °C for 3 h with a temperature ramp rate of 12 °C/min.

Table 1. Deposition conditions of the films.

	Condition	Nb layer thickness (nm)	Sn layer thickness (nm)	Nb:Sn	Total thickness (nm)
Films with different stoichiometry	1	10	10	1	900
	2	20	10	2	900
	3	30	10	3	900
	4	40	10	4	900
Films with fixed stoichiometry	5	10	5	2	1200
	6	20	10	2	1200
	7	50	25	2	1200
	8	200	100	2	1200

2.2. Characterization

X-ray diffraction (XRD) patterns of the films were obtained from Rigaku Miniflex II X-ray diffractometer with Cu-Kα radiation. The crystallite sizes of the films were measured from the peak broadening using Scherrer's equation [9]. The film microstructure was observed by field-emission scanning electron microscopic (SEM) images (FESEM S-4700, Hitachi, Japan). A Noran 6 energy dispersive X-ray spectroscopy (EDS) detector connected to a Jeol JSM 6060 LV scanning electron microscope system was used to quantitatively measure the surface composition. EDS was performed using 15 kV accelerating voltage on a surface covering 1.2 mm^2 area from each condition and the composition was measured on 5 different areas. The surface roughness of the samples were measured from the AFM images obtained by a Digital Instrument Dimension 3100 Atomic Force Microscope (AFM) in tapping mode. Root mean square (RMS) roughness was obtained over 5 × 5 μm scan area. Superconducting T_c's were measured by four-point probe resistance measurement down to cryogenic temperature. T_c was calculated from the mean value of T^{90} and T^{10} (temperature when the resistance is 90% and 10% of the final resistance before transition respectively) and the superconducting transition width ΔT_c was calculated from the difference of T^{90} and T^{10}.

3. Results

3.1. Film with different stoichiometry

3.1.1. Structural properties. The Sn composition of as-deposited and annealed films are shown in Table 2. The Sn composition was reduced with increasing Nb layer thickness on as-deposited films. About 43.6% Sn was observed on films with condition 1, where both Nb and Sn layer thickness was maintained at 10 nm. All annealed films showed Sn loss after annealing. Large amount of Sn loss was observed on films with condition 1. The Sn composition changed from ~43.6% to ~23.8%. This large amount of Sn loss occurred due to sublimation of Sn during annealing. All annealed films showed Sn composition ~20–23%. The composition of the as-deposited films was ~16% in condition 4, however, the annealed film had ~20% Sn. Annealed film showed more Sn due to the uniform distribution of Sn throughout the film after annealing.

Table 2. Summary of structural, morphological, and superconducting data.

Condition	At. % Sn as-deposited	At. % Sn annealed	RMS Roughness (nm)	T_c (K)	ΔT_c (K)
1	43.6	23.8	66.2±24.3	17.93	0.02
2	28.4	23.0	31.8±3.2	17.84	0.03
3	20.3	21.4	31.6±0.9	17.56	0.09
4	16.0	20.2	26.8±1.2	17.54	0.11
5	29.5	23.1	28.2±2.9	17.82	0.02
6	28.9	21.7	31.5±1.2	17.83	0.01
7	27.2	22.4	41.0±5.5	17.83	0.02
8	20.0	21.4	64.7±3.2	17.84	0.03

Figure 1 shows the SEM and AFM images of the films. For condition 1, where the Nb and Sn layer thicknesses were 10 nm, several clusters separated by large voids were observed. The grains were not visible on the surface of this film. However, for the other three conditions, the films had surfaces with uniformly distributed grains showing some voids. The film grown with condition 4, where the Nb:Sn thickness ratio was 4:1, showed smaller grains than that for other conditions. The surface roughness for condition 1 was higher due to the film having clusters of larger heights and deeper voids. The RMS roughness for this condition is 66.2±24.3 nm. The large standard deviation originated from the irregular distribution of the clusters and voids. Films with the other three conditions had smoother surfaces. The average roughness was measured to be 31.8±3.2, 31.6±0.9, and 26.8±1.2 nm for conditions 2, 3, and 4, respectively.

The XRD patterns of the films with different Nb thicknesses (10, 20, 30, 40 nm) and constant Sn thickness of 10 nm after annealing at 950 °C for 3 h are compared in Figure 2 (a). All annealed films were polycrystalline Nb_3Sn films showing (200), (210), (222), (320), (321), (400), (420), (421) diffraction orders. The crystallite size and lattice parameter corresponding to Nb_3Sn (210) diffraction peak of the films as a function of Nb:Sn thickness ratio are shown in Figure 1 (b). The crystallite size decreased with increasing Nb layer thickness, which is in agreement with the grain size observed in SEM and AFM. Lattice parameters calculated from the *d* value of XRD peak are less than lattice parameter of bulk Nb_3Sn (5.290 Å). This is because the Sn composition in the film is less than ideal condition of 25%.

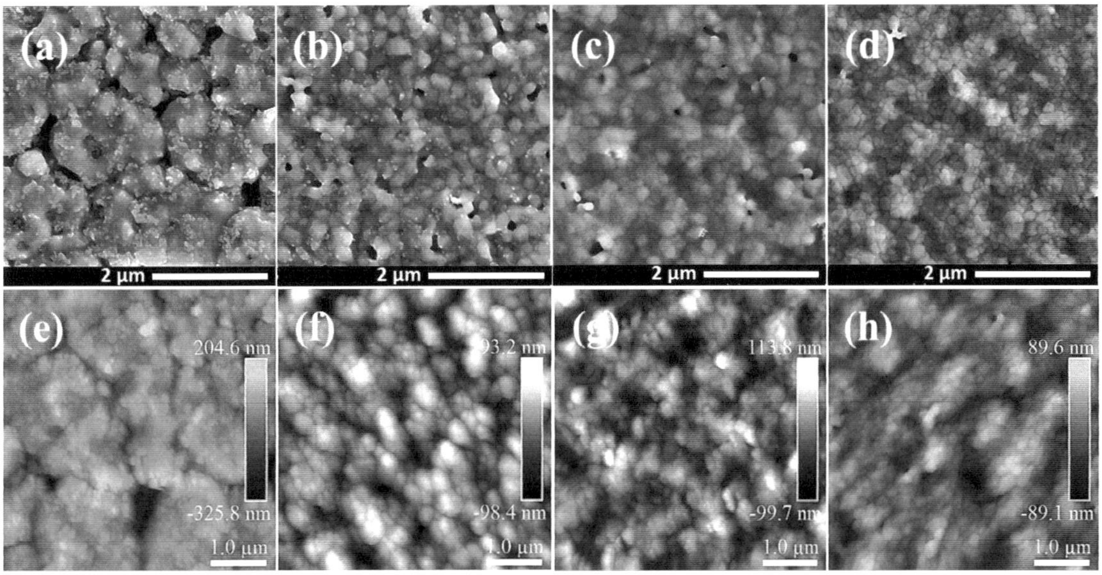

Figure 1. SEM and AFM images of the films: (a) and (e) condition 1, (b) and (f) condition 2, (c) and (g) condition 3, (d) and (h) condition 4.

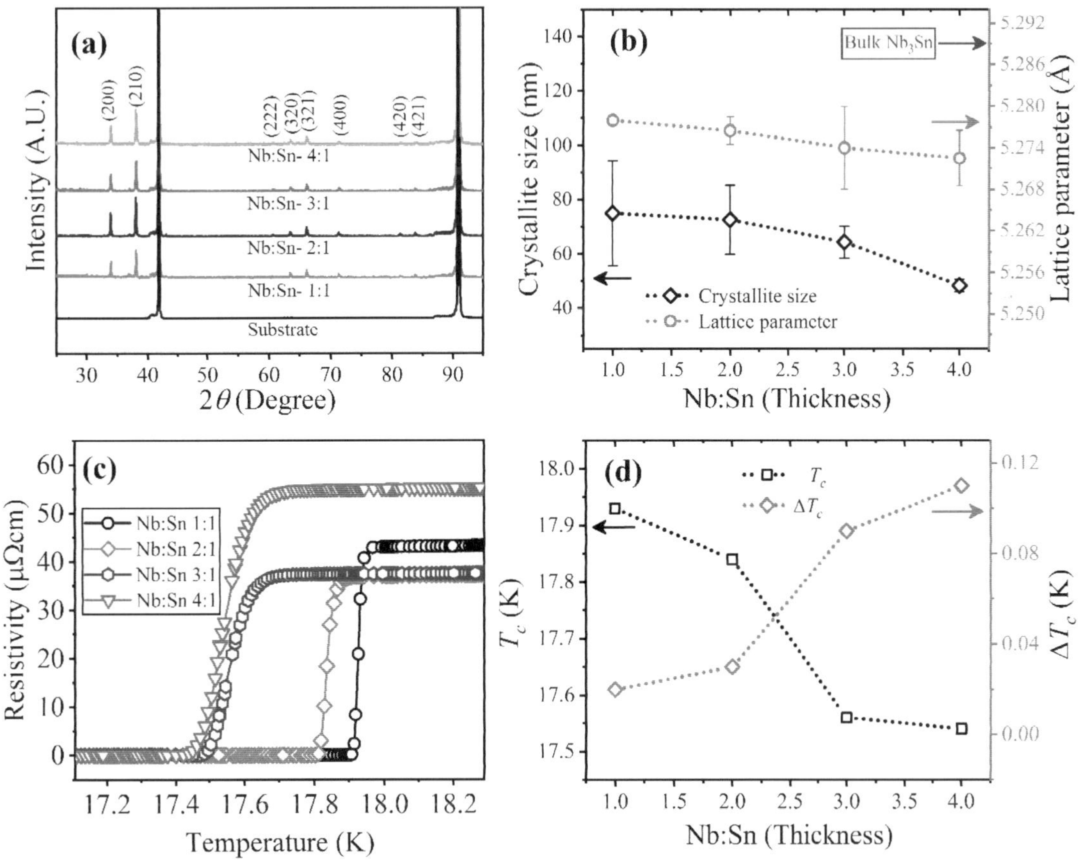

Figure 2. (a) X-ray diffraction patterns of the films with conditions 1–4, (b) crystallite size and lattice parameters of the films as a function of Nb:Sn film thickness, (c) resistivity versus temperature of the films, (d) corresponding T_c and ΔT_c for different Nb:Sn layer thicknesses.

3.1.2. Superconducting properties. The resistivity versus temperature is shown in Figure 2 (c). The calculated T_c, and ΔT_c from the graph are shown in Table 2 and Figure 2 (d). All films exhibited good superconducting properties. The highest critical temperature is observed on the film fabricated with condition 1, where the Nb and Sn thicknesses were the same. Better T_c at this condition was obtained due to higher Sn composition on the annealed films. It has been reported that T_c of Nb_3Sn is dependent on the Sn composition of the films [1]. The transition width also became wider with increasing Nb:Sn.

3.2. Films with similar stoichiometry

3.2.1. Structural properties. All annealed films (conditions 4–8) had Sn composition in the range of 21–23%. Figure 3 shows the SEM and AFM images of annealed films of different thicknesses. These films showed uniformly distributed grains with some voids, similar to those fabricated under conditions 1–4. However, for films fabricated under condition 8 (Figure 3 (d) and (h)), the presence of deep voids was not observed. Surface of the film for this condition shows small grains aggregating into clusters. The color pattern on the AFM image of the surface in Figure 3 (h) confirmed that the clusters have different heights, which caused a relatively rough surface. RMS roughness of the films are shown in Table 2. The surface became increasingly rough with increased layer thickness. The lowest roughness of 28.2±2.9 nm was observed for films of condition 5 (Sn layer thickness 5 nm), whereas films of condition 8 (Sn layer thickness 100 nm) had the highest surface roughness of 64.7±3.2 nm.

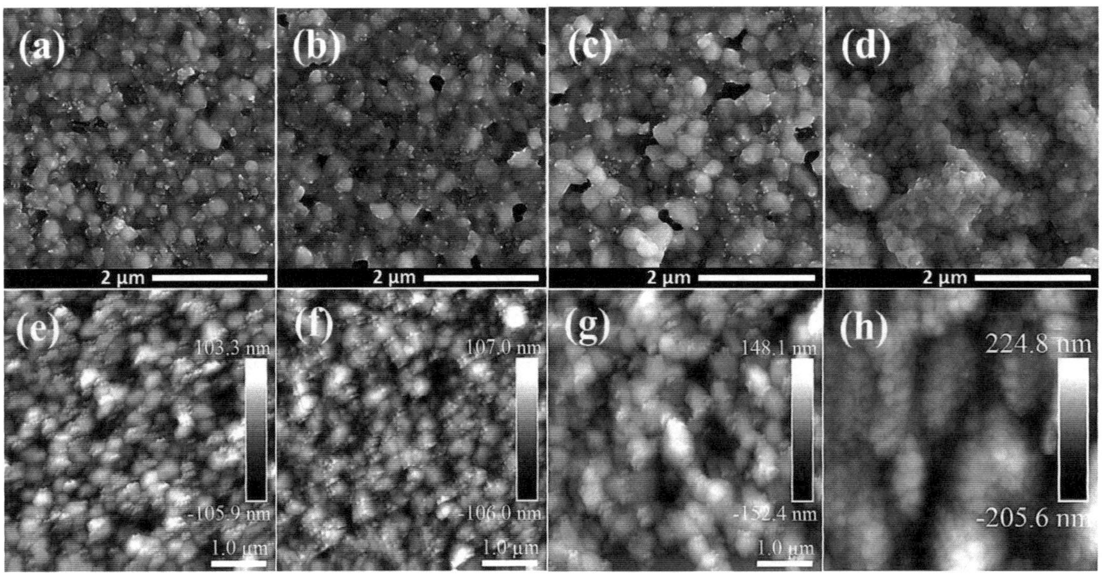

Figure 3. SEM and AFM images of the films: (a) and (e) condition 5, (b) and (f) condition 6, (c) and (g) condition 7, (d) and (h) condition 8.

Figure 4 (a) shows the XRD patterns of annealed films with condition 5–8. Similar to the films fabricated under condition 1–4, these films also exhibit diffraction peaks corresponding to Nb_3Sn (200), (210), (222), (320), (321), (400), (420), (421) orders. The lattice parameter and crystallite size of (210) reflection is shown in Figure 4 (b). The lattice parameter and crystallite size correspond to the film fabricated under condition 2 is also plotted. The lattice parameter of the films showed little variation due to the almost similar Sn composition of the films. The crystallite size of the films also did not vary significantly with increased layer thickness. Films for condition 5 had crystallites with an average size of 72.1±5.2 nm, whereas the films for condition 8 had crystallites with an average size of 68.5±2.5 nm.

3.2.2. Superconducting properties. The surface resistivity of the films for different Sn layer thicknesses is shown in Figure 4 (c) and the corresponding T_c and ΔT_c data are shown in Table 2 and Figure 4 (d), respectively. The T_c and ΔT_c of all films are close. Slight increased T_c is observed on the film for condition 8, however, the ΔT_c also increased.

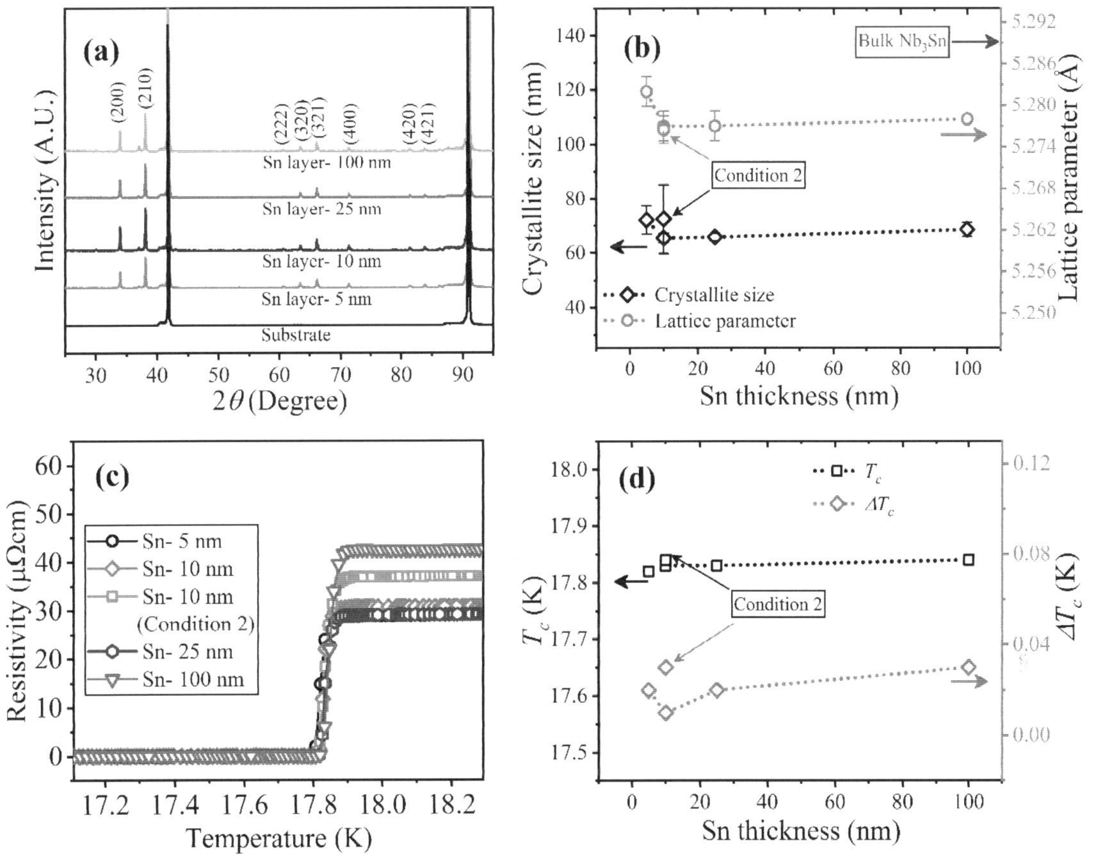

Figure 4. (a) X-ray diffraction patterns of the films with condition 5–8, (b) crystallite size and lattice parameters of the films as a function of Sn film thickness, (c) resistivity vs temperature curve of the films, (d) corresponding T_c and ΔT_c as a function of Sn layer thickness with condition 5–8.

4. Discussion

Sn composition plays an important role on the superconducting properties of Nb_3Sn. In SRF cavities, Nb_3Sn films coated by conventional Sn vapor diffusion method resulted to Sn deficient patchy regions which degraded the performance of SRF cavities [10, 11]. Therefore, it is important to maintain the stoichiometry of the films after annealing. We have experienced Sn loss for almost all deposition conditions after annealing. This Sn loss resulted from the Sn diffusion and evaporation from the surface. The Sn composition for annealed films obtained from EDS in conditions 3, 4 and 8 are higher than that of as-deposited films. This is due to the limitation of the spatial resolution of EDS. For 15 keV electron, the Anderson-Hasler equation [12] gives the X-ray transmission fraction of 0.68 μm for Nb-Lα lines and 0.76 μm for Sn-Lα lines, which is shorter than the thicknesses of the deposited films. Because of the short X-ray escape depth, layer ordering and the individual layer thickness affects the apparent EDS composition. Films, where the top layer is thick Nb, such as films with condition 3, 4, and 8, appear Nb-rich in EDS measurements. The annealed film showed more Sn due to the uniform distribution of Sn throughout the film after annealing.

The critical temperature of Nb_3Sn varies from 6 to 18.3 K when the Sn composition varies from 17 to 26% [1]. According to the critical temperature versus atomic Sn content plot reported by Godeke [1], the T_c of films with 22% Sn should be ~10–12 K whereas our fabricated films had T_c of ~17.83 K. One possible explanation of higher T_c on our fabricated films is Sn content variation in Nb_3Sn. Depth profile of the films can provide the distribution of Sn content throughout the films. Though we reported ToF-SIMS depth profile of the films coated on Nb substrates before [13], we have not observed it for films on sapphire substrates. Off-stoichiometric Nb_3Sn films with high T_c were observed previously on the films grown on sapphire substrate by magnetron sputtering from a single target. Kampwirth *et al.* reported Nb_3Sn film on sapphire substrate with superconducting transition range of 17–17.5 K where atomic Sn composition was ~22% [14]. Recently, Ilyina *et al.* reported Nb_3Sn films with a T_c of 17.5 K on Al_2O_3 substrate whereas films of same condition on copper substrates resulted to reduced T_c of 14.7 K [7].

The surface morphology of the films is affected by the coating parameters. It is possible to change the surface roughness varying the thickness of Nb and Sn layers. The grain sizes of the films were also varied when Nb:Sn ratio was varied. The cross-section image of these films could give more detailed idea about the grain structure near the surface as well as the interface. The technique has been used on vapor-diffused samples [15], but not yet on the present films.

5. Conclusion

Nb_3Sn films were fabricated by sputtering multilayers of Nb and Sn with varied layer thicknesses followed by annealing. Throughout annealing, evaporation of Sn from the surface and growth of Nb_3Sn by diffusion of Sn into Nb surface occur simultaneously. Sn loss is the outcome of the competition of these two processes. Sn loss depends on annealing parameters (annealing temperature and annealing time). For annealing at 950 °C for 3 h, Sn composition were 20–23% for all coating conditions. The surface morphology and the roughness of the films varied with the thickness of the layers. For different Nb:Sn thicknesses, the grain size and surface roughness decreased with increasing thickness ratio. For different layer thicknesses with constant thickness ratio, the surface roughness increased with increasing thickness of the layers, whereas the grain size did not vary significantly. All films showed good superconducting properties with a superconducting critical temperature T_c up to 17.93 K.

Acknowledgments

This material is based upon work supported by the U.S. Department of Energy, Office of Science, Office of Nuclear Physics under Contract No. DE-AC05-06OR23177. The authors acknowledge Michael J. Kelley, Gianluigi Ciovati for their suggestions and Joshua Spradlin for his help with Tc measurement.

References

[1] A Godeke, "A review of the properties of Nb_3Sn and their variation with A15 composition, morphology and strain state," *Superconductor Science and Technology*, vol. 19, no. 8, pp. R68-R80, 2006.

[2] X. Xu, "A review and prospects for Nb_3Sn superconductor development," *Superconductor Science and Technology*, vol. 30, no. 9, p. 093001, 2017.

[3] S. Posen and D. L. Hall, "Nb_3Sn superconducting radiofrequency cavities: fabrication, results, properties, and prospects," *Superconductor Science and Technology*, vol. 30, no. 3, p. 033004, 2017.

[4] A.-M. Valente-Feliciano, "Superconducting RF materials other than bulk niobium: a review," *Superconductor Science and Technology*, vol. 29, no. 11, p. 113002, 2016.

[5] U. Pudasaini et al., "Recent Results from Nb_3Sn Single Cell Cavities Coated at Jefferson Lab", in *Proc. 19th International Conference on RF Superconductivity, (SRF'19)*, Dresden, Germany, paper MOP018.

[6] C. Wu, R. Kampwirth, and J. Hafstrom, "High-rate magnetron sputtering of high T_c Nb_3Sn films," *Journal of Vacuum Science and Technology*, vol. 14, no. 1, pp. 134-137, 1977.

[7] E. Ilyina et al., "Development of sputtered Nb_3Sn films on copper substrates for superconducting radiofrequency applications," *Superconductor Science and Technology*, vol. 32, no. 3, p. 035002, 2019.

[8] M. N. Sayeed, U. Pudasaini, C. E. Reece, G. Eremeev, and H. E. Elsayed-Ali, "Structural and superconducting properties of Nb_3Sn films grown by multilayer sequential magnetron sputtering," *Journal of Alloys and Compounds*, vol. 800, pp. 272-278, 2019.

[9] B. Cullity, *Elements of X-Ray Diffraction 2nd edition*, Addision-Wesley Pub. Co, Inc., CA USA, vol. 197, p. 356, 1978.

[10] J. Lee, S. Posen, Z. Mao, Y. Trenikhina, K. He, D.L. Hall, M. Liepe, and D.N. Seidman, "Atomic-scale analyses of Nb_3Sn on Nb prepared by vapor diffusion for superconducting radiofrequency cavity applications: a correlative study", *Superconductor Science and Technology*, 32(2), p.024001, 2018.

[11] Y. Trenikhina, S. Posen, A. Romanenko, M. Sardela, J.M. Zuo, D.L. Hall, and M. Liepe, "Performance-defining properties of Nb3Sn coating in SRF cavities", *Superconductor Science and Technology*, 31(1), p.015004, 2017.

[12] C.A. Anderson, M.F. Hasler, in X-ray Optics and Microanalysis, *Fourth International Congress on X-ray Optics and Microanalysis, Orsay*, 1965, eds. by R. Castaing, P. Deschamps, K. Philibert (Herman, Paris, 1966), p. 310.

[13] M.N. Sayeed, H. Elsayed-Ali, G.V. Eremeev, M.J. Kelley, U. Pudasaini, and C.E. Reece, "Magnetron Sputtering of Nb3Sn for SRF Cavities", in *Proc. 9th Int. Particle Accelerator Conf. (IPAC'18)*, Vancouver, BC, Canada, Apr. 4,, pp. 3946-3949, doi:10.18429/JACoW-IPAC2018-THPAL129.

[14] R. Kampwirth, J. Hafstrom, and C. Wu. "Application of high rate magnetron sputtering to the fabrication of A-15 compounds." *IEEE Transactions on Magnetics* 13.1 : 315-318, 1977.

[15] J. Tuggle, U. Pudasaini, J. Angle, G. Eremeev, C.E. Reece, F.A. Stevie, and M.J. Kelley, "Electron backscatter diffraction of Nb_3Sn coated niobium: Revealing structure as a function of depth", *Journal of Vacuum Science & Technology B*, 37(5), 052901, 2019.

IOP Conf. Series: Materials Science and Engineering **756** (2020) 012015 doi:10.1088/1757-899X/756/1/012015

Pressure-induced critical current reduction in impregnated Nb_3Sn Rutherford cables for use in future accelerator magnets

P. Gao[1], M. Dhallé[1] and H. H. J. ten Kate[1,2]

[1] University of Twente, Faculty of Science & Technology, Energy, Materials and Systems (EMS), P.O. Box 217, 7500 AE Enschede, The Netherlands

[2] CERN, European Organization for Nuclear Research, 1211 Geneva 23, Switzerland

p.gao@utwente.nl

Abstract. The measured critical current reduction in Nb_3Sn Rutherford cables under magnet-relevant transverse pressure levels is analyzed in terms of the strain state of the filaments inside their strands. Several straightforward mechanical 2D FE models of the cables' cross-section are used to translate the stress that is applied to the surface of the impregnated cables into a strain distribution on its strands. The resulting critical current reduction of the cable is then estimated from the average deviatoric strain in the strands' filamentary zone, using the well-established strain scaling relations obtained for isolated strands. This allows to identify the main factors that influence the pressure response of impregnated Nb_3Sn accelerator cables. The analysis is presented for state-of-the-art cable samples that were measured at the University of Twente and shows how especially stiff and incompressible resins reduces the deviatoric strain in the filamentary zone of the cable strands, but also how relatively small alignment errors can lead to stress concentrations that reduce the critical current density significantly.

1. Introduction

Future circular accelerators such as HL-LHC, FCC or SppC call for bending magnets that generate fields of 16 T or above [1, 2] and will therefore have to rely on Nb_3Sn or HTS technology. This paper focusses on the mechanical stress tolerance of Nb_3Sn Rutherford cables for such magnets. 16 T magnet lay-outs ranging from 'classical' cos-theta [3, 4] over block [5] and common-coil [6] to canted-cos-theta [7] have been proposed, but all involve unprecedented peak transverse pressure levels in the range of about 140 to 200 MPa on their cables, which operate at non-Cu current densities J_c of about 1500 A/mm^2.

The superconducting properties of Nb_3Sn are well-documented [8] and a-priori understood [9] to change under mechanical strain ε, leading to an intrinsic and reversible J_c reduction at lower strain levels [10] and an irreversible microstructural J_c degradation due to filament cracking when an 'irreversible strain limit' is exceeded [11]. The strain dependence of J_c in the reversible regime is often described as:

$$J_c(T,B,\varepsilon) = \frac{C_1}{B} s(\varepsilon_{dev})(1-t^{1.52})(1-t^2)b^p(1-b)^q\,, \qquad (1)$$

with t and b the reduced temperature T/T_c and the reduced magnetic field B/B_{c2}, respectively; C_1, p and q are material-specific fitting parameters and $s(\varepsilon_{dev})$ the strain function [8]. The 'deviatoric strain' ε_{dev} is the square root of the second strain invariant [12], which essentially describes the distortion of the crystal lattice.

To minimize such distortion, and the corresponding reversible J_c reduction, Nb_3Sn strands [13, 14] and cables [15, 16] can be resin-impregnated, which causes a distribution of transverse stresses more evenly and leads to a strongly improved stress tolerance [17]. In this paper we use a number of straightforward FE models to illustrate semi-quantitatively how the resin properties, winding pack composition and detailed layout of the cable confinement influence the transverse pressure tolerance of impregnated Rutherford cables.

2. 2D mechanical models and results

2.1. Influence of the epoxy properties on the deviatoric strain

The isotropic 2D two-strand plane-strain COMSOL model presented in Figure 1 is a simplified simulation of the 'hair-pin' sample holder that is used at the University of Twente in transverse pressure experiments on accelerator-relevant superconducting cables [18]. The strands are initially modelled as homogenous Nb_3Sn cylinders with a diameter of 1 mm that are fully surrounded by an epoxy resin. This uniform-strand assumption will be relaxed later. According to the material properties of the CTD-101K and the composite of CTD-101K + glass-sleeve [19]. The Young's modulus and Poisson's ratio of resin are assumed to be varied in the ranges of 1 to 40 GPa and 0.1 to 0.4, respectively. A normal pressure of 100 MPa is applied to the top of the impregnated sample over a section of 1.2×46 mm^2 with a stainless-steel anvil. The total thickness of the impregnated stack is 2.6 mm, with a distance of 0.1 mm in-between the two strands. The assumed mechanical properties of various materials at 4.2 K are presented in Table 1. The bottom edge of the epoxy is fixed, the left and right sides of the stack and of the anvil are modelled as 'sliding boundaries', which implies that no displacement in horizontal direction can occur at these sides.

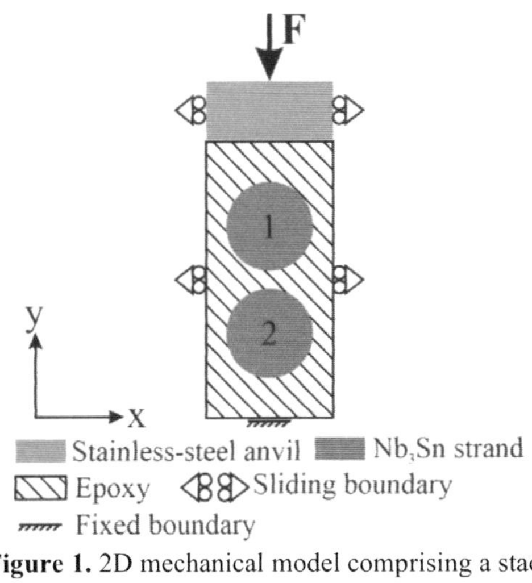

Figure 1. 2D mechanical model comprising a stack of two Nb_3Sn strands.

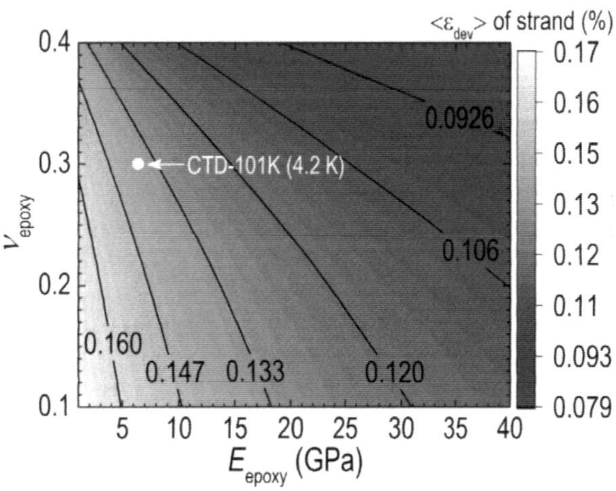

Figure 2. Variation of average deviatoric strain acting on strand-1 with respect to the Young's modulus and the Poisson's ratio of the epoxy resin.

Table 1: MECHANICAL PROPERTIES OF VARIOUS MATERIALS USED IN THE 2-STRAND MODEL.

$T = 4.2$ K	E (GPa)	ν
Nb_3Sn	75 [20]	0.36 [21]
Epoxy	1 to 40	0.1 to 0.4
Stainless steel 316L	207 [20]	0.284 [20]

Because of the symmetry in the model, only the average deviatoric strain $<\varepsilon_{dev}>$ acting on strand-1 needs to be calculated and is presented in Figure 2 as a function of the Young's modulus E_{epoxy} and the Poisson's ratio ν_{epoxy} of the epoxy resin. With increasing Young's modulus and Poisson's ratio, $<\varepsilon_{dev}>$ decreases. This can be understood intuitively: as the epoxy becomes stiffer (larger E_{epoxy}) it carries a larger fraction of the applied force and the stress exerted on the strand becomes smaller, causing a corresponding reduction of the strain in the strand. A larger Poisson's ratio of the epoxy resin on the other hand increases its bulk modulus [12], Equation (2), and thus renders the stress distribution more hydrostatic:

$$k_{epoxy} = \frac{E_{epoxy}}{3(1-2v_{epoxy})}. \quad (2)$$

Correspondingly, the x- and y-strains inside the strand become more similar, which also results in a reduction of $<\varepsilon_{dev}>$. It can thus be concluded that using an epoxy resin with either a large Young's modulus or a large Poisson's ratio reduces the deviatoric strain component acting on the strand and hence increases its critical current. In the transverse pressure experiments on Nb_3Sn Rutherford cables [17], the epoxy resin CTD-101K [22] is used, with a reported Young's modulus E_{epoxy} of 6.44 GPa and a Poisson's ratio v_{epoxy} of 0.3 at 4.2 K [19]. These values are indicated by the dot in Figure 2.

2.2. *Influence of glass reinforcement on the deviatoric strain*

Vacuum impregnation with epoxy resin is commonly applied in accelerator magnets to enhance the coils' mechanical properties. For large-scale magnets, the epoxy resin CTD-101K is often used due to its long pot life of 60 h and low viscosity 400 mPa.s at 40 °C [20]. In addition, a woven glass sleeve is often applied around the cables to ensure adequate electrical turn-to-turn isolation. As added advantages, compared to the pure epoxy resin the combination of CTD-101K and glass fibers has a thermal contraction coefficient that is closer to the one of the Nb_3Sn cables, as well as a larger Young's modulus. The orthotropic mechanical properties of the glass-resin composite are presented in Table 2.

Table 2. MECHANICAL PROPERTIES OF CTD-101K WITH AN S-2 GLASS-FIBER SLEEVE [19].

T = 4.2 K	E_x / E_y / E_z (GPa)	v_x / v_y / v_z	G_x / G_y / G_z (GPa)
CTD-101K & glass-fiber	39.4 / 19.7 / 32.9	0.21 / 0.12 / 0.22	9 / 4.9 / 9

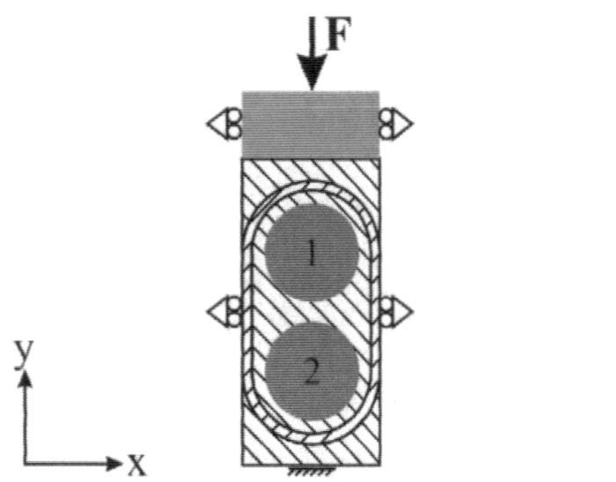

Figure 3. 2-strand model including epoxy resin & woven glass sleeve.

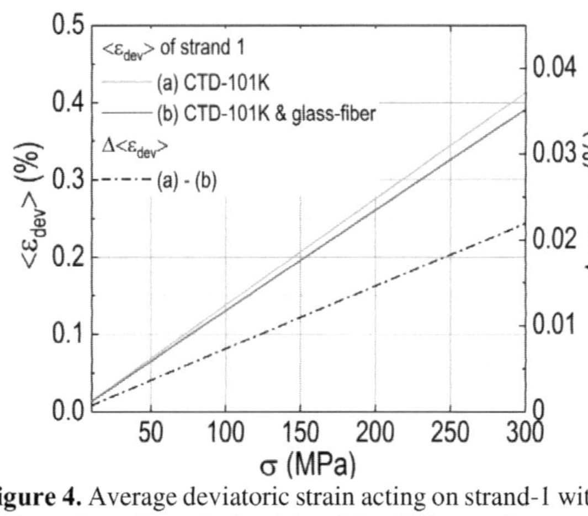

Figure 4. Average deviatoric strain acting on strand-1 with and without glass sleeve (left axis) and the difference between both (right axis) plotted against the applied pressure.

We can adapt the simple cable model of Figure 1 to include the glass and thus to estimate the change in the average deviatoric strain in the strand due to the presence of the sleeve. The resulting 2D mechanical model is presented in Figure 3. It has the same size and boundary constraints as the 2-strand model of Figure 1, but this time the medium enclosing the strands has three different regions: the pure epoxy resin outside the sleeve, the resin mixed with glass fibers and the pure resin inside the sleeve. The thickness of the sleeve is taken 50 μm, and the minimum distance between glass-fiber and strand is also

50 μm. As before, transverse pressure is applied to the top surface of the impregnated sample stack through the anvil.

The average deviatoric strain $<\varepsilon_{dev}>$ acting on strand-1 with and without the sleeve is plotted against applied pressure in Figure 4. The difference in deviatoric strain with and without glass is proportional to the applied pressure, with the strands in the sample impregnated with CTD-101K mixed with glass-fiber experiencing a 5% lower average deviatoric strain compared to the sample impregnated with pure CTD-101K.

2.3. Influence of the confinement geometry on the deviatoric strain (effect of the alignment impregnation)

As reported elsewhere [17], a first series of critical current versus transverse pressure measurements on Rutherford cables for the Short Model Coil (SMC) demonstrator magnet program at CERN (comprising 18 Restacked-Rod-Processed (RRP) or Powder-In-Tube (PIT) Nb_3Sn strands) yielded an irreversible pressure limit of 170 ± 25 MPa for the RRP cable and 125 ± 10 MPa for the PIT cable, which was considerably lower than expected. A possible reason for these low values was a suspected misalignment between the impregnated cable and the pressure anvil. The measurements were repeated on a second set of cable samples with an additional 'alignment impregnation' step [18, 23] to improve the interface parallelism. As a result, the measured transverse pressure tolerance of these two cable types improved significantly, to an irreversible pressure limit larger than 190 ± 3 and 150 ± 3 MPa, respectively.

To estimate how significant the effect of misalignment is on the deviatoric strain in the strands of the measured Nb_3Sn cables, as well as to understand better how the alignment impregnation improves the transverse stress tolerance of the cables, a 2D mechanical model for impregnated cables was developed, as shown in Figure 3. Just like in the previous models, a constant force is applied to the stainless steel anvil while the left and right boundaries of both the anvil and the impregnated cable sample are defined as sliding. Also like in the models above, the strands are still represented by uniform Nb_3Sn cylinders. The input variable d_s in the COMSOL simulations is the vertical displacement of the anvil, while g_s is the initial vertical gap that is left on the "non-touching" edge of the cable when d_s is zero (i.e. when the anvil first comes into contact with the sample at the other edge). θ is the misalignment angle between the anvil and the sample surface.

The various geometrical parameters of the cable model of Figure 5 are summarized in Table 3. The 1 mm diameter strands are surrounded by pure CTD-101K epoxy resin. The minimum distance $d_{c.i}$ between strands and the glass sleeve is 50 μm. Within one layer of strands, neighboring strands are separated by a distance d_a of 50 μm, while the vertical distance d_c between the strands in different layers is 100 μm. This structure is enveloped in a glass sleeve with a thickness $t_{c.g}$ of 50 μm. An additional 100 μm thick pure epoxy resin layer is assumed outside the glass at the top and the bottom of the resulting sample stack. Nb_3Sn, epoxy resin and glass-fiber properties are taken to be the same as in the previous models, as listed in Tables 1 and 2.

Figure 6 shows the calculated average deviatoric strain acting on the strands in the top layer of the cable for a perfectly aligned experiment ($\theta = 0°$) and for an experiment with a small misalignment angle ($\theta = 0.2°$). In both cases, the force applied by the anvil is 20 kN, corresponding to a nominal pressure of 40 MPa. Strikingly, in the misaligned case the deviatoric strain in strand 9, the strand situated directly below the 'first contact' area of the anvil, is about 3 to 3.5 times higher than in the 'perfect' experiment. As we move from the right to the left, from strand 9 to strand 3 in Figure 5, the deviatoric strain decreases approximately linearly. For the left-most strands 1 and 2 the strain becomes less position-dependent. These observations can be understood in terms of the system's geometry, as sketched in Figure 7. For a non-zero misalignment and relatively low anvil displacement d_s, only part of the anvil is in contact with the sample, with an "overall" strain $\varepsilon_y(x)$ that decreases linearly from right to left over the width of the contact area w_c. As long as $d_s < g_s$, w_c will be smaller than the sample width and a gap remains between the anvil and the sample on the left-hand side.

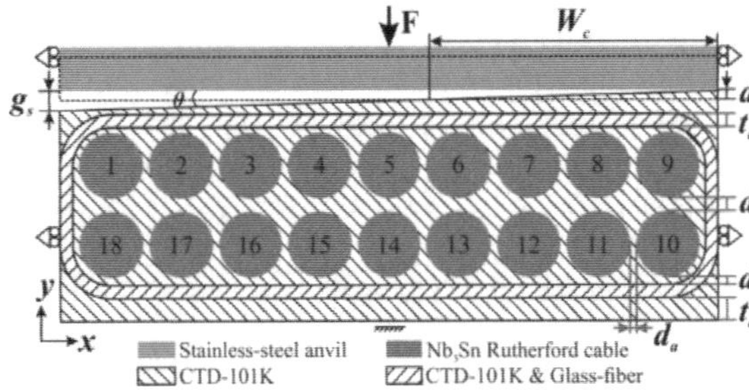

Figure 5. 2D mechanical model of the cable under transverse pressure that includes a misaligned interface.

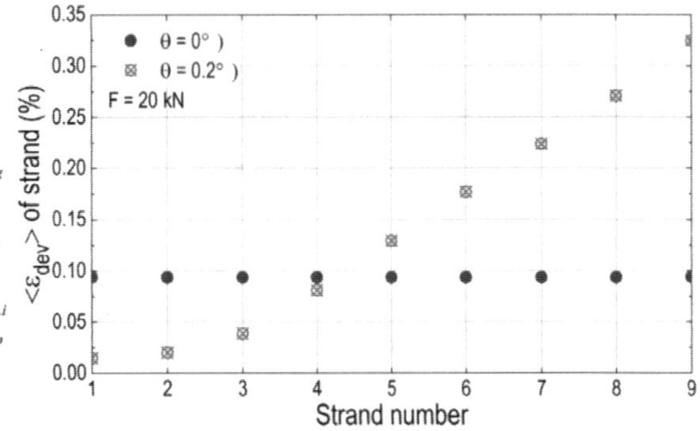

Figure 6. Average deviatoric strain in strands 1 to 9 of the top cable face in Figure 5. The blue solid symbols are modelled assuming perfect parallelism between cable and pressure anvil, the red crossed ones with a small misalignment angle of 0.2°.

Table 3. DIMENSIONAL PARAMETERS OF THE 2D MECHANICAL MODEL OF THE CABLE ARCHITECTURE.

Parameter	Description	Unit	Value
r	Radius of the strand	mm	0.5
$t_{c.o}$	Thickness of the outer CTD-101K layer	mm	0.1
$t_{c.g}$	Thickness of CTD-101K and glass-fiber layer	μm	50
$d_{c.i}$	Distance between strand and glass-fiber sleeve	μm	50
d_a	Distance between neighboring strands in the same layer	μm	50
d_c	Distance between neighboring strands in different layers	mm	0.1

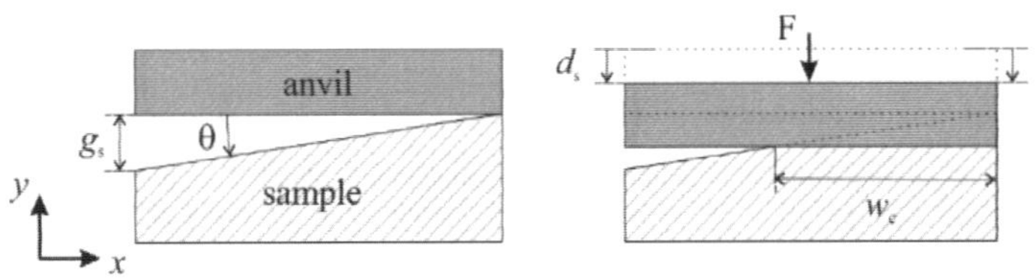

Figure 7. Schematic geometry of the compressed sample in the case of a non-zero misalignment angle θ.

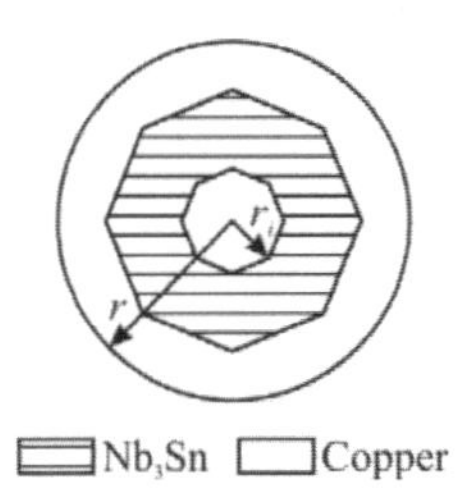

Figure 8. Layout of the refined strand model.

In reality, the strain concentration shown in Figure 6 can be expected to be even larger on a filament level. As discussed above, the models presented so far estimate the deviatoric strain in *homogeneous* Nb_3Sn cylindrical 'strands', not taking into account the internal strand composition. More detailed FE modelling by other authors shows that further stress- and strain- concentrations occur within the filamentary zone of the strands [21, 24, 25]. However, also without resorting to such detailed level of modelling, we can get an impression of the magnitude of this effect from a simplified model. The strain calculation with the cable model of Figure 5 was repeated, but this time replacing the homogeneous Nb_3Sn cylinders by strands modelled as in Figure 8. In this simple refinement, the 1 mm diameter strand is modelled as an octagonal uniform Nb_3Sn zone surrounded by a cylindrical copper matrix, and containing a central copper core. The Cu to non-Cu ratio of 1.22 is taken from the SMC-RRP strand. The diameter of the central copper core r_i is 200 μm. The Young's modulus and Possion's ratio of the copper are set to 137 GPa and 0.35 [21], respectively. Running the calculations of the model in Figure 2 again and comparing the average deviatoric strain in the filamentary zone of this refined strand model with the one in the homogeneous strand, reveals a further increase of $<\varepsilon_{dev}>$ with about 33%, as shown in Figure 9.

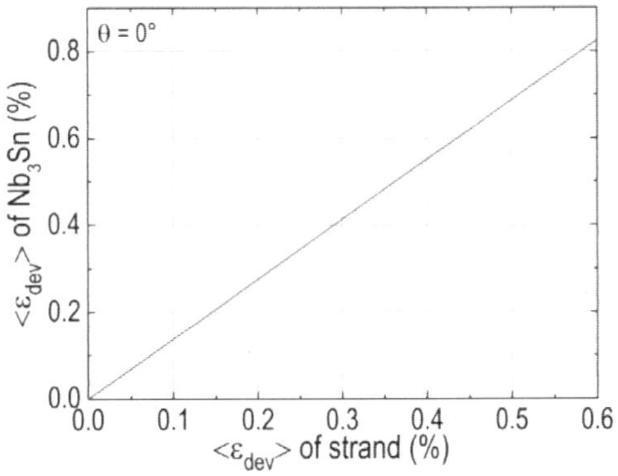

Figure 9. Average deviatoric strain in the filamentary zone of the refined strand model, plotted against the average deviatoric strain in the uniform strand model.

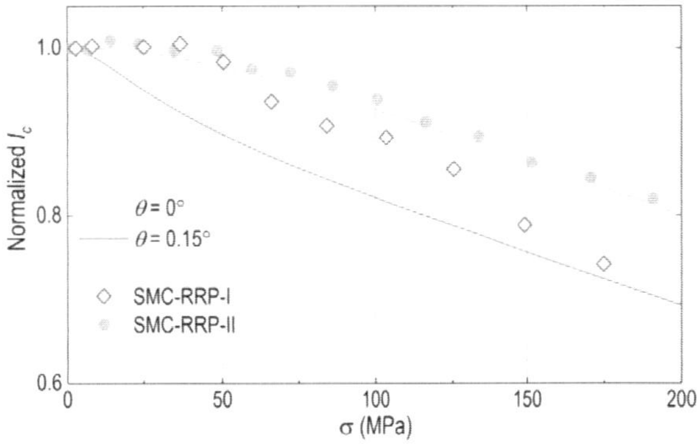

Figure 10. Normalized critical current versus transverse stress, illustrating the influence of a small misalignment angle on the pressure-dependence of the critical current.

The model outcomes presented in Figures 6 and 9 can now be combined with the strain scaling relation for the critical current density $J_c(T, B, \varepsilon)$ presented in the introduction, Equation (1) is applied to obtain a rough estimate of the effect of sample-anvil misalignment on the critical current measured in the press experiments [17]. Once more, we take an approach as simple as possible: the calculated average deviatoric strain in the filamentary zone of model strand 9 is inserted in Equation (1), using the characteristic scaling-law parameters given in Table 4, and it is assumed to dominate the overall pressure response of the critical current of the whole cable. Note that this is a rather crude approximation, neglecting possible effects of current redistribution among filaments or among strands and assuming that the magnetic peak-field location coincides with the region of maximum stress concentration. The result is shown as the solid lines in Figure 10. The top blue line is calculated for perfect alignment, when θ is 0°, while the bottom green curve is obtained with a misalignment angle θ of 0.15°. Strikingly, even this small misalignment causes a 13% lower I_c value at a transverse pressure of 200 MPa compared to the perfectly aligned case.

Table 4. SCALING-LAW PARAMETERS IN EQUATION (1) USED TO CALCULATE THE EFFECT OF STRAIN ON I_c [13].

Parameter	Description	Unit	Value
$B_{c2}(4.2\text{ K}, 0)$	Upper critical magnetic field at 4.2 K and zero strain	T	26.38
$T_{c2}(0,0)$	Critical temperature at zero field and strain	K	16.8
B	Applied magnetic field	T	11.63
T	Operation temperature	K	4.2
C_l	Scaling constant	AT/mm^2	46300
C_{dev}	Strain constant	-	1000
P	Low-field exponent of the pinning force	-	0.5
q	High-field exponent of the pinning force	-	2
e_{0d}	Residual strain component	%	0.09

The symbols in Figure 10 represent measurements on SMC-RRP cables, reported more extensively in [17]. Here we suffice to say that the only difference between the two cable experiments is the alignment impregnation that was applied to cable sample SMC-RRP-II but not to the identical cable SMC-RRP-I. The shape of the modeled $I_c(\sigma)$ curve when θ is 0.15° differs from the exact trend in the SMC-RRP-I data, but this is not too surprising given the crudeness of the mechanical and electrical models. However, at higher pressures the magnitude of the extra I_c reduction comparing the $\theta = 0.15°$ with the $\theta = 0°$ simulation is similar to the measured I_c difference between the cables SMC-RRP-I (without alignment impregnation) and SMC-RRP-II (with alignment impregnation). This suggests that the alignment impregnation is indeed effective in avoiding strain concentrations in the cable samples.

To investigate the effect of the alignment impregnation further, the calculations on the cable model of Figure 5 are repeated once more, but this time filling the gap between the anvil and the sample with a mixture of Stycast 2850FT and glass fiber, as shown in Figure 11. Note that this mixture is also used in actual alignment impregnation described in [17]. No literature data on the mechanical properties of a Stycast-glass mixture at 4.2 K were found. Instead, we used the data published in [26], reporting a Young's modulus of pure Stycast at 77 K in the range 22 to 29 GPa. To account for the lower temperature of 4.2 K and for the glass layers, we assume a high-end value of 30 GPa. Since also no published data were found for the Poisson's ratio, two extreme values are used: 0.1 and 0.4.

The average deviatoric strain values in the strands of this model are presented in Figure 12, including for comparison also the results without alignment impregnation previously shown in Figure 6. Both models were simulating a misalignment angle θ of 0.2°. Clearly, also in the simulations, the alignment impregnation virtually eliminates the effect of the misalignment angle, reducing the variation of the deviatoric strain between the strands to less than 1% and yielding a $<\varepsilon_{dev}>$ value that is near-identical to the one found for $\theta = 0°$.

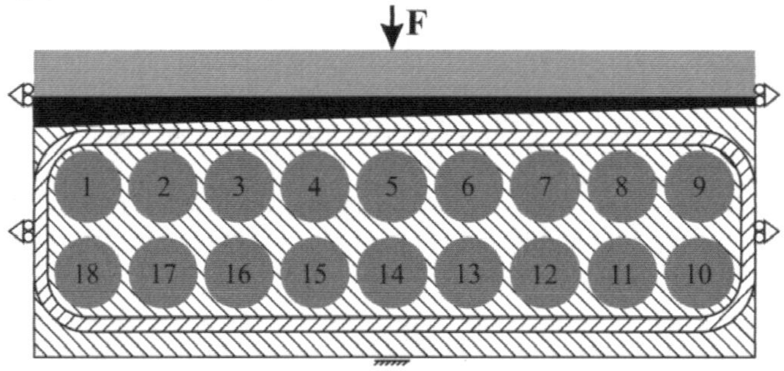

Figure 11. 2D mechanical model of the cable under transverse pressure, but now with and extra alignment impregnation to fill the gap between cable surface and anvil.

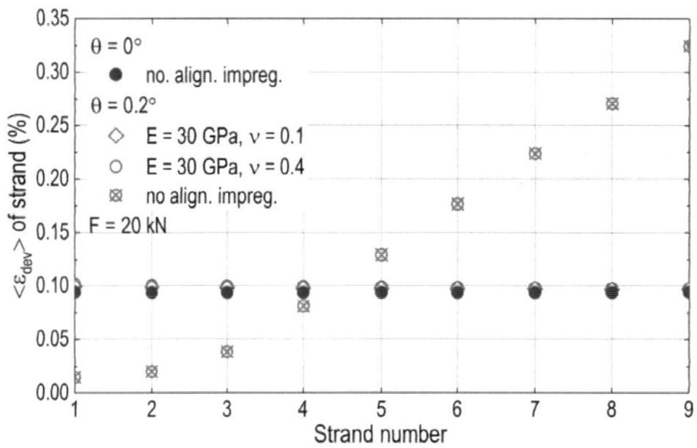

Figure 12. Average deviatoric strain in the strands 1-9 of the top cable face in Figure 5, with and without misalignment, and with and without alignment impregnation.

3. Conclusion

Simplified 2D FE modelling of impregnated and transversely loaded Nb_3Sn Rutherford cables shows that the average deviatoric strain acting on their strands decreases with increasing Young's modulus of the epoxy resin, as the resin carries an increasing part of the load. Also a larger Poisson's ratio of the resin, and therefore a larger bulk modulus, renders the stress on the strand more hydrostatic and thus reduces the deviatoric strain. Since the deviatoric strain of Nb_3Sn filaments has a direct impact on their critical current density, the use of a stiff and relatively incompressible epoxy resin significantly improves the pressure tolerance of the cables. The mechanical effect of an insulating glass sleeve around the cable, on the other hand, is relatively small.

The same type of mechanical model was also used to quantify the effect of a possible misalignment between the cable surface and the pressure anvil used in the cable experiments. For a typical Nb_3Sn cable layout, even a misalignment angle as small as 0.2° can cause a strain concentration by a factor 3 in the strands at the side of the cable that comes into contact with the anvil first. It should be noted that similar stress- and strain concentrations due to imperfect cable stacking may very well occur also in the winding pack of real magnets, which the experiment is designed to mimic. The alignment impregnation significantly improves the strain homogeneity in the transverse press experiments, rendering it essentially equal to the strain modelled for the perfectly aligned situation.

4. References

[1] Schoerling D *et al.* 2019 *IEEE Trans. Appl. Supercond.* **29**(5)
[2] Xu Q *et al.* 2016 *IEEE Trans. Appl. Supercond.* **26**(4)
[3] Marinozzi V, Bellomo G, Caiffi B, Fabbricatore P, Farinon S, Ricci A M, Sorbi M and Statera M 2018 *IEEE Trans. Appl. Supercond.* **28**(3)
[4] Kashikhin V V, Andreev N, Barzi E, Novitski I and Zlobin A V 2015 *Materials Science and Engineering, 101(conference 1)*
[5] Lorin C, Segreti M and Durante M 2018 *IEEE Trans. Appl. Supercond.* **28**(3)
[6] Toral F, Munilla J and Salmi T 2018 *IEEE Trans. Appl. Supercond.* **28**(3)
[7] Auchmann B, Brouwer L, Caspi S, Gao J, Montenero G, Negrazus M, Rolando G and Sanfilippo S 2018 *IEEE Trans. Appl. Supercond.* **28**(3)
[8] Bottura L and Bordini B 2009 *IEEE trans. Appl. Supercond.* **19**(3)
[9] Godeke A, Hellman F, Ten Kate H H J and Mentink M G T 2018 *Supercond. Sci. Technol.* **31**(10)
[10] Ten Haken B and Ten Kate H H J 1993 *Fusion Engineering and Design.* **20**
[11] Jewell M C, Lee P J and Larbalestier D C 2003 *Supercond. Sci. Technol.* **16**(9)
[12] Sadd M 2014 *Elasticity: theory, applications and numerics* (Academic Press) ISBN:978-0-12-408136-9
[13] Ten Haken B 1994 PhD dissertation, *University of Twente, The Netherlands*
[14] Calzolaio C, Mondonico G, Ballarino A, Bordini B, Bottura L, Oberli L and Senatore C 2015 *Supercond. Sci. Technol.* **28**(5)
[15] Van Oort J M 2000 PhD dissertation, *University of Twente, The Netherlands*
[16] Vallone G, Bordini B and Ferracin P 2018 *IEEE Trans. Appl. Supercond.* **28**(4)
[17] Gao P, Dhallé M, Bordini B, Ballarino A and Ten Kate H H J, to be published
[18] Gao P, Wessel W A J, Dhallé M, Otten S, Kario A, Van Nugteren J, Kirby G, Bottura L and Ten Kate H H J 2019 *Supercond. Sci. Technol.* **32**(5)
[19] Murtomaki J S, Van Nugteren J, Kirby G, Rossi L, Ruuskanen J and Stenvall A 2017 *IEEE Trans. Appl. Supercond.* **27**(4)
[20] Bauer P, Rajainmaki H and Salpietro E 2007 *Inside document of CERN*
[21] Calzolaio C, Mondonico G, Ballarino A, Bordini B, Bottura L, Oberli L and Senatore C 2015 *Supercond. Sci. Technol.* **28**(5)
[22] CTD-101K 2014 *Composite Technology Development, Lafayette, CO, USA*
[23] Otten S, Dhallé M, Gao P, Wessel W, Kario A, Kling A and Goldacker W 2015 *Supercond. Sci. Technol.* **28**(6)
[24] Daly M, Loffler C H, Smekens D, Fontenla A T, De Frutos O S, Guinchard M and Savary F 2018 *IEEE Trans. Appl. Supercond.* **28**(3)
[25] Zhai Y, D'Hauthuille L, Barth C and Senatore C 2016 *IEEE Trans. Appl. Supercond.* **26**(4)
[26] Kirby G A *et al.* 2016 *IEEE Trans. Appl. Supercond.*, **26**(3)

Acknowledgments
The work presented in this paper was done in the frame of the EuroCirCol project, partially funded by EU's H2020 framework program under grant agreement no. 654305.

IOP Conf. Series: Materials Science and Engineering **756** (2020) 012016 doi:10.1088/1757-899X/756/1/012016

Trial Manufacturing of Jelly-Rolled Nb/Al Monofilamentary Wire with Very Small Diameter below 50 microns

Akihiro Kikuchi[1], Yasuo Iijima[1], Ataru Ichinose[2], Masatoshi Kawano[3], Masaru Yamamoto[3], Motoyoshi Kimura[3], Jun Nagamatsu[4], Masato Otsubo[4], Kazuto Hirata[1], Shigeki Nimori[1] and Kiyosumi Tsuchiya[5]

[1] National Institute for Materials Science, Tsukuba, Ibaraki 305-0047, Japan.

[2] Central Research Institute of Electric Power Industry, Yokosuka, Kanagawa 240-0196, Japan.

[3] Meiko Futaba Co., Ltd., Chuo, Yamanashi 409-3842, Japan.

[4] Yuki Precision Co., Ltd., Chigasaki, Kanagawa 253-0084, Japan.

[5] High Energy Accelerator Research Organization, Tsukuba, Ibaraki 305-0801, Japan.

E-mail: KIKUCHI.Akihiro@nims.go.jp

Abstract. Jelly-rolled Nb/Al composite monofilamentary wires having an outer diameter of 50 microns and more than 100 meters in length were successfully fabricated. Moreover, some of these thin wires were additional drawn to 30 microns in outer diameter in this study. We recognized again that jelly-rolled Nb/Al composite monofilamentary wires have excellent cold workability and drawability. The transmission electron microscopic observation revealed that the niobium and aluminum laminations retained a nano-scaled fiber composite structure in the longitudinal direction like the microarchitecture of Japanese bamboo, even in the very thin wire with an outer diameter of 30 μm. The thickness of the aluminum sheet is approximately 70 nm, which is almost comparable to the value calculated from the reduction ratio. The critical temperature of 16.2 K (on set) and 15.7 K (off set) were obtained after the diffusion reaction at 850 °C for 10 h. The transport critical currents in liquid helium (4.2 K) under external magnetic fields of 1 T and 8 T are 7.2 A and 0.7 A, corresponding to non-Cu critical current densities of 15,000 and 1,500 A/mm^2, respectively. These values are greater than those of industrial Nb-Ti and MgB_2 commercial multifilamentary wires.

Introduction

In general, practical metallic superconducting wires, such as Nb-Ti, Nb_3Sn and others, have a fine multifilamentary configuration. The predominant reasons are: a) a reduction of electromagnetic losses, b) improvement of intrinsic and thermal stabilities, and c) reduction of mechanical strain [1]. In particular for the A15 compounds, the promotion of chemical reaction through shortening of the diffusion distance should be added to the above list. According to these understandings, we have been developing the Nb_3Al monolith wires having multifilament configuration. However, several R&D issues, such as the wire breakage of multifilamentary precursors, costly Cu stabilizer fabrication and low mechanical irreversible strain, still remain [2]. On the other hand, we are considering a different approach to fabricate Nb_3Al superconducting wires which would have low electromagnetic losses, good intrinsic and thermal stabilities, and excellent flexibility for winding through the React & Wind method. The anticipated fabrication cost would be lower. The conductor would consist of a multiple very fine single strand wires braided into a cylindrical pattern. An example of such a conductor braid geometry is known as the "BNL Braid" [3]. In this paper, as the first step for our new approach for Nb_3Al conductors, we manufactured Jelly-Rolled Nb/Al monofilamentary precursor wires of diameter below 50 μm. The drawing workability, microstructure changes, and superconducting properties after heat treatment of very fine Nb_3Al monofilamentary wires were investigated.

1. Jelly-rolled Nb/Al laminated billet assembly and hydrostatic extrusion

In 2017, some jelly-roll (JR) production facilities, such as special metal fixtures for manual jelly rolling, a precision sheet cutting machine, and an automatic tightening and spot welding machine, have been moved from SH Copper Products (now Hitachi Metals Neomaterial, Ltd.) to NIMS. Therefore, at present, jelly-rolled Nb/Al laminated billets is being fabricated in-house at NIMS. Figure 1 is a picture of the jelly-rolled billet assembly room established in 2017 at NIMS. In addition, Figures 2, 3 and 4 are pictures for precision sheet cutting, manual jelly rolling and automatic JR tightening and spot welding, respectively.

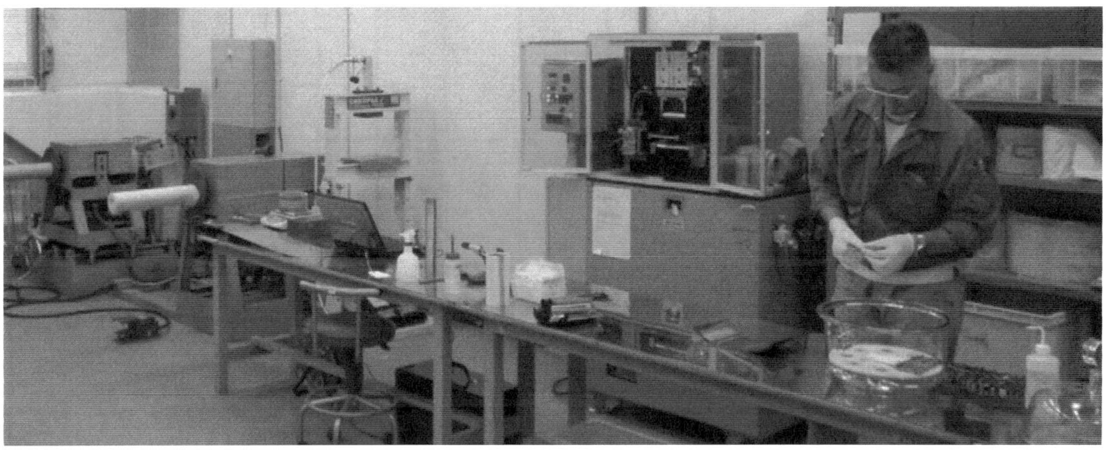

Figure 1. Photograph of the jelly-rolled billet assembly room established in 2017 at NIMS.

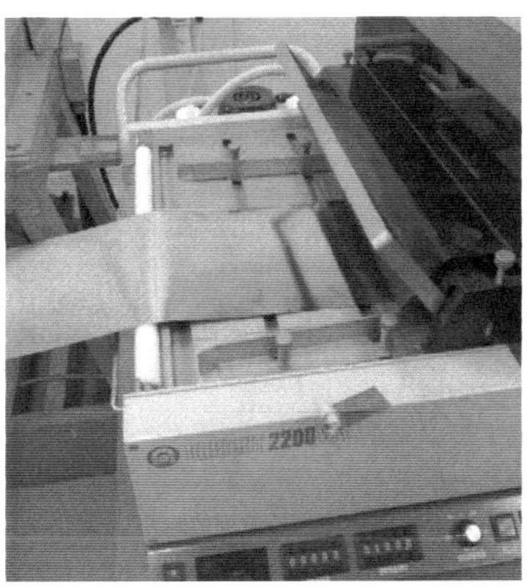

Figure 2. Photograph of precision sheet cutting machine.

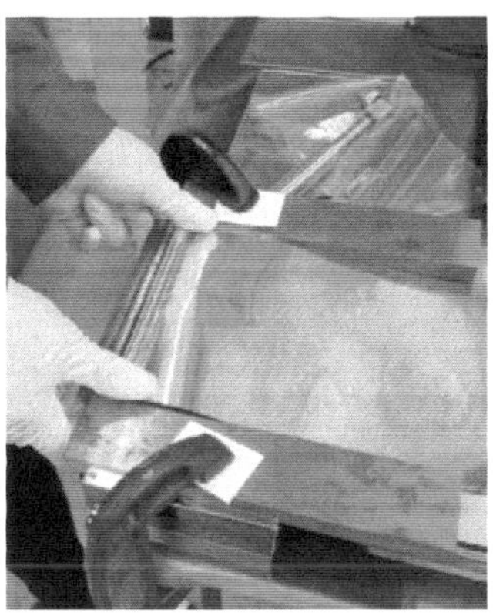

Figure 3. Photograph of manual jelly rolling.

Figure 4. Photograph of automatic JR tightening and spot welding machine.

A pure niobium sheet of nominal dimensions 0.1 mm in thickness and 150 mm wide, and a pure aluminum sheet of 0.3 mm thickness and 150 mm width, were used as starting materials for the jelly-rolled Nb/Al laminated billet. Those sheets were tightly overlapped several tens of turn together around a pure tantalum rod with a diameter of 2.0 mm and 150 mm in length. Note that the final five turns were overlapped by only the niobium sheet, which would act as a diffusion barrier. Then, this Nb/Al laminated composite was inserted into an annealed OFC tube with tip OFC and rear stainless steel (SUS) plugs. The outer and inner diameters of the OFC tube are 12.3 mm and 10.1 mm, respectively. These jelly-rolled Nb/Al laminated billets were extruded at room temperature by using a vertical hydrostatic extruder having 100 tons of maximum load. The hydrostatic extrusion was performed at Hitachi Metals Neomaterial, Ltd. The liquid medium used for applying hydrostatic pressure at room temperature was castor oil. The diameter of the billet decreased from 12.3 mm to 5.0 mm, which corresponds to an extrusion ratio of 6.05. Figure 5 are transverse cross-section images of the jelly-rolled Nb/Al laminated composites taken by an optical microscope (Nikon: ECLIPSE-LV150), before and after the hydrostatic extrusion. These cross-sections keep similar configurations, and thus the hydrostatic extrusion at room temperature resulted in a normal smooth metal flow. The Cu/non-Cu ratio of the present billet is 0.5.

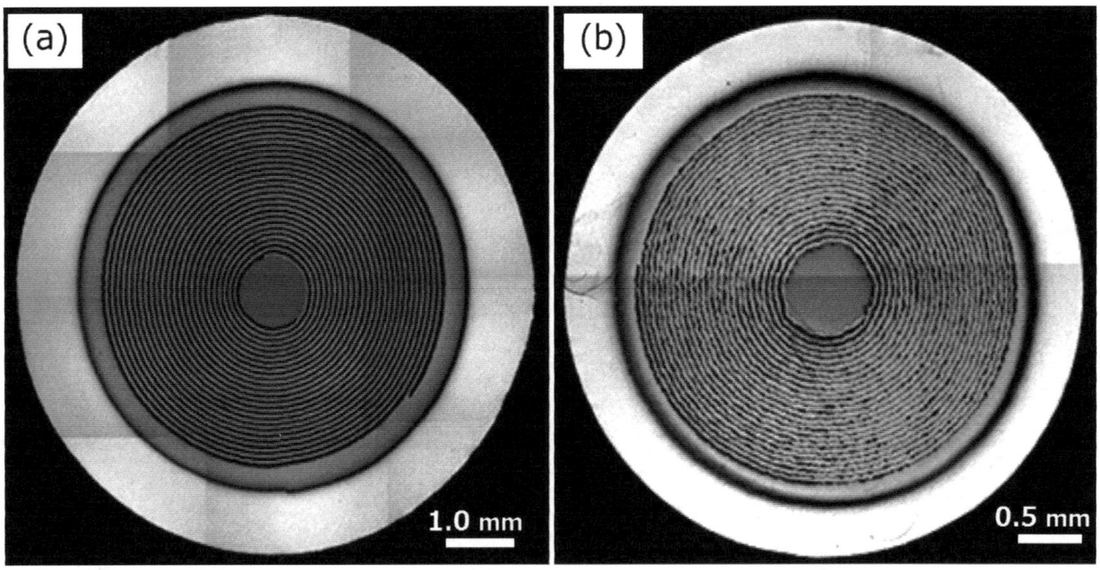

Figure 5. Comparison of transverse cross-sections of the jelly-rolled Nb/Al laminated composites (a) before and (b) after the hydrostatic extrusion at room temperature. The outer diameters of (a) and (b) are 12.3 mm and 5.0 mm, respectively.

2. Cold die drawing of jelly-rolled Nb/Al laminated precursors

After the hydrostatic extrusion at room temperature, we have tried an area reduction of the jelly-rolled Nb/Al laminated monofilamentary precursor only using cold die drawing without multifilament bundling. Intermediate annealing was not performed. In NIMS, the outer diameter was drawn to 0.56 mm from 5.0 mm by using a single-head drawing machine with general hard steel metal dies and a lubricant of a rapeseed oil. An area reduction ratio between dies was consistently about 11 % and the drawing speed was approximately 5.0 m/min. No wire breakages happened at this stage. Then, 0.56 mm wires were taken to Meiko Futaba Co., Ltd. for drawing down the wire diameter from 0.56 mm to 50 microns using a wet-type continuous drawing machine with diamond dies. These are mass-production facilities for pure electro-Cu wires. The area reduction ratio between dies was 10-20 %

IOP Conf. Series: Materials Science and Engineering **756** (2020) 012016 doi:10.1088/1757-899X/756/1/012016

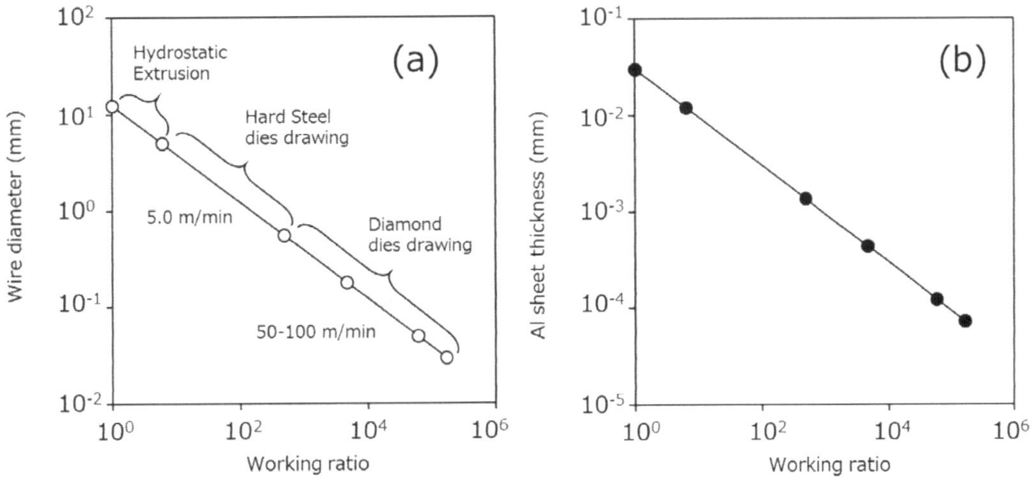

Figure 6. (a) Jelly-rolled wire diameter versus working ratio, and (b) The aluminum sheet thickness in the jelly-rolled Nb/Al composite versus working ratio.

and the drawing speed was 50-100 m/min. No wire breakages happened down to a diameter of 0.18 mm, but several wire breakages happened between 0.18 mm and 0.05 mm. Eventually a jelly-rolled Nb/Al laminated monofiamentary wire with a diameter of 50 μm and 128 meters in length was successfully fabricated. The superconducting filament diameter without Cu is below 50 microns. A part of the 50 micron wire could be demonstrated additionally to draw down to 30 μm in outer diameter. Figure 6 is a calculated scheme of the cold work in this study: (a) shows the relationship between the wire diameter and the working ratio and (b) shows the relationship between the aluminum sheet thickness and the working ratio.

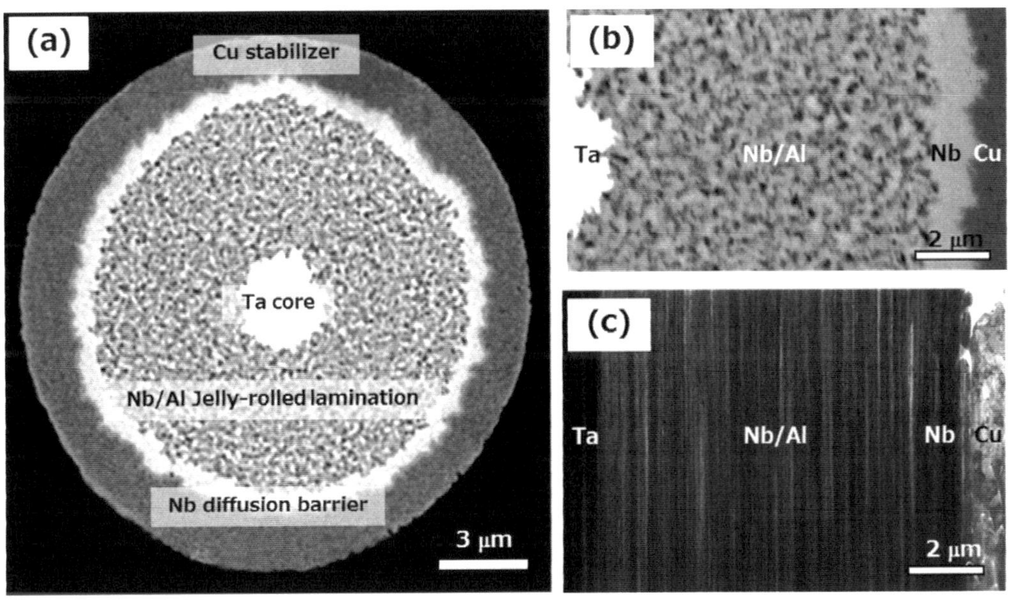

Figure 7. (a) Whole transverse cross-section image taken by SEM on the jelly-rolled Nb/Al laminated composites wire with a diameter of 30 μm. (b) Enlarged transverse cross-section image of (a). (c) Longitudinal cross section image taken by STEM of (a).

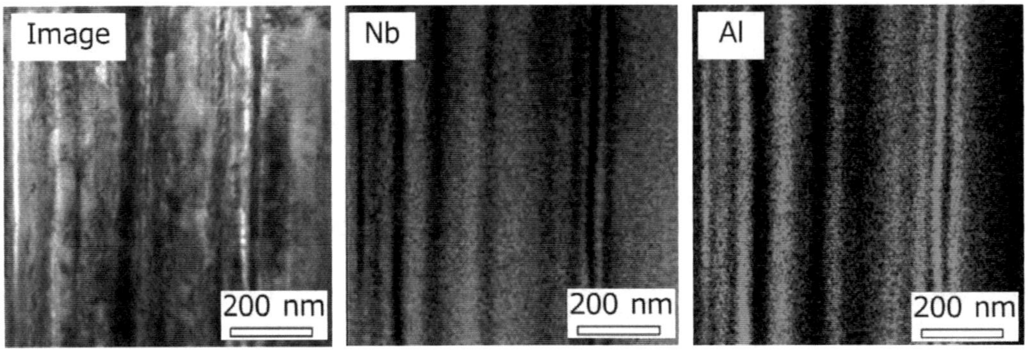

Figure 8. (a) Enlarged longitudinal cross-section image taken by TEM on the jelly-rolled Nb/Al laminated composite wire 30 μm in diameter. (b) EDX element mapping of niobium and (c) EDX element mapping of aluminum.

Figure 7 are microstructure images of the cross-sections of the jelly-rolled Nb/Al laminated composites wire with 30 μm in diameter. Figure 7 (a) and (b) show back-scattered images of a whole and an enlarged vertical cross-sections taken by a scanning electron microscope (Hitachi; TM5050). Figure 7 (c) is a scanning transmission electron microscope (JEOL; JEM-2100F) image of the longitudinal cross-section. Focused ion beam milling (FIB) was used for the scanning transmission electron microscopy (STEM) sample preparation. According to the vertical cross-section image of Figure 7 (a) and (b), the Nb/Al laminated structure seems to be almost broken and to mechanically mix together well comparison to the cross-section just after extrusion. But, Nb/Al lamination retains a nano-scaled fiber composite structure in the longitudinal direction as shown in Figure 7 (c). These fiber architectures are expected to maintain a flexibility and tenacity like Japanese bamboo, thus the very thin wire could be drawn even to below 30 μm in diameter. Figure 8 show an enlarged TEM image in the longitudinal direction of the 30 micron wire and the element mappings of niobium and aluminum by energy dispersive X-ray Spectroscopy (EDX). The thickness of the aluminum fibers is approximately 50-70 nm, which is comparable to the value calculated from the reduction ratio as shown in Figure 6 (b).

3. Critical temperature, critical current and non-Cu critical current density

Figure 9 shows the resistivity versus temperature curves for the 30 micron and 50 micron wires. This measurement was performed by a DC four probe method with 1 mA of constant current. The distance between the voltage taps is 10 mm. Measurement samples were set at an appropriate stable temperature position from the surface of liquid helium and the temperature was controlled by small coil heater with a ramping rate of 4.0 K/h. Both wire samples were diffusion-reacted at 850 °C for 10 h in an electric tube furnace with a high dynamic vacuum atmosphere of 10^{-4} Pa. The critical temperature, T_c of the 30 micron wire was slightly higher than that of the 50 micron wire. Its onset value is 16.2 K and offset value is 15.7 K. These value are lower than that (17.6 K) of the rapid heating/quenching and transforming (RHQT) processed Nb_3Al multifilamentary wires [2]. The normal state resistivity, ρ_n just before a superconducting transition of the 30 micron wire was larger than that of the 50 micron wire.

Figure 10 shows the transport critical current, I_c as a function of external magnetic field in liquid helium (4.2 K). The closed circles and the open circles are results for the 50 micron and 30 micron wires, respectively. The heat treatment conditions of the wire samples are the same as for Figure 8, i. e. 800 °C for 10 h. The transport I_c (4.2 K) of the 50 micron wire was about two times larger than that of the 30 micron wires, and it exceeded 10 A at 4.2 K and 1 T. However, the vertical cross-sectional area of the 50 micron wire is 2.8 times larger than that of the 30 micron wire, therefore, the non-Cu critical current density of the 30 micron wire becomes larger than that of the 50 micron wire.

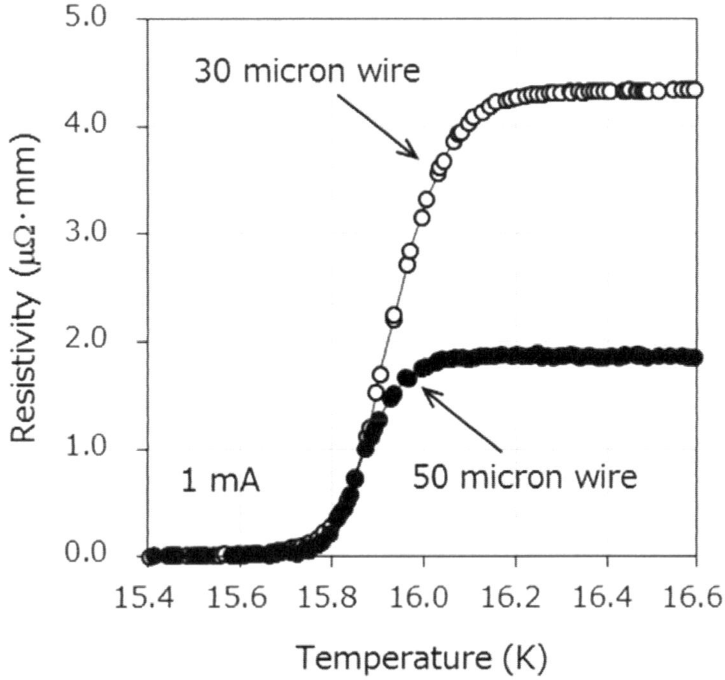

Figure 9. The resistivity versus temperature curves for 30 micron wire (open circles) and 50 micron wire (closed circles), respectively.

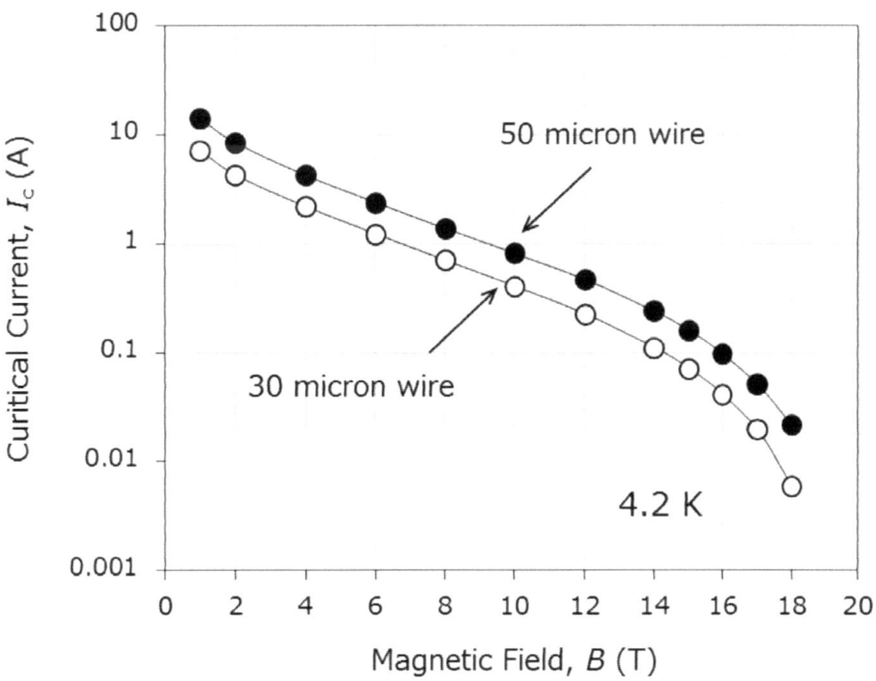

Figure 10. The transport critical current versus external magnetic field curves for 30 micron Nb_3Al wire (open circles) and 50 micron Nb_3Al wire (closed circles), respectively.

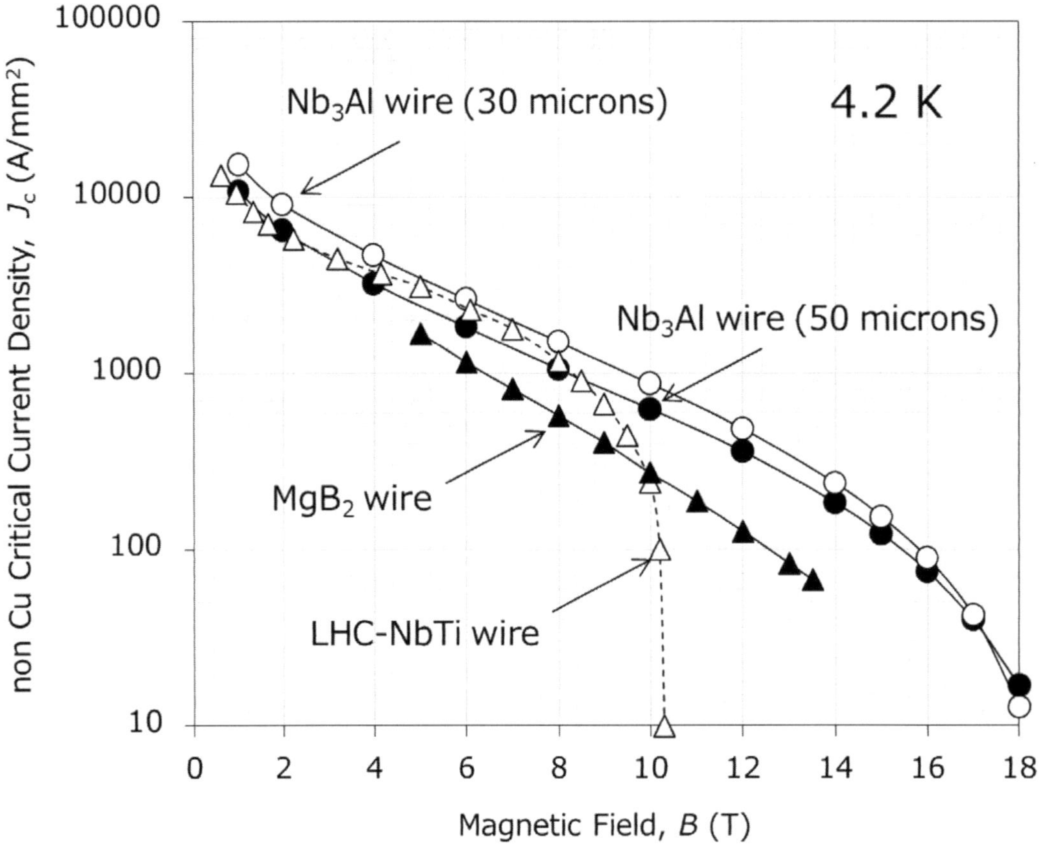

Figure 11. Comparison of the non-Cu critical current density as a function of external magnetic field for (a) 30 micron Nb_3Al wire, (b) 50 micron Nb_3Al wire, (c) industrial Nb-Ti wire used for the LHC [4] and (d) MgB_2 wires produced by Hyper Tech Research in the USA [5].

The diffusion distance between niobium and aluminum becomes smaller on decreasing the wire diameter, and thus the synthesis of A15-Nb_3Al phases may improve in the 30 micron wire. Figure 11 is the comparison of the non-Cu critical current density as a function of external magnet field for the present Nb_3Al wires in comparison to commercial Nb-Ti and MgB_2 multifilamentary wires. The non-Cu J_c of the present 30 micron Nb_3Al wire is exceeding that of the Nb-Ti wire and is much larger than that of MgB_2 wire. The heat treatment condition has not been optimized yet, and it can be expected to much improve the J_c in near future. Besides, the RHQT process [2] should be of interest for obtaining good superconducting properties as well.

4. Conclusions

In this study, we have recognized that jelly-rolled Nb/Al composite monofiamentary wires have excellent cold workability and drawability, and very thin Nb_3Al wires have been demonstrated. The reduction of the wire diameter is very effective to reduce the bending moment. In addition, it is well known that Nb_3Al wires show excellent strain tolerance, therefore, very thin Nb_3Al wires may keep the flexibility without performance degradations after heat treatment. This is promising for wire braiding and cabling, as well as for a coil application through the React & Wind method. Because the heat treatment condition for Nb_3Al thin wires has not been optimized yet, the present critical current density is not so large in comparison to industrial Nb_3Sn wires. In addition, in order to increase the

IOP Conf. Series: Materials Science and Engineering **756** (2020) 012016 doi:10.1088/1757-899X/756/1/012016

transport current capacity, a braid conductor using several hundreds of Nb_3Al fine monofiamentary wires will be demonstrated in near future. Furthermore, it will be needed to study the relationship between a braiding compaction factor and the critical current density, in order to achieve comparable the engineering J_c (J_e) to conventional monolith wire with multifilamentary architectures.

5. References

[1] M. N. Wilson 1983 *Superconducting Magnets* (Oxford: Oxford University Press) p. 289

[2] A. Kikuchi 2012 *J. Cryo. Super. Soc. Jpn.* **47** 503

[3] A. D. Mcinturff 1979 *Treatise on Materials Science and Technology* vol 14, ed. Thomas Luhman and David Dew-Hughes (London: Academic Press) p. 132

[4] T. Boutboul, S. Le Naour, D. Leroy, L. Oberli and V. Previtali 2006 *IEEE Transactions on Applied Supercond.* **16** 1184

[5] G. Z. Li, M. D. Sumption, J. B. Zwayer, M. A. Susner, M. A. Rindfleisch, C. J. Thong, M. J. Tomsic, and E. W. Collings 2013 *Supercond. Sci. Technol.* **26** 095007

Acknowledgments

The author would like to thank our technical staff, Mr. M. Chikatsu, Mr. H. Takigawa, Ms. E. Tsuchiya and Ms. N. Shinohara for their great help with this study. In addition, we also received some technical support for sample preparations and T_c measurements from Mr. K. Yamada, who was on a traineeship at NIMS and also a graduate student at the Sophia University.

A study on the extent of Ag protrusions in different TiO_2-coated Bi-2212 wires

S I Hossain[1,2,4], J Jiang[1], U P Trociewitz[1], J Lu[1], E Bosque[1], Y Kim[1], D C Larbalestier[1,2,3] and E Hellstrom[1,2,3]

[1] National High Magnetic Field Laboratory, Florida State University, Tallahassee, Florida, FL 32310, USA
[2] Materials Science and Engineering, Florida State University, Tallahassee, Florida, FL 32310, USA
[3]Department of Mechanical Engineering, FAMU-FSU College of Engineering, Florida State University, Tallahassee, Florida, FL 32310, USA
[4]Corresponding author. Email:sh15c@my.fsu.edu

Abstract. We report on the extent of Ag protrusions into the TiO_2 insulation layer on 9 different Bi-2212 wires after overpressure heat treatment. These wires were made with different powders and had different diameters and geometries, including aspected and twisted wires. To replicate coil heat treatments, we also studied whether increased time spent in the melt state affects the protrusions. We found that Ag protrusions are not universal and increasing the time in the melt state does not affect the protrusions.

1. Introduction

$Bi_2Sr_2CaCu_2O_x$ (Bi-2212) high temperature superconductor offers very high critical current density [1] and its irreversibility field is much higher than Nb_3Sn [2]. This high critical current density can be achieved in both round and aspected wires, which are multifilamentary and twistable. The round wire geometry of Bi-2212 enables employing the cabling technology developed for earlier generation Nb-Ti and Nb_3Sn superconductors, while twistability and multifilamentary geometry reduce AC loss. Also, the magnetic shielding current in Bi-2212 is much lower compared to REBCO coated conductors, which results in much reduced field uniformity errors [3]. Significant advances have been made in recent years that have put Bi-2212 on the verge of commercial applications and it is considered as a very attractive option for high field NMR and general purpose research magnets [4-10].

One of the key issues facing Bi-2212 superconducting magnet technology is the development of a suitable insulation material. In a wind-and-react coil fabrication process, the insulation has to be applied to the wire before coil winding, meaning that it has to withstand a heat treatment with a maximum temperature of around 890° C. The TiO_2 insulation developed in collaboration between the National High Magnetic Field Laboratory (NHMFL) and nGimat meets this criterion while offering a high winding current density due to its relative thinness compared to other available insulation materials [11-14]. However, while dip-coated TiO_2 insulation worked well in short samples, when used in a large coil it had severe electrical shorting after the overpressure heat treatment (OPHT). We found that this quiet unexpected shorting was caused by Ag protrusions through the TiO_2 layer. In our earlier work [15], we identified the combination of Ag- 0.2 wt.% Mg alloy (Ag(Mg)) sheath and overpressure heat treatment (OPHT) as the underlying reasons behind these Ag protrusions.

In this study, we wanted to investigate if all Ag(Mg) sheathed Bi-2212 wires with TiO_2 insulation have protrusions after OPHT. To do this we used 9 different wires produced between 2011 and 2018.

The diameter of the wires used varied from 0.8 to 1.3 mm and both round and aspected wires were used. We used the standard OPHT schedule (figure 1) and also heat treated some samples with increased time in the melt state to simulate the longer melt time that coils often seen compared to short samples due to their larger thermal mass.

2. Experimental procedure

The wires used in this study were fabricated by Bruker OST using the powder-in-tube method. Powders from three different manufacturers, Nexans, nGimat, and MetaMateria, were used. The nominal composition of all the powders was similar to $Bi_{2.17}Sr_{1.74}Ca_{0.89}Cu_{2.00}O_x$. The wires were all double restacked, with a pure Ag matrix in contact with the Bi-2212 and a Ag(Mg) outer sheath enclosing the restacked bundles. All the wires, except for one, were dip coated in-house at the NHMFL with a TiO_2 slurry. nGimat dip coated the other wire. Table 1 contains detailed information about the wires. Variation in coating thickness was observed between different wires and also along the circumference of the same wire.

Table 1: Brief description of the wires used in this study and results.

Wire No.	Powder[b]	Archi-tecture	Shape[c]	Dia or LxW (mm)	Insulation[d]	Twist (Y/N)	Results[e]
Pmm110106-2	Nex	85x18	R	1.2	nGi	N	NP, B
Pmm130125	Nex	121x18	A	1.38x0.98	I-H	N	NP, B
Pmm140328-2	Nex	55x18	R	1	I-H	Y	NP, B
Pmm140606[a]	Nex	55x18	R	1	I-H	Y	FP
Pmm151103	Nex	121x18	R	1	I-H	Y	NP, B
Pmm150227[a]	Nex	121x18	R	1.3	I-H	Y	EP
Pmm160301[a]	Meta	85x18	R	1	I-H	N	NP
Pmm180410[a]	nGi	85x18	R	1	I-H	N	NP
Pmm181004	nGi	55x18	R	0.9	I-H	N	NP, B

[a]Wire used for extended time in the melt study -- [b]Nex = Nexans; Meta = MetaMateria; nGi = nGimat -- [c]R = round; A = aspected -- [d]nGi = nGimat; I-H = in-house -- [e]EP = extensive protrusions; FP = few protrusions; NP = no protrusions; B = blob

8 cm long samples were used with each end hermetically sealed with a Ag-bead using a torch before the heat treatment. All the wires first went through a burnout heat treatment with a maximum temperature of 450° C in flowing oxygen to remove the organic binders and plasticizers in the TiO_2 slurry. The heating rate was 10° C/h to avoid ignition of the organics. After that, the wires underwent the OPHT processing heat treatment at 50 bar overpressure (1 bar O_2 + 49 bar Ar). Two slightly different OPHT schedules were used. In one, the standard OPHT was carried out (figure 1) in which the wires took about 1 hr to cool down from 888° C (the maximum temperature) to 878° C. For the other OPHT, this cooling time was extended to 3 hrs.

Longitudinal sections of heat treated wires were mounted in conductive epoxy using a mounting press. The wires were dry polished using SiC papers up to 800 grit, followed by final polishing in an automatic vibratory polisher (Buehler Vibromet 2) using a suspension of 50 nm alumina powder in methanol. A Zeiss 1540EsB SEM was used to observe the microstructure of the TiO_2- Ag(Mg) interface. About 6 cm of this interface was examined for each wire.

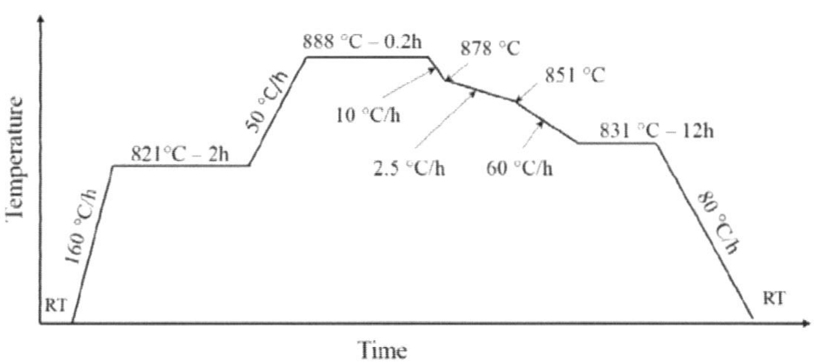

Figure 1. Standard heat treatment schedule for Bi-2212. For increased time in the melt samples, the cooling rate between 888 °C and 878 °C was reduced to 3.3 °C/h. The figure is not to the scale.

3. Results

3.1. . Standard OPHTed wires:

3.1.1. Pmm150227. This is the wire that was used for the large coil that shorted. Similar to the coil, the short sample showed a lot of Ag protrusions into the insulation layer (figure 2a, 2b) which varied in size, some almost penetrating through the whole 20 μm insulation layer. The distribution of the Ag protrusions was not homogenous; there were areas where they were significantly more abundant. The wire also showed reduced roundness, which is a common feature for twisted wires. In addition to the large Ag protrusions, there were many Ag particles distributed in the portion of the insulation close to the wire surface. We suspect that these discrete Ag regions are actually part of the large Ag protrusions.

3.1.2. Pmm110106. The TiO_2 on this wire was dip coated by nGimat and the composition of the TiO_2 slurry was possibly different as the insulation was brown instead of the usual bright white colour of in-house coated wires. Almost no Ag protrusions were observed in this wire (figure 2c). There were some Ag "blobs" on the wire surface (5 or 6 blobs along the 6 cm of interface that was examined). The size of the blobs was less than 5 μm and none of them penetrated the 12 μm thick insulation layer. These Ag blobs are a common occurrence in the Ag(Mg)-sheathed wires and can be seen in bare wires too. We refer to them as blobs because their shape and size are distinctly different from the Ag protrusions. Also, the insulation had very few of the Ag particles that we saw in Pmm150227 (figure 2a, 2b).

3.1.3. Pmm130125. This wire was rolled into an aspected geometry from a round wire. The insulation thickness was around 14 μm. The microstructure (figure 2d) was similar to Pmm110106, with only 5-6 Ag blobs along 6 cm of interface. The blobs were mostly smaller than 5 μm, with one reaching about 8 μm.

3.1.4 Pmm140328. The insulation thickness for this wire was about 16 μm. Its microstructure (figure 2e) was similar to Pmm110106 with very few Ag blobs.

3.1.5. Pmm140606. The wire lacked roundness like most of other twisted wires. The insulation thickness, which was on average around 15 μm, also varied due to the unevenness of the wire surface and was considerably thinner at some places. This wire and Pmm150227 were the only wires that had Ag protrusions (figure 2f) that were as large as the insulation layer thickness. However, the number of protrusions was significantly less compared to Pmm150227, with only 10-12 protrusions observed along the 6 cm of interface examined. Like Pmm150227, there were randomly distributed Ag particles in the insulation layer in contact with the Ag.

3.1.6. Pmm151103. This wire showed the least number of Ag blobs (figure 2g), only 1-2 that were all less than 5 μm in size. Surprisingly, this twisted wire was fairly round.

3.1.7. Pmm160301. The microstructure of this wire (figure 2h), which was made with MetaMateria powder, was quite clean and there were almost no Ag protrusions observed in the microstructure checked.

3.1.8. Pmm180410. This wire showed very little to no sign of Ag protrusion (figure 2i).

3.1.9. Pmm181004. The microstructure (figure 2j) was similar to Pmm110106, with only 5-6 blobs that were around 5 μm in size.

3.2. Increased time in the melt

We choose 4 wires to study the effect of increased time in the melt. These were the two wires that showed significant protrusions after the heat treatment (Pmm150227 and Pmm140606), which were made with Nexans powder, and one made with nGimat and one with MetaMateria powder, Pmm180410 and Pmm160301, respectively, which did not show any protrusions. No significant change in the protrusions were observed with longer time in the melt for any of the wires.

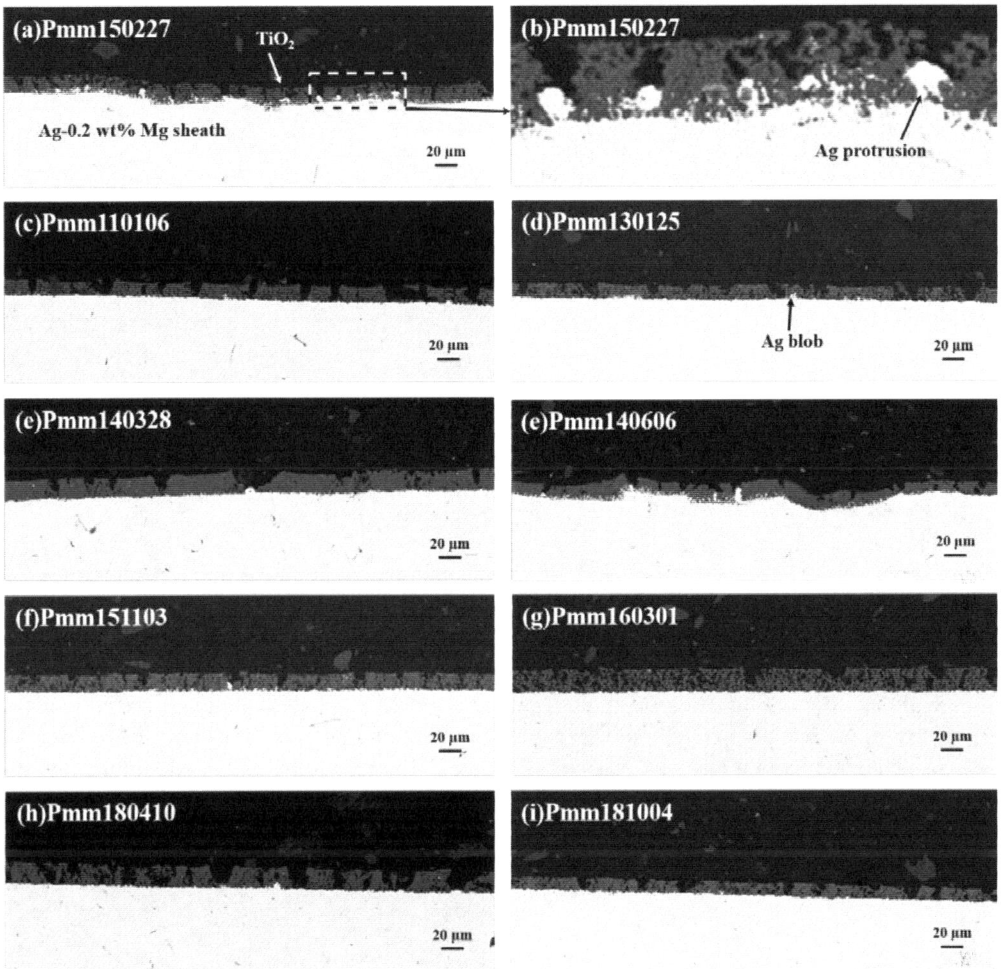

Figure 2. Longitudinal cross sections of different wires that underwent standard OPHT. A section of (a) is magnified in (b) for better demonstration of Ag protrusion. In (d), an arrow points to a Ag "blob".

4. Discussion

The wires used in our earlier study [15] showed extensive protrusions, leading to our concern that all Bi-2212 wires with a TiO_2 insulation layer would form protrusions after OPHT, thus defeating the essential purpose of TiO_2 as an insulation. An important observation of this subsequent study is that the broad range of wires studied displayed a wide range of behavior, ranging from extensive protrusions (Pmm150227) to essentially no protrusions. There is no apparent correlation between the wire properties listed in Table 1 and whether the wire had protrusions. For example, the two wires that had extensive protrusions were twisted, but the two other twisted wires did not have extensive protrusions. Unsatisfactory though it is, it presently appears that the only way to tell if a wire will form protrusions is to test it.

One positive result is that we did not find that increasing the time in the melt affected the development or extent of protrusions. This relives one of our concerns for continued use of TiO_2 for coil insulation, since coils tend to spend a longer time in the melt than optimised short samples of low thermal mass.

In our earlier study [15], we found one curious result which was that Ag protrusions were devoid of the MgO particles that provide precipitation hardening of the sheath. We did not do TEM studies in the current investigation, but suspect that the protrusions in wire Pmm140606 were also devoid of MgO. Our working hypothesis was that the Mg or MgO is mobile during the highest temperature portion of the heat treatment and that the thermodynamic stability of MgO is greater at the surface where it has access to 1 bar O_2. Protrusions can then be viewed as a form of extrusion of the purer and weaker Ag into the interstices in the TiO_2 layer. The variability of this result suggests either that the quality of the Ag(Mg) alloy used in the sheath matters or perhaps that the quality of the oxidation matters.

As part of this work, we also heat treated wire Pmm150227 at 1 bar pressure and found that it did not show any Ag protrusions. This confirms our observation [15] that OPHT is required to form the protrusions. A working hypothesis is that the hydrostatic pressure of the gas in the furnace can actually produce a non-hydrostatic response in the Ag sheath if there are local variations in the strength of the wire caused by variations in the local MgO concentration. Thus 50 bar can provide a driving force for protrusion formation, while 1 bar total pressure cannot.

We changed the insulation strategy we use for Bi-2212 coils after the extensive protrusions in wire Pmm150227 electrically shorted the large coil. Now we use two insulation layers: first the dip-coated TiO_2 layer in direct contact with the Ag(Mg) sheath and then an alumino-silicate braid that prevents shorting if protrusions occur. This alumino-silicate braid does however react with Ag, meaning that we do require the TiO_2 layer to physically separate the braid from the Ag(Mg) sheath. Unfortunately the alumino-silicate braid is much thicker (150 μm) than our preferred 20-30 μm TiO_2 layer thickness, meaning that we do lose winding current density. This study suggests that it is worth further effort to understand how protrusions form in some wires and not others. We should point out however, that the braid helps epoxy penetrate the coil and strengthens it after curing, meaning that we are not yet ready to go back to single layer TiO_2 insulation.

5. Conclusion

We examined the extent of Ag protrusions in 9 different Bi-2212 wires coated with TiO_2. Our findings showed that the extent of Ag protrusions varies strongly among different wires and indeed that most of the wires did not show Ag protrusions. Samples that spent 3 hours longer in the melt state did not show any change in the overall amount or size of Ag protrusions. We speculate that Ag protrusion are related to variations in the particular Ag(Mg) alloy used or its oxidation heat treatment, allowing us optimism that we may be able to control and avoid protrusions with better understanding of the state of the Ag(Mg) sheath.

Acknowledgment

The work is supported by the US DOE Office of High Energy Physics under grant number DE-SC0010421 and by the NHMFL, which is supported by NSF under Award Number DMR-1644779, and by the State of Florida, and is amplified by the U.S. Magnet Development Program (MDP).

References:

[1] Jiang J, Bradford G, Hossain SI, Brown MD, Cooper J, Miller E, Huang Y, Miao H, Parrell JA, White M, Hunt A, Sengupta S, Revur R, Shen T, Kametani F, Trociewitz U, Hellstrom E, Larbalestier D 2019 High-Performance Bi-2212 Round Wires Made With Recent Powders. *IEEE Transactions on Applied Superconductivity*. **29**(5):1-5.

[2] Chen B, Halperin WP, Guptasarma P, Hinks DG, Mitrović VF, Reyes AP and Kuhns PL 2007 Two-dimensional vortices in superconductors. *Nature physics*. **3**(4):239-42.

[3] Hwang YJ, Hahn S, Lee S, Jang JY, Han JH, Lee H, Kim J, Yeom H, Yoon S, Kim K, Ahn MC 2016 A study on mitigation of screening current induced field with a 3-T 100-mm conduction-cooled metallic cladding REBCO magnet. *IEEE Transactions on Applied Superconductivity*. **27**(4):1-5.

[4] Larbalestier DC, Jiang J, Trociewitz UA, Kametani F, Scheuerlein C, Dalban-Canassy M, Matras M, Chen P, Craig NC, Lee PJ and Hellstrom EE 2014 Isotropic round-wire multifilament cuprate superconductor for generation of magnetic fields above 30 T. *Nature materials*. **13**(4):375-81.

[5] Hasegawa T, Ohtani N, Koizumi T, Aoki Y, Nagaya S, Hirano N, Motowidlo L, Sokolowski RS, Scanlan RM, Dietderich DR, Hanai S 2001 Improvement of superconducting properties of Bi-

2212 round wire and primary test results of large capacity Rutherford cable. *IEEE transactions on applied superconductivity*. **11**(1):3034-37.

[6] Weijers HW, Trociewitz UP, Marken K, Meinesz M, Miao H and Schwartz J 2004 The generation of 25.05 T using a 5.11 T $Bi_2Sr_2CaCu_2O_x$ superconducting insert magnet. *Superconductor Science and Technology*. **17**(4):636-44.

[7] Miao H, Marken KR, Meinesz M, Czabaj B and Hong S 2005 Development of round multifilament Bi-2212/Ag wires for high field magnet applications. *IEEE Transactions on applied superconductivity*. **15**(2):2554-57.

[8] Huang Y, Miao H, Hong S and Parrell JA 2013 Bi-2212 round wire development for high field applications. *IEEE Transactions on Applied Superconductivity*. **24**(3):1-5.

[9] Godeke A, Cheng DW, Dietderich DR, Hannaford CR, Prestemon SO, Sabbi G, Wang XR, Hikichi Y, Nishioka J and Hasegawa T 2009 -and-react Bi-2212 accelerator magnet technology. *IEEE Transactions on Applied Superconductivity*. **19**(3):2228-31.

[10] Zhang K, Higley H, Ye L, Gourlay S, Prestemon S, Shen T, Bosque E, English C, Jiang J, Kim Y, Lu J 2018 Tripled critical current in racetrack coils made of Bi-2212 Rutherford cables with overpressure processing and leakage control. *Superconductor Science and Technology*. **31**(10):105009.

[11] Lu J, McGuire D, Kandel H, Xin Y, Chen P, Jiang J, Trociewitz U, Hellstrom E and Larbalestier DC 2016 Ceramic insulation of $Bi_2Sr_2CaCu_2O_{8-x}$ round wire for high-field magnet applications. *IEEE Transactions on Applied Superconductivity*. **26**(4):1-5.

[12] Kandel H, Lu J, Jiang J, Chen P, Matras M, Craig N, Trociewitz UP, Hellstrom EE and Larbalestier DC 2015 Development of TiO_2 electrical insulation coating on Ag-alloy sheathed $Bi_2Sr_2CaCu_2O_{8-x}$ round-wire. *Superconductor Science and Technology*. **28**(3):035010.

[13] Chen P, Trociewitz UP, Dalban-Canassy M, Jiang J, Hellstrom EE and Larbalestier DC 2013 Performance of titanium oxide–polymer insulation in superconducting coils made of Bi-2212/Ag-alloy round wire. *Superconductor Science and Technology*. **26**(7):075009.

[14] Ishmael S, Luo H, White M, Hunte F, Liu XT, Mandzy N, Muth JF, Naderi G, Ye L, Hunt AT and Schwartz J 2013 Enhanced Quench Propagation in $Bi_2Sr_2CaCu_2O_x$ and $YBa_2Cu_3O_{7-x}$ Coils via a Nanoscale Doped-Titania-Based Thermally Conducting Electrical Insulator. *IEEE Transactions on Applied Superconductivity*. **23**(5):7201311.

[15] Hossain I, Jiang J, Matras M, Trociewitz UP, Lu J, Kametani F, Larbalestier D, Hellstrom E 2017 Effect of sheath material and reaction overpressure on Ag protrusions into the TiO_2 insulation coating of Bi-2212 round wire. *IOP Conference Series: Materials Science and Engineering*. **279**(1):012021.

The Role of CHPD and AIMI processing on enhancing J_C and transverse connectivity of *in-situ* MgB_2 strand

F Wan[1], M D Sumption[1], M A Rindfleisch[2], and E W Collings[1]

[1]Center for Superconductor and Magnetic Materials, Department of Materials Science and Engineering, The Ohio State University, Columbus, OH 43210, USA.
Corresponding author e-mail: wan.108@osu.edu
[2]Hyper Tech Research Inc, Columbus, OH 43228, USA

Abstract. Research into *in-situ* MgB_2 strand has been focused on improvements in J_C through reduction of porosity. Both of cold-high-pressure-densification (CHPD) and advanced-internal-magnesium-infiltration (AIMI) techniques can effectively remove the voids in *in-situ* MgB_2 strands. This study shows the nature of the reduced porosity for *in-situ* MgB_2 strands lies on increases in transverse grain connectivity as well as longitudinal connectivity. The CHPD method bi-axially applying 1.0 GPa and 1.5 GPa yielded 4.2 K $J_{CM\|}$s of 9.6×10^4 A/cm^2 and 8.5×10^4 A/cm^2 at 5 T, respectively, with compared with 6.0×10^4 A/cm^2 for typical powder-in-tube (PIT) *in-situ* strand. Moreover, AIMI-processed monofilamentary MgB_2 strand obtained even higher J_Cs and transverse grain connectivity than the CHPD strands.

1. Introduction

MgB_2 superconducting strands are promising to the practical magnetic application due to its high transition temperature T_C (39 K) [1], high coherence length [2, 3], and low anisotropy of upper critical fields (B_{c2}) [3–6]. The powder-in-tube (PIT) *in-situ* MgB_2 strands were fabricated by filling a mixture of Mg and B powder into a non-reactive metallic tube and then being cold-worked into wires or tapes. The PIT strands have large amount of voids elongated along longitudinal strand axis which were left behind by molten Mg powders after heat treatment [7]. The present of voids tends to limit the number of the continuous current path in the *in-situ* MgB_2 strand and therefore suppress the current-carrying-capacity of the strand.

The CHPD technique can effectively increase the transport properties of PIT *in-situ* MgB_2 strands by eliminating the pores [8–11]. The pre-reacted powder-in-tube composite was bi-axially cold-densified at room temperature to increase the mass density of the Mg + B mixture. In this case, the cold-densified MgB_2 strand can obtain higher grain connectivity after heat treatment. Additionally, the Mg reactive – liquid – infiltration (RLI) process, initiated by Giunchi *et al* [12], also has the ability to eliminate the pores and produce a dense MgB_2 layers in *in-situ* MgB_2 strands. For RLI process, a Mg rod is inserted axially into a boron-filled metallic tube. After wire drawing the heat treatment (H. T.), MgB_2 layer is formed through the reactive diffusion of Mg into B layer. Since Mg is totally separated with precursor B layer before H. T., the RLI process can totally eliminate the "Mg-site porosity" from MgB_2 layer but induce the formation of a big hole at the central region of the strand [13]. Furthermore, since the molar volume of MgB_2 (17.46 cm^3/mol) is twice as that of two B atoms (9.18 cm^3/mol), the volume expansion associated with the reactive transformation from 2B to MgB_2 enables even better connections between MgB_2 grains during heat treatment [13]. Our group named our RLI-processed strands as advanced – internal – magnesium – infiltration (AIMI) strands due to their optimized strand architecture and high J_Cs [13].

The previous researches mostly focused on investigating the effect of CHPD and AIMI techniques on the transport properties along longitudinal strand axis of *in-situ* MgB_2 strands, such as transport J_Cs and longitudinal grain connectivity. However, Shi and Susner pointed out that high anisotropic grain connectivity exists in PIT *in-situ* strands, which was resulted from the elongated voids between elongated MgB_2 stringers [7]. Therefore longitudinal transport properties are different with the transport properties along transverse strand axis for MgB_2 superconducting wires. In continuing along these lines, the transport properties along transverse strand axis were investigated for the 2.0 mol% C-doped *in-situ* MgB_2 strands in this study. The anisotropic connectivity of the PIT strands results in the differences between perpendicular magnetic J_C ($J_{CM\perp}$) and parallel J_C ($J_{CM||}$). The influence of aspect ratio (S = length/diameter) on transverse and longitudinal J_C were investigated for the PIT strand (P00). The CHPD-processed strands were P10 (1.0 GPa cold-pressing) and P15 (1.5 GPa cold-pressing). According to the previous results of our group, the CHPD technique increased the transport J_C of the monofilamentary PIT *in-situ* MgB_2 strand from 3.0×10^4 A/cm^2 to 3.6×10^4 A/cm^2 at 4.2 K and 10 T due to decreased porosity [11] and AIMI-processed MgB_2 strands attained the 4.2 K, 10 T transport layer J_Cs of $1.0 \sim 1.5 \times 10^5$ A/cm^2 [13-16]. In this study, we compared the $J_{CM||}$s and transverse connectivity of the CHPD- and AIMI-processed strands with those of the P00 strand at 4.2 K and 20 K. The relationship between porosity and transverse flux pinning force density $F_{p||}$, which is $J_{CM||} \times B$, for the *in-situ* MgB_2 strands was also discussed.

2. Experimental

2.1. Sample preparation

A series of pre-reacted powder-in-tube (PIT) *in-situ* strands, typically 0.834 mm diameter, with a Nb barrier and a Cu outer sheath were fabricated by Hyper Tech Research, Inc. (HTR). Two PIT strands (designated P10 and P15) were bi-axially densified with 1.0 GPa and 1.5 GPa at room temperature, respectively. The other strand (designed A00) manufactured through AIMI technique were also provided by HTR. The AIMI-processed strand, with 0.55 mm diameter, has a Nb barrier and a Monel outer sheath. The powders used for the present strands were 2 mol% C–doped amorphous B (10 – 100 nm) from Specialty Materials Inc. (SMI). The cold–densified PIT strands were heat-treated at 675 ^oC for 1 h and the AIMI strand was heat – treated at 625 ^oC for 16 h. The specification and heat treatment (H. T.) conditions of the strands are presented in table 1.

2.2. Transport and Magnetic Measurements

The transport I_C (I_{CT}) test was conducted in perpendicular magnetic field up to 13 T in a pool of liquid Helium at 4.2 K on the MgB_2 strands with a total length of 50 mm and a gauge length of 5 mm. The electric criterion used for determining I_{CT}s is 1.0 μV/cm. The magnetizations versus perpendicular and parallel magnetic fields (M – H) loops were measured by a Quantum Design Model 6000 Physical Property Measuring System (PPMS) for all strands with a sample length of 3 – 5 mm.

Table 1. Strand specification and H. T. conditions

Strand Name	Strand Type	CHPD (GPa)	H. T. (h/^{o}C)
P00	PIT	0.0	1/675
P10	PIT	1.0	1/675
P15	PIT	1.5	1/675
A00	AIMI	0.0	16/625

3. Results

The J_{CT}s of the PIT strands were the transport critical current normalized by MgB_2 core area. As shown in Figure 1(a) the MgB_2 core of the typical PIT *in-situ* strand is a solid cylinder. Figure1(b)-(d) shows the shape of MgB_2 cores for the CHPD- and AIMI-processed strands are cuboid and hollow cylinder, respectively. The J_{CT} of the AIMI–processed strands, which is also named as transport layer J_C, were calculated by dividing the I_{CT} by the area of annulus MgB_2 layer. Values of magnetic J_C ($J_{CM\perp}$ and $J_{CM\|}$) for the MgB_2 strands were extracted from the full $M - H$ loops heights ΔM, using the standard Bean model equations [7, 17]:

For the PIT *in–situ* wire P00:

$$\text{Perpendicular Magnetic } J_C\text{: } J_{CM\perp} = \frac{3\pi\Delta M}{8R_0} \tag{1}$$

$$\text{Parallel Magnetic } J_C\text{: } J_{CM\|} = \frac{3\Delta M}{2R_0} \tag{2}$$

Here R_0 is the radius of the cylinder MgB_2 core in the PIT wire.

For the densified wires P10 and P15:

$$\text{Parallel Magnetic } J_C\text{: } J_{CM\|} = \frac{2\Delta M}{b(1-\frac{b}{3a})} \tag{3}$$

Here a, b are both lengths of the transverse cross – sectional area of cuboid MgB_2 core, a > b.

IOP Conf. Series: Materials Science and Engineering **756** (2020) 012018 doi:10.1088/1757-899X/756/1/012018

For the AIMI wire A00:

$$\text{Parallel Magnetic } J_C\text{: } J_{CM\|} = \frac{3\Delta M}{2}\frac{R_0^2 - R_i^2}{R_0^3 - R_i^3} \tag{4}$$

Here R_i is the inner diameter of the annulus MgB_2 layer and R_0 is the outer diameter of the annulus MgB_2 layer.

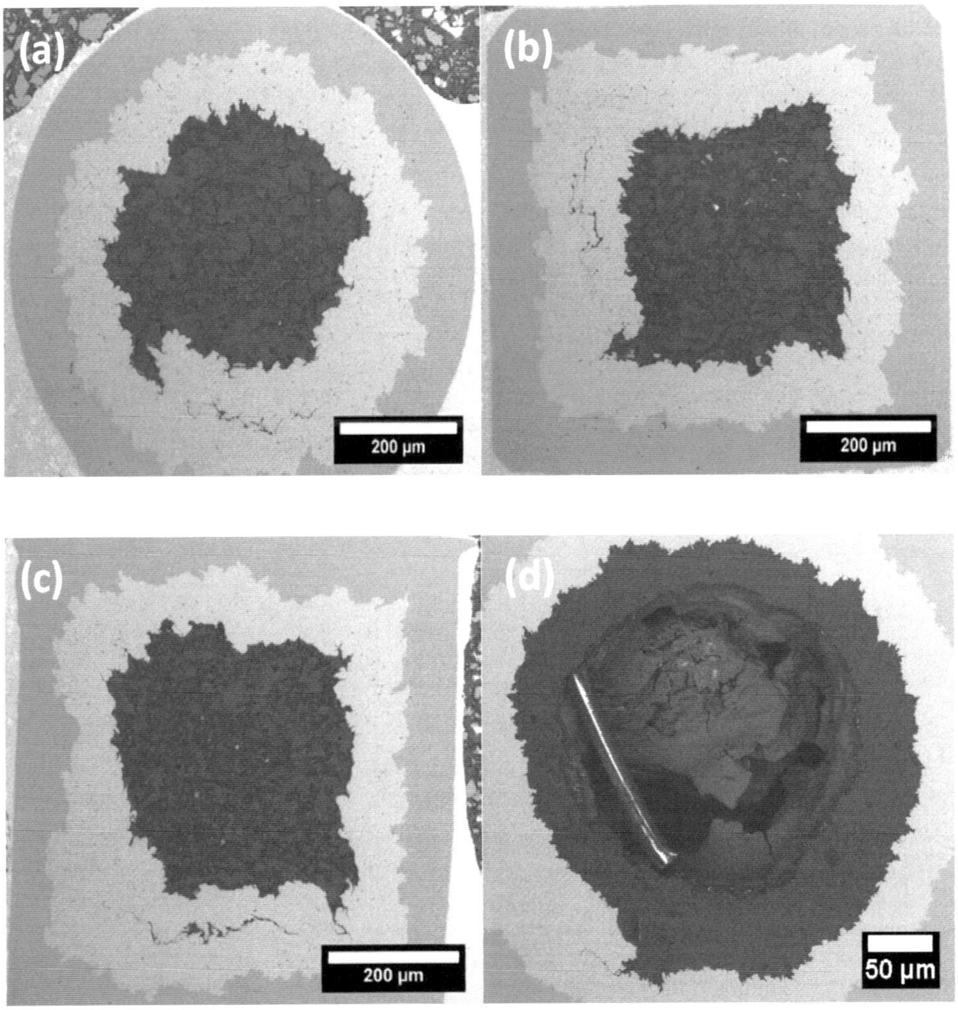

Figure 1. (a) Back scattered SEM images of (a) strand P00, (b) strand P10 (1.0 GPa cold pressure), (c) strand P15 (1.5 GPa cold pressure), and (d) AIMI strand A00.

IOP Conf. Series: Materials Science and Engineering **756** (2020) 012018 doi:10.1088/1757-899X/756/1/012018

3.1. Transport and magnetic critical current densities, J_{CT} and J_{CM}

Figure 2(a) shows the J_{CT} and $J_{CM\_\perp}$ versus B at 4.2 and 20 K for the strand P00. It can be seen that $J_{CM_\perp}$s agree with J_{CT} at low fields, whereas the bifurcation of J_{CT} and $J_{CM\_\perp}$ happened at high fields. Moreover, $J_{CM_\perp}$s were greatly affected by the aspect ratio S, especially at high magnetic fields. The relationships among J_{CT}, $J_{CM\_\perp}$, and aspect ratio were fully discussed in ref [7] and [18]. Therefore, the values of $J_{CM\_\perp}$ are not only determined by the intrinsic properties but also the extrinsic properties of the *in-situ* MgB_2 strands. As shown in Figure 2(b) $J_{CM||}$s were independent on the aspect ratio S. Based on this result, we can know the values of $J_{CM||}$s as well as J_{CT}s are merely determined by the intrinsic properties of the *in-situ* MgB_2 strands. The MgB_2 macrostructure of the reacted PIT *in-situ* strand is characterized by elongated polycrystalline MgB_2 fibers and elongated pores partially separating the fibers [7, 11, and 18]. The J_{CT}s and $J_{CM||}$s can represent the longitudinal and transverse connectivity of the MgB_2 fibers, respectively. In summary, we can investigate the effect of CHPD and AIMI technique on the current–carrying–capacity of the *in–situ* MgB_2 strands through $J_{CM||}$s as well as J_{CT}.

Figure 3 shows the J_{CT} and $J_{CM||}$ versus B at 4.2 and 20 K for all strands. The CHPD technique significantly enhanced $J_{CM||}$s of PIT *in-situ* strands at 4.2 K. Strand P10 and P15 attained 9.5 x 10^4 A/cm^2 and 8.5 x 10^4 A/cm^2 at 4.2 K and 5 T, with respect to 6.0 x 10^4 A/cm^2 for strand P00. 4.2 K J_{CT}s were slightly enhanced by the CHPD technique. Therefore, the increases in J_{CT} and $J_{CM||}$ indicated that the both longitudinal and transverse connections between MgB_2 fibers were enhanced. With the formation of high density MgB_2 layer, strand A00 attained 4.2 K, 10 T J_{CT} of 9.4 x 10^4 A/cm^2, which is 180% higher than those of CHPD-processed strands. On the other hand, the AIMI strand obtained 5 T $J_{CM||}$s of 3.1 x 10^5 A/cm^2 at 4.2 K and 1.7 x 10^4 A/cm^2 at 20 K, which are about 240% higher than those of CHPD strands.

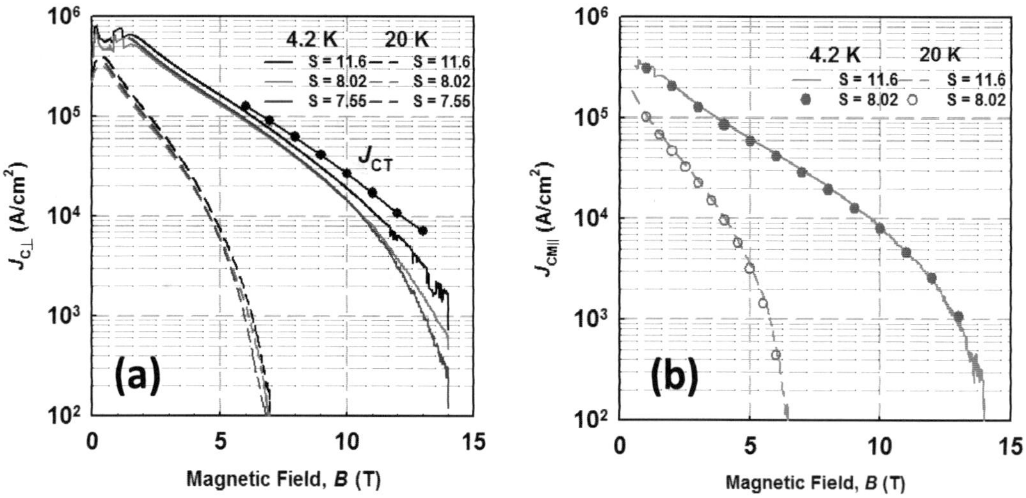

Figure 2. (a) J_{CT} and $J_{CM\_\perp}$ versus B at 4.2 and 20 K (b) $J_{CM||}$ versus B at 4.2 and 20 K for strand P00

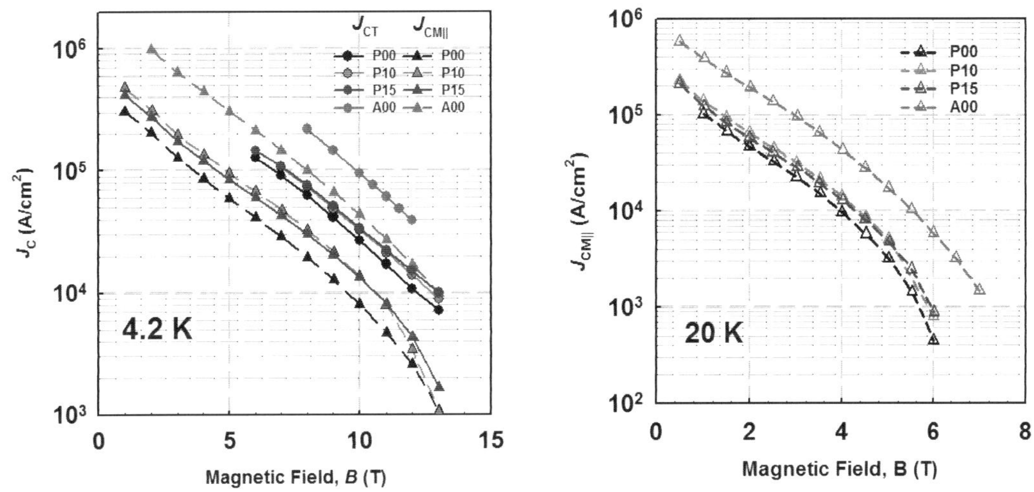

Figure 3. J_{CT} and $J_{CM||}$ versus B at 4.2 and 20 K for all strands

3.2. Transverse Flux Pinning, Porosity and Transverse Grain Connectivity

Figure 4 shows the $F_{p||}/F_{p,max||}$ versus B/B_{irr} for all strands, where $F_{p||} = J_{CM||} \times B$. It can be seen that the peak pinning occurred at $b = B/B_{irr||}$ close to 0.2 at 4.2 and 20 K, which is in agreement with the Dew-Hughes/Kramer model [19, 20]. In other words, the dominant pinning centers for the all strands are also grain boundaries for the direction along transverse strand axis. According to the previous work [11], the densified wires have decreased porosities, the values of porosity are ~ 50% (p = 0) and ~ 30 % (p = 1.0 or 1.5 GPa). Since the Mg–site porosity was totally eliminated for AIMI–processed strands, the porosity of the AIMI strand is close to 0. Therefore, both CHPD and AIMI processes enable the resulting MgB_2 phases to be denser and more connected. For the discussion, it can be concluded that the enhanced J_{CT} and $J_{CM||}$ in densified wires and AIMI wire is correlated with lower porosity (higher grain connectivity).

The connectivity K defined by Rowell [21] can represent the grain connectivity of MgB_2 strands. The connectivity K is calculated by the equation:

$$K = \frac{\Delta\rho_{SC}}{\Delta\rho} \tag{5}$$

Here $\Delta\rho$ is the difference between the sample's resistivity at 300 K and the sample's resistivity at 40 K and $\Delta\rho_{SC}$ is the resistivity difference for an ideal single crystal. However, transverse connectivity $K_{||}$ is difficult to be determined for the MgB_2 wires. It has been reported that the maximum flux pinning force densities of fully–connected MgB_2 superconductor, where grain connectivity K = 100 %, are estimated to be 90 GN/m^3 at 4.2 K and 22 GN/m^3 at 20 K [22]. Therefore, we can roughly estimate the transverse grain connectivities $K_{||}$s of the *in-situ* MgB_2 strands by normalizing the $F_{p,max||}$ with the $F_{p,max}$ of fully-connected MgB_2. Table 2 shows the transverse maximum flux pinning force densities $F_{p,max||}$ at 4.2 K and 20 K and estimated transverse grain connectivities for all the strands. The connectivity of 5% is achieved by the 2 mol% C-doped PIT strand, P00. The cold–densification increases the connectivity of PIT strand by 20 %. The AIMI strand A00 obtained the highest transverse grain connectivities of 20 %.

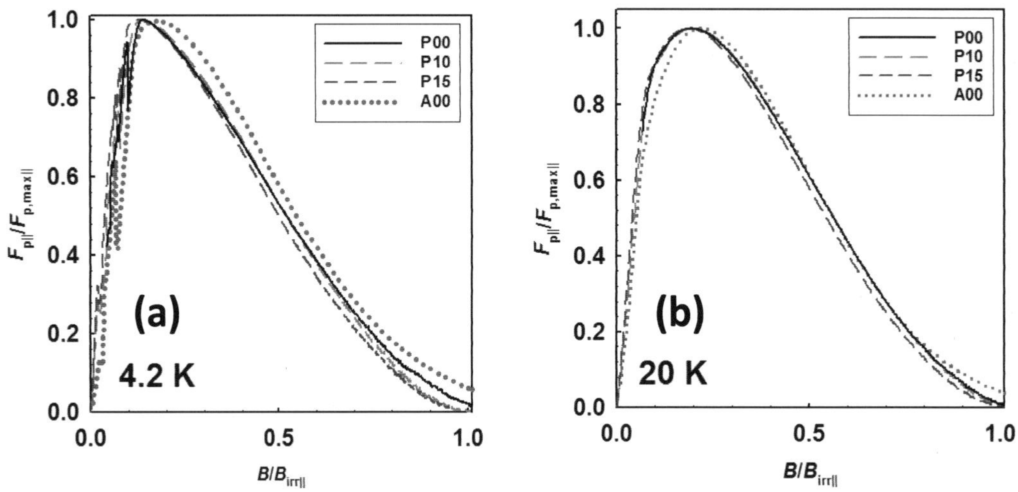

Figure 4. (a) $F_{p||}/F_{p,max||}$ versus B/B_{irr} for all strands at 4.2 K, (b) $F_{p||}/F_{p,max||}$ versus B/B_{irr} for all strands at 20K

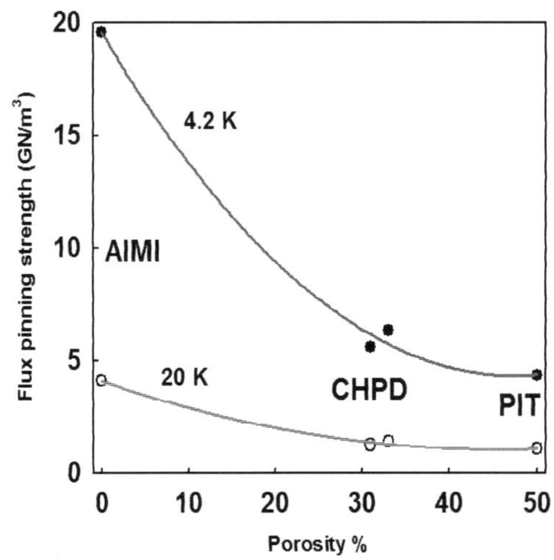

Figure 5. Relationship between porosity and transverse flux pinning force densities for *in–situ* MgB_2 strands

IOP Conf. Series: Materials Science and Engineering **756** (2020) 012018 doi:10.1088/1757-899X/756/1/012018

Table 2. Transverse flux pinning force densities and estimated connectivity of the strands.

Strand Name	$F_{p,max||}$, 4.2 K (GN/m^3)	$F_{p,max||}$, 20 K (GN/m^3)	Transverse Connectivity, $K_{||}$ %
P00	4.3	1.1	5.0
P10	6.3	1.4	6.0
P15	5.6	1.3	6.0
A00	19.6	4.1	20

4. Conclusion

$J_{CM||}$s is merely determined by the intrinsic properties of *in-situ* MgB_2 strands, so the current-carrying capacity of *in-situ* MgB_2 strands can be represented by $J_{CM||}$ as well as J_{CT}. The CHPD of 1.0 GPa and 1.5 GPa enhanced the 4.2 K, 5 T $J_{CM||}$ from 6.0×10^4 A/cm^2 to 9.6×10^4 A/cm^2 and 8.5×10^4 A/cm^2, respectively. AIMI strand attained the highest J_{CT} and $J_{CM||}$ at 4.2 and 20 K due to the formation of a high dense MgB_2 layer. By eliminating the voids in *in-situ* MgB_2 strands through CHPD and AIMI technique, better connections between MgB_2 grains along transverse strand axis can be obtained in *in–situ* MgB_2 strands

5. References

[1] Nagamatsu J, Nakagawa N, Muranaka T, Zenitani Y, and Akimitsu J 2001 *Nature* ***410*** 63-64
[2] Finnemore D K, Ostenson J E, Bud'ko S L, Lapertot G and Canfield PC 2001 *Rhys. Rev. Lett* **86** 2420-2422
[3] Vinod K, Abhilash Kumar R G and Syamaprasad U 2007 *Supercond. Sci. Technol* **20** R1-R3
[4] Eltsev Y, Lee S, Nakao K, Chikumoto N, Tajima S, Koshizuka N and Murakami M 2002 *Phys. Rev. B* **65** 140501
[5] Eisterer M, Zehetmayer M and Weber H W 2003 *Phys. Rev. Lett* **90** 247002
[6] De Liam O F, Ribeiro R A, Avila M A, Cardoso C A and Coelho A A 2001 *Phys. Rev. Lett* **86** 5974
[7] Shi Z X, Susner M A, Majoros M, Sumption M D, Peng X, Rindfleisch M, Tomsic M J and Collings E W 2010 *Supercond. Sci. Technol* **23** 045018
[8] Flukiger R, Hossain M S A and Senatore C 2009 *Supercond. Sci. Technol* **22** 085002
[9] Hossain M S A, Senatore C, Flukiger R, Rindfleisch M A, Tomsic M J, Kim J H and Dou S X 2009 *Supercond. Sci. Technol* **22** 095004
[10] Hossain M S A et al 2014 *Supercond. Sci. Technol* **27** 095016
[11] Wan F, Sumption M D, Rindfleisch M A and Collings E W 2017 *IOP Conf. Series: Materials Science and Engineering* **279** 012024
[12] Giunchi G, Ripamonti G, Perini E, Cavallin T and Bassani E 2007 *IEEE TRANSACTIONS ON APPLIED SUPERCONDUCTIVITY* **17** 2761
[13] Li G Z, Sumption M D, Susner M A, Yang Y, Reddy K M, Rindfeisch M A, Tomsic M J, Thong C J and Collings E W 2012 *Supercond. Sci. Technol* **25** 115023
[14] Ye S J, Matsumoto A, Zhang Y C and Kumakura H 2014 *Supercond. Sci. Technol* **27** 085012
[15] Li G Z, Sumption M D, Zwayer J B, Susner M A, Rindfleisch M A, Thong C J, Tomsic M J and Collings E W 2013 *Supercond. Sci. Technol* **26** 095007
[16] Kumakura H, Hur J, Togano K, Matsumoto A, Wada H and Kimura K 2011 *IEEE Trans. Appl. Supercond.* **21** 2643
[17] Sumption M D, Peng X, Lee X, Wu X and Collings E W 2004 *Cryogenics* **44** 711-725
[18] Susner M A, Daniels T W, Sumption M D, Rindfleisch M A, Thong C J and Collings E W 2012 *Supercond. Sci. Technol* **25** 065002
[19] Dew-Hughes D 1974 *Philos. Mag.* **30** 293-305
[20] Kramer E J 1973 *J. Appl. Phys.* **44** 1360-70
[21] Rowell J 2003 *Supercond. Sci. Technol* **16** R17-27
[22] Matsushita T, Kiuchi M, Yamamoto A, Shimoyama J and Kishio K 2008 *Physica C* **468** 1833

IOP Conf. Series: Materials Science and Engineering **756** (2020) 012019 doi:10.1088/1757-899X/756/1/012019

Increased flux pinning force and critical current density in MgB_2 by nano-La_2O_3 doping

Danlu Zhang[1], Fang Wan[1], Michael D. Sumption[1], Edward W. Collings[1], CJ Thong[2], Matt Rindfleisch[2], Mike Tomsic[2]

1 CSMM, Materials Science and Engineering, The Ohio State University, Columbus, OH 43220, USA.
2 Hyper Tech Research Inc., Columbus, OH 43228, USA.

zhang.5952@osu.edu

Abstract. MgB_2 superconducting wires and bulks with nano-La_2O_3 addition have been studied. A series of MgB_2 superconducting bulk samples with nano-La_2O_3 addition levels of 0, 5, 7, 18wt% were prepared. AC resistivity data showed slight increases of Bc_2 and unchanged B_{irr} for the bulk samples with doping levels lower than 7 wt% and decreased critical fields for the heavily doped (18 wt%) bulk. X-ray diffraction (XRD) showed the presence of LaB_6 in the nano-La_2O_3 doped MgB_2 bulk samples and decreased MgB_2 grain size in nano-La_2O_3 doped bulks. Monocore powder-in-tube (PIT) MgB_2 wires without and with 5 wt% nano-La_2O_3 addition (P-05) were prepared for transport property measurement. 2mol%C-doped *Specialty Materials Inc.* (SMI) boron powder was used for wire P-05 and previously prepared control wires (control wires were made without the addition of nano-La_2O_3 powder, W-00 and P2). Low field magnetic properties were obtained from magnetization loop (M–H), transport critical current density (J_c) was measured at 4.2 K for the nano-La_2O_3 doped PIT wire (P-05) and the control samples (P2 and W-00). The transport critical current density J_c(B) of P-05 at 4.2 K and 8 T (6.0×10^4 A/cm^2) was twice that of the control wire. The critical magnetic fields (Bc_2 and B_{irr}) of P-05 and the control sample P2 were compared. The critical fields of P-05 were slightly less than those of P2. Kramer-Dew-Hughes plots indicated a change from surface pinning to a mixture of volume pinning and surface pinning. It is shown that enhancement of P-05's transport properties is due to additional flux pinning by the fine-size rare-earth borides rather than enhanced Bc_2 or B_{irr}.

1. Introduction

Since the discovery of MgB_2 superconductors in 2001 [1], substantial improvement on the material has been achieved in terms of critical field, transport property, wire manufacture processes. Due to the relatively high T_c (39 K) in MgB_2 and the shortage of liquid helium worldwide, MgB_2 is particularly useful for helium-free MgB_2 MRI magnets [2] and is expected to replace Nb-based MRI magnets in the future. To reach these goals, improvements in the critical current density (J_c) of MgB_2 conductors are necessary.

Numerous approaches have been taken to produce MgB_2 materials with high transport properties: (1) chemical doping of MgB_2 wires [3-5] ; (2) cold pressing [6-7]; (3) hot isostatic pressing (HIP) [8-9]; (4) the introduction of the internal magnesium diffusion (IMD) method to address porosity and connectivity issues [10-12]. So far, the best "non-barrier" transport J_c values were obtained by the advanced internal magnesium infiltration (AIMI) approach, with the addition of C and Dy_2O_3 (1.07×10^5 A/cm^2) at 10 T, 4.2 K, [13]. The AIMI technique has the benefit of forming dense MgB_2 layers with improved longitudinal connectivity compared to the extrinsic conventional powder-in-tube (PIT) method, which usually produces wires with randomly connected MgB_2 fibers. On the other hand, the PIT method is favored as a simple and inexpensive approach to studying wire properties in response to chemical doping.

Many chemicals have been added to MgB_2 in the past 18 years to study their effects on the resultant transport and other superconducting properties, such as upper critical field (Bc_2) [14-15], and irreversibility field (B_{irr}) [16-17]. Generally, improvements in J_c have been attributed to the improvements of either Bc_2 or B_{irr}. Till now, C has been shown to be the most effective doping element for the enhancement of Bc_2 [18] in MgB_2 materials. Many MgB_2 wires with high transport properties were doped with C or C-containing materials. In addition to C and C containing materials, doping with Dy_2O_3 has shown to increase both J_c and B_{irr} [19-20]. In particular, Chen *et al.* [19] and Li *et al.*[20] both demonstrated the effectiveness of a combination of flux pinning by Dy_2O_3 and carrier scattering by C. The effects on the transport J_c of rare earth oxide additions such as Pr_6O_{11} [21-22], CeO_2 [23], Eu_2O_3 [24] have also been studied.

This paper describes the effect of adding nano-La_2O_3 dopants on superconducting and structural properties in MgB_2 superconductors. La_2O_3 was chosen because some rare-earth oxide additions have improved J_c and Bc_2/B_{irr} [19-24] in MgB_2 superconductors. Besides, La_2O_3 doping has been previously studied for MgB_2 tapes and nano LaB_6 flux pinning centers as well as increase of J_c in the doped tapes were observed. Therefore, the effects of La_2O_3 doping in MgB_2 wires and bulks were studied here to fully explore the doping effect as well as the optimum doping level. Consequently, four different doping levels (0, 5, 7,18 wt%) were chosen for the bulks and 5 wt% was chosen to make PIT wires based on the property measured for bulks with the same doping level. Transport J_c in the 5 wt% nano-La_2O_3 doped monocore PIT-processed MgB_2 wire increased to twice the value of the control sample at 4.2 K, 8 T.

2. Experimental

2.1. Sample preparation

2.1.1. Bulk samples. Four bulk samples B-00 (0 wt% La_2O_3 added), B-05 (5 wt% La_2O_3 added), B-07 (7 wt% La_2O_3 added) and B-18 (18 wt% La_2O_3 added) were made. The precursor powders used were 2 mol% C doped SMI boron powder, Mg powder (325 mesh) from Alfa Aesar, and nano-level (10-100 nm) La_2O_3 powder manufactured by US Research Nanomaterials Inc. The mole ratio of Mg to B powder used was 1: 2. The powder was mixed inside a glove box, and then transferred to a hydraulic press for densification at 5000 psi (or 34.5 MPa). The resulting pellets were heat treated at 850 °C for 30 minutes in flowing Ar, followed by furnace cooling.

2.1.2. Wire Samples. The wire samples were prepared by Hyper Tech Research using their well-known "continuous tube filling and forming" (CFTT) powder-in-tube (PIT) process [25] followed by wire drawing to 0.83 mm in diameter. Selected for controls were two samples from previous studies. Designated P2 [26] and W-00 [27] precursor powders were 2 mol% C doped B from SMI and Mg powder as in the undoped bulk. The sample specifications are listed in Table 1.

Table 1. Info on bulk and PIT wire samples

Sample Form	Sample Name	Wt% La_2O_3	Heat Treatment	Crystallite size (nm)
Bulk	B-00	0	30min/850°C	20.6
Bulk	B-05	5	30min/850°C	19.8
Bulk	B-07	7	30min/850°C	7.6
Bulk	B-18	18	30min/850°C	17

Wire	P2*	0	20min/675°C	-
Wire	W-00*	0	60min/650°C	-
Wire	P-05	5	60min/650°C	-

*Previously made/published [26][27]
Both referred to as "Control"

2.2. *Measurements*

2.2.1. X- Ray diffraction (XRD) measurement. The powdered bulk samples were scanned on a Rigaku Miniflex 600 XRD machine at a scan rate of 5 deg / min. Phases and peaks were studied with the help of PDXL software.

2.2.2. AC resistivity. The AC resistivity vs temperature measurements on the bulk samples were performed on a Quantum Design Model 6000 Physical Property Measuring System (PPMS). The bulk samples were cut into rectangular prisms and four-point probe measurements were used to obtain sample resistance of 5 K to 45 K. The usual "10% normal resistivity" and "90% normal resistivity" rules were used to obtain B_{irr} and Bc_2, respectively.

2.2.3. Transport J_c measurement. The transport J_c measurements were carried out in transverse magnetic fields of up to 12 T in a liquid He bath at 4.2 K on samples 50 mm long with a gauge length of 5 mm and an electric field criterion of 1 μV/cm. Measurements were also done using a gauge length of 4 mm. The layer J_c of the CTFF PIT strands were calculated using the critical transport currents divided by the MgB_2 layer area.

2.2.4. Magnetic J_c measurement. Magnetization-Magnetic field (M-H) loops were measured under VSM mode in a Quantum Design Model 6000 Physical Property Measuring System (PPMS) on sample P-05 3.2 mm long. Magnetic J_{cm} at 4.2 K was extracted from M-H loop at a ramp rate of 10 mT/s in transverse fields of up to 9 T. Based on Bean's model [28], the 4.2 K magnetic J_{cm} for the superconducting CTFF monocore strand and was derived from:

$$J_{cm} = \frac{3\cdot\pi\cdot\Delta M}{4\cdot d},$$

where d is the diameter of the MgB_2 layer inside the wire and ΔM is the full height of the M-H loop at a certain field. The magnetic J_{cm} played a role in the pinning force calculation (see section 3.4).

3. Results and Discussion

3.1. *Bulk sample's magnetic and structure properties*

Bc_2 and B_{irr} values were obtained for MgB_2 bulks doped with nano-La_2O_3 and the control bulk sample (B-00) at temperatures of 16 K ~ 36 K, Figure 1. B_{irr} stayed unchanged for B-05 and B-07. Bc_2 increased by 0.2 T with the addition of 5 wt% La_2O_3 and increased by 0.4 T with addition of 7 wt% La_2O_3. However, both Bc_2 and B_{irr} decreased significantly in response to doping with 18 wt% La_2O_3. Based on the data in Figure 1, it can be seen that T_c of the doped bulks did not change compared to the control bulk sample. This is similar to the unchanged T_c observation on Dy_2O_3 doped MgB_2 bulks [29], meaning no significant atomic substitution occurred on MgB_2 host lattice sites.

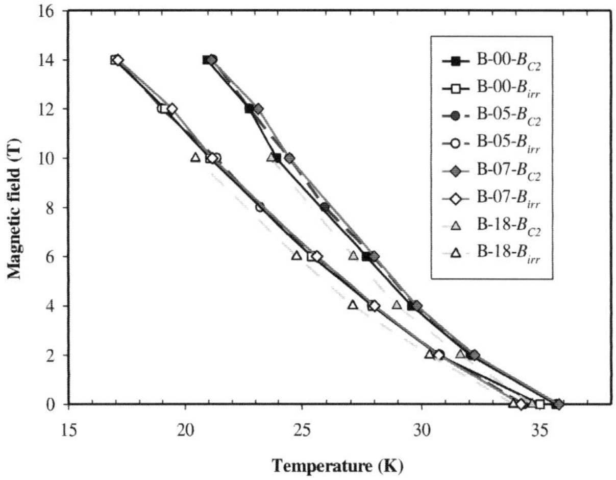

Figure 1. Critical magnetic fields as a function of temperatures for undoped bulk sample (B-00) and doped samples (B-05, B-07, B-18).

Phases and peaks in the MgB_2 bulk samples have been studied with the aid of XRD. The patterns shown in Figure 2 indicates the presence of LaB_6 in all doped bulk samples (shown by green arrows), presumably there are pinning centers. The MgB_2 peaks of B-05 and B-07, which is consistent with the minor increases of Bc_2 seen in Figure 1. The lattice shifts are smaller than those observed by Gao *et al.* [30] in MgB_2 tapes with acetone and La_2O_3 additions. Usually MgB_2 peak shifts observed in XRD patterns are caused by atomic substitution and/or strain. The shifts seen in Figure 3 is probably due to the lattice strain generated by the nanoparticles (such as LaB_6) since T_c was shown to be unchanged in doped samples. MgB_2 grain size was estimated based on XRD data using William-Hall method, Table 1. As the La_2O_3 doping level increases from 0 wt% to 7 wt%, MgB_2 grain size decreased from 20.6 nm to 7.6 nm. The grain size of MgB_2 was 17 nm in the heavily doped sample. Yuan [27] observed same trend in his Dy_2O_3 doped MgB_2 bulks [27]. He attributed this grain size reduction to secondary phase DyB_6 nanoparticles. Here, the reduction of grain size in La_2O_3 doped MgB_2 bulks might be due to the secondary phase LaB_6, which inhibits the grain growth and pins fluxons.

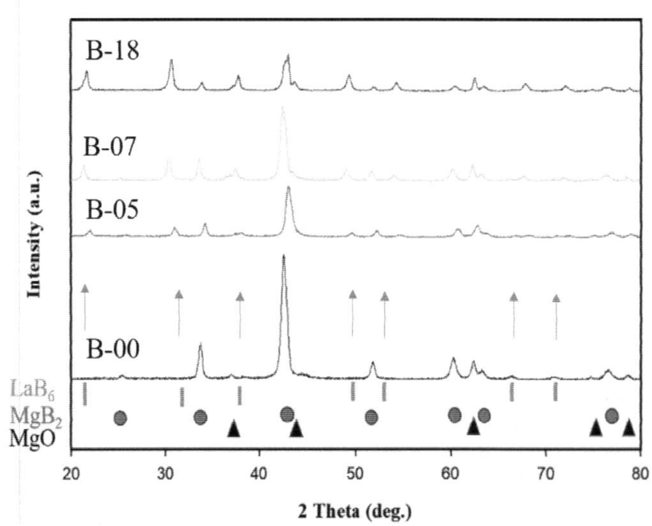

Figure 2. XRD patterns for undoped MgB_2 bulk and MgB_2 bulks doped with nano-La_2O_3.

3.2. Wire strength, transport and magnetic properties
This paper also describes the effect of 5 wt% nano-La_2O_3 addition to MgB_2 PIT wires. Two previously-made control samples (P2 [26] and W-00 [27]) were used for reference. These two wires were interchangeably used not only because the compositions are the same and the heat treatment procedures are very similar, but also because measured data from two controls combined can offer complete data for comparison purpose in this paper. Though they were manufactured from different batches, these mono PIT wires (with the use of 2mol%C-doped B powder) usually possess consistent properties. The transport properties of PIT MgB_2 wire P-05 and undoped wires P2 and W-00 were measured at 4.2 K in transverse magnetic fields up to 12 T, Figure 3. It can be seen that La_2O_3 doping resulted in a significant increase in J_c. In particular, at 8 T and 4.2 K, the J_c of P-05 is 6.0 $\times 10^4$ A/cm^2, about twice that of the control wire P2. It is important to keep in mind that AIMI approach usually produces MgB_2 wires with higher J_c compared to conventional PIT method for wires with same composition and heat treatment procedure because of the enhanced connectivity. Thus, AIMI-produced wires are not discussed here.

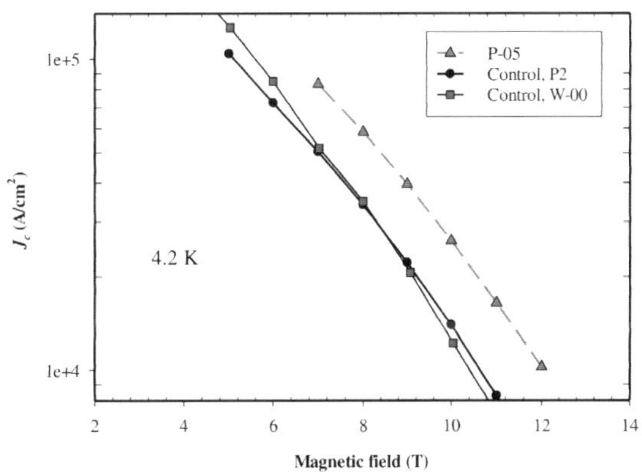

Figure 3. Critical current density (J_c) as a function of magnetic field for control wire and the La_2O_3 doped sample P-05 a 4.2 K.

3.3. Wire sample: Critical fields and flux pinning

In order to enquire into the factors responsible for the enhancement of J_c, the critical fields (Bc_2 and B_{irr}) were measured via resistivity approach in the PPMS in Figure 4. The T_c of P-05 decreased with the addition of La_2O_3 and its Bc_2 decreased by about 1 T at lower temperatures and very insignificantly at higher temperatures. Likewise, its B_{irr} decreased by around 0.6 T at lower temperatures and by a smaller amount at higher temperatures. A decrease in the Kramer field, B_k (a surrogate for B_{irr}) of 1.0 T at 4.2 K (Figure 5) confirms the above result. The relatively larger Bc_2 values in the undoped wires could be due to the existence of nano-level electron scattering centers (e.g., grain boundaries, point defects, volume defects and secondary phases). Since these two wires shown on Figure 4 are manufactured at different heat treatment conditions, it is very likely that fine secondary impurity phases such as MgO, B_2O_3 were introduced in the undoped wire P2 and therefore enhanced Bc_2 to a slight degree.

Two common mechanisms for enhancing J_c in MgB_2 wires are to enhance B_{irr} or enhance Bc_2. Here, we propose another mechanism for enhanced J_c in MgB_2 wires with nano-La_2O_3 doping. It is well known that enhanced transport J_c by way of C doping in MgB_2 wire can be attributed to enhanced B_{c2} by way of enhanced scattering effect [16, 19]. The increased transport J_c in Dy_2O_3-doped MgB_2 wire was attributed to an increase in B_{irr} [19-20]. With regard to La_2O_3 doped wire samples, as mentioned above, the 4.2 K, 8 T, J_c of P-05 was about twice that of the undoped control wire. But this increase is accompanied by no changes or decreases in B_{c2} and B_{irr} (1 T and 0.6 T, respectively, at lower temperatures), Figure 4. Clearly, a mechanism other than critical field is responsible for the increase in J_c; flux pinning is the obvious choice.

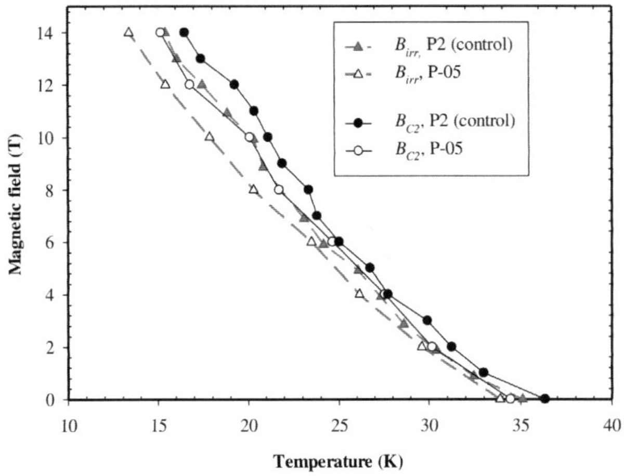

Figure 4. Upper critical field and irreversibility field for the control sample and nano-La_2O_3 added PIT MgB_2 wire P-05.

To further study the phenomena, Kramer plot has been made, A Kramer Plot, $J_c^{0.5}B^{0.25}$ vs. B, was shown in Figure 5. The irreversibility fields based on Kramer model [31], B_k, was taken at the x-axis intercepts of linear fittings (evenly spaced dashed lines) on Figure 5. B_k of the doped wire has a lower value than the control wire value, which agrees with the critical field analysis done above on the wire samples.

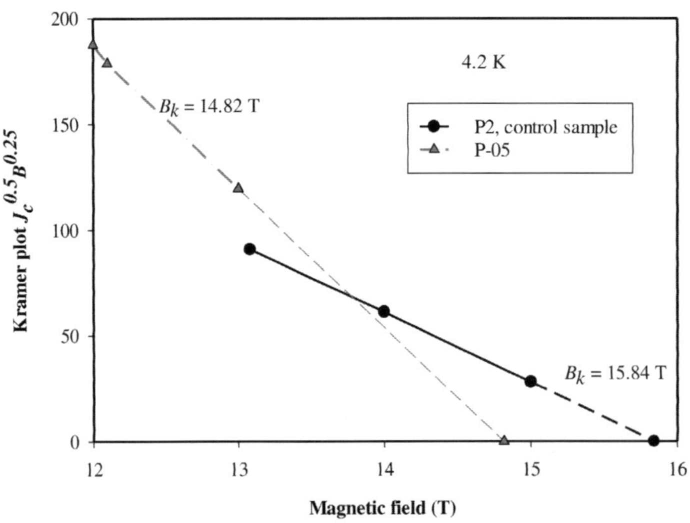

Figure 5. Kramer plot ($J_c^{0.5}B^{0.25}$ vs. B) for control wire sample and P-05 at 4.2 K. The irreversibility fields based on the Kramer model [31], B_k, were taken as the extrapolated x- axis intercepts of linear fittings (evenly spaced dashed lines).

3.4. Flux pinning in response to La_2O_3 doping

Using data derived from the M-H loop, the normalized flux pinning force density $F_p/F_{p,max}$ was plotted against normalized magnetic field $b=B/B_k$ at 20 K to avoid flux jump effect, Figure 6(a). B_k here was derived based on magnetic measurement conducted at 20 K in PPMS. Figure 6(b) showed the J_c data masured by transport and magnetic methods. Apparently, magnetic measured J_c agrees well with the transport data for P-05. Based on Dew-Hughes's analysis [32] about $f_p \propto b^{1/2}(1-b)^2$, the curve for the control sample which peaks at $B/B_k= 0.2$ is indicative of grain boundary pinning. In contrast, the curve for P-05 wire peaked at around $b = 0.168$. This phenomenon has been observed in Dy_2O_3 doped samples [29] and Yang *et al.* [29] stated that two reasons can be responsible for this behaviour: (1) The doped sample might contain a set of local B_ks instead of one distinct value, which can lead to an artificial error in the estimation of the peak positions; (2) The deviation from $b_{peak} = 0.2$ might be due to other pinning mechanisms (e.g., normal volume pinning in which f_p maximizes at b < 0.2 due to anisotropy [33]) in association with the GB pinning. The presence of LaB_6 peaks in all the XRD patterns of the doped bulk samples strongly suggests that LaB_6, formed by reaction with the B powder, would also be present in the doped wires and as such would be responsible for the observed volume pinning.

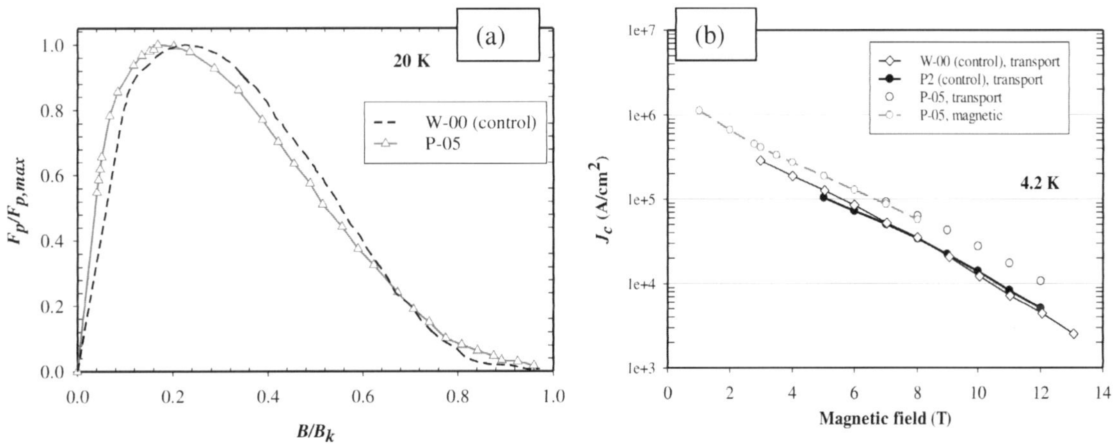

Figure 6. (a, left) Normalized flux pinning force as a function of reduced field for control sample and the doped wire, P05, based on the magnetic data; (b, right) Critical current density measured for wire P-05 using both magnetic measurement and transport measurement.

4. Concluding Discussion

Four MgB_2 bulk samples with 0-18 wt% nano-La_2O_3 additions and a monocore PIT MgB_2 wire with 5 wt% nano-La_2O_3 addition were prepared for measurements of J_c, B_{irr}, Bc_2 and magnetic and transport J_c (see Table 1). Resistive measurements of B_{irr} and Bc_2 in response to La_2O_3 doping were made in the PPMS on the bulk and wire samples. For the bulks B_{irr} remained unchanged in B-05 and B-07 but decreased in B-18; Bc_2 increased by 0.2 T in B-05 and 0.4 T in B-07 but also decreased in B-18. For the wires (controls and P-05) B_{irr} decreased by ~0.6 T at low temperatures and a smaller amount at high temperatures. Bc_2 decreased by 1 T at low temperatures (resistive results and Kramer plot) and insignificantly at high temperatures. The 4.2 K, 8 T, J_c of P-05 was about twice that of the undoped control. There is no correlation between changes in the critical fields and J_c. On the other hand, the occurrence of LaB_6 in the XRD patterns and the shapes of the normalized flux pinning curves (after Dew-Hughes) indicate that the increase in J_c is a consequence of increased flux pinning in the doped wire.

5. References

[1] Nagamatsu J, Nakagawa N, Muranaka T, Zenitani Y and Akimitsu J 2001 *Nature* **410** 63
[2] Zhang D, Kovacs C, Rochester J, Majoros M, Wan F, Sumption M D, Collings E W, Rindfleisch M, Panik D, D Doll D and Avonce R 2018 *Supercond. Sci. Technol.* **31** 085013
[3] Matsumoto A, Kumakura H, Kitaguchi H and Hatakeyama H 2003 *Supercond. Sci. Technol.* **16** 926
[4] Ma Y, Zhang X, Xu A, Li X, Xiao L, Nishijima G, Awaji S, Watanabe K, Jiao Y, Xiao L and Bai X 2005 *Supercond. Sci. Technol.* **19** 133
[5] Woźniak M, Hopkins S C, Gajda D and Glowacki B A 2012 *Phys. Procedia* **36** 1594-8
[6] Hossain M S, Senatore C, Flükiger R, Rindfleisch M A, Tomsic M J, Kim J H and Dou S X 2009 *Supercond. Sci. Technol.* **22** 095004
[7] Wan F, Sumption M D, Rindfleisch M A and Collings E W 2017 *IOP Conference Series: Mater ials Science and Engineering* Vol. **279**, No. 1, p. 012024
[8] Gajda D, Morawski A, Zaleski A, Cetner T, Małecka M, Presz A, Rindfleisch M, Tomsic M, Tho -ng C J and Surdacki P 2013 *Supercond. Sci. Technol.* **26** 115002
[9] Prikhna T, Gawalek W, Savchuk Y, Soldatov A, Sokolovsky V, Eisterer M, Weber H W, Noudem Serga M, Turkevich V and Tompsic M 2011 *J SUPERCOND NOV MAGN* **24** 137-50
[10] Li G Z, Sumption M D, Zwayer J B, Susner M A, Rindfleisch M A, Thong C J, Tomsic M J and Collings E W 2013 *Supercond. Sci. Technol.* **26** 095007
[11] Ye S J, Matsumoto A, Togano K and Kumakura H 2011 *Physica C* **471** 1133-6
[12] Kováč P, Hušek I, Melišek T, Kopera L and Kulich M 2016 *Supercond. Sci. Technol.* **29** 10LT01
[13] Li G Z, Sumption M D, Susner M A, Yang Y, Reddy K M, Rindfleisch M A, Tomsic M J, Thong CJ and Collings E W 2012 *Supercond. Sci. Technol.* **25** 115023
[14] Patnaik S, Gurevich A, Bu S D, Kaushik S D, Choi J, Eom C B, Larbalestier D C 2004 *Phys. Rev. B* **70** 064503
[15] Angst M, Bud'ko S L, Wilke R H and Canfield P C 2005 *Phys. Rev. B* **71** 144512
[16] Sumption M D, Bhatia M, Dou S X, Rindfliesch M, Tomsic M, Arda L, Ozdemir M, Hascicek Y and Collings E W 2004 *Supercond. Sci. Technol.* **17** 1180
[17] Wang J, Bugoslavsky Y, Berenov A, Cowey L, Caplin A D, Cohen L F, MacManus Driscoll J L, Cooley L D, Song X and Larbalestier D C 2002 *Appl. Phys. Lett.* **81** 2026-8
[18] Dai W, Ferrando V, Pogrebnyakov A V, Wilke R H, Chen K, Weng X, Redwing J, Bark C W, Eom C B, Zhu Y and Voyles P M 2011 *Supercond. Sci. Technol.* **24** 125014
[19] Chen S K, Wei M, MacManus-Driscoll J L 2006 *Appl. Phys. Lett.* **88** 192512
[20] Li G Z, Sumption M D, Rindfleisch M A, Thong C J, Tomsic M J and Collings E W 2014 *Appl. Phys. Lett.* **105** 112603
[21] Pan X F, Shen T M, Li G, Cheng C H and Zhao Y 2007 *Phys. Status Solidi A.* **204** 1555-60
[22] Ojha N, Malik VK, Bernhard C and Varma G D 2010 *Phys. Status Solidi A.* **207** 175-82
[23] Pan X F, Cheng C H and Zhao Y 2011 *J SUPERCOND NOV MAGN* **24** 1611-6
[24] Ojha N, Malik V K, Bernhard C and Varma G D 2009 *Physica C* **469** 846-51
[25] Tomsic M, Rindfleisch M, Yue J, McFadden K, Doll D, Phillips J, Sumption M D, Bhatia M, Bohnenstiehl S and Collings E W 2007 *Physica C* **456** 203-8
[26] Li G Z, Yang Y, Susner M A, Sumption M D, Collings E W 2011 *Supercond. Sci. Technol.* **25** 025001
[27] Yang Y 2016 Influence of Chemical Doping on Microstructures and Superconducting Properties of MgB_2 Wires and Bulk Samples (Doctoral dissertation, The Ohio State University)
[28] Sumption M D, Peng X, Lee E, Wu X and Collings E W 2004 *Cryogenics* **44** 711
[29] Yang Y, Sumption M D and Collings E W 2016 *Sci. Rep.* **6** 29306
[30] Gao Z, Wei Y M, Wang D, Z Xian-Ping, Satoshi A and Kazuo W 2010 *Chin. Phys. Lett.* **27** 117401
[31] E. J. Kramer 1973 *J. Appl. Phys.* **44** 1360
[32] Dew-Hughes D 1974 *Phil. Mag.* **30** 293-305
[33] Eisterer M 2008 *Phys. Rev. B.* **77** 144524

Structure and properties of MgB_2 bulks: *ab-initio* simulations compared to experiment

V V Romaka[1], T A Prikhna[2,9], M Eisterer[3], A P Shapovalov[2], W Goldacker[4], V E Moshchil[2], A Kozyrev[2], M Rindfleisch[5], M Tompsic[5], E E Hellstrom[6], G E Grechnev[7], V Sh Zhang[8], F Yang[8], Ch Li[8]

[1] Lviv Polytechnic National University, 12 Bandera Str., Lviv, 79013, Ukraine
[2] V. Bakul Institute for Superhard Materials of the National Academy of Sciences of Ukraine, 2, Avtozavodskaya Str., Kiev 07074, Ukraine
[3] Atominstitut, TU Wien, Stadionallee 2, 1020 Vienna, Austria
[4] Karlsruhe Institute of Technology (KIT), 76344 Eggenstein, Germany
[5] Hyper Tech Research, Inc., 539 Industrial Mile Road, Columbus, OH, USA
[6] Colledge of Engineering, University of Wisconsin-Madison, 247 ASC - Shaw Building, 2031 East Paul Dirac Dr, Tallahassee, FL 323, USA
[7] B. Verkin Institute for Low Temperature Physics of the National Academy of Sciences of Ukraine, Kharkov 61103, Ukraine
[8] Superconducting Materials Research Center, Northwest Institute for Nonferrous Metal Research, 710016, Xi'an, China

[9]Author to whom any correspondence should be addressed, e-mail: prikhna@ukr.net

Abstract. Analysis of XRD patterns by Rietveld refinement has been shown that the main phase of superconducting MgB_2-based bulk materials (with high level of superconducting characteristics) has AlB_2 type structure and near $MgB_{1.8-1.68}O_{0.2-0.32}$ stoichiometry. The materials demonstrated the critical current densities up to 0.9 – 0.4 MA/cm^2 j_c (at 0 - 1 T, 20 K); up to 15 T B_{c2} (at 22.5 K) and B_{irr} (at 18 K). The *ab-initio* simulation confirmed (1) benefits in binding energy and enthalpy of formation if stoichiometry of the solid solution is near $MgB_{1.75}O_{0.25}$; (2) energetic advantage in case if impurity oxygen present only in each second boron plane of MgB_2 cell while the first boron plane of the same cell stays pristine and location of substituted oxygen atoms in the nearby positions. Besides, the results of *ab-initio* modeling allow explanation of the tendency towards segregation of O-impurity in MgB_2 structure during synthesis or sintering, and formation of Mg-B-O inclusions or nanolayers (with MgO type of structure) which effect pinning. Calculated transition temperatures, T_c, for $MgB_{1.75}O_{0.25}$ occurred to be 23.3 K, while for MgB_2 it was 21.13 K only. Experimental T_c of the bulk materials was 35.7-38.2 K.

1. Introduction

MgB_2 is oxygen-free superconducting compound with hexagonal AlB_2 type of crystallographic lattice. It contains close packed Mg layers and graphene like B layers alternate along the c axis of the hexagonal unit cell. MgB_2 shows a complex bonding structure consisting of ionic interlayer bonding and covalent/metallic mixed in-plane bonding. Existence of two energy gaps was shown for MgB_2 [1, 2]. The XRD study demonstrates that the MgB_2-based materials usually contain MgB_2 phase and some admixture MgO phase even if special atmosphere is used during preparation for avoiding oxidation, see for, example [3, 4], what can be explained by high affinity of Mg towards oxygen. Besides, the presence of other admixture phases such as MgB_4 (orthorhombic), MgH_2 and BN, etc. can be observed on XRD patterns [5]. But in MgB_2-based materials structures can be as well present dispersed inclusions of higher magnesium borides, MgB_x, with x>4 (as SEM and Auger study witnessed), which cannot be reveled by XRD on the background of MgB_2 reflexes. The unit cells of higher borides such as MgB_7 (orthorhombic), MgB_{12} (orthorhombic), MgB_{12} (trigonal), MgB_{17} (trigonal), MgB_{20} (trigonal) and Mg_2B_{25} (trigonal) are rather complicated, containing many atoms. The large number of atoms in their unit cells and low symmetries of the cells are resulting in many 'reflecting planes' which essentially reduce the intensities of the XRD reflections, as compared to those of the MgB_2 (i.e. from a simple hexagonal unit cell). Because of this the dispersed inclusions of the higher magnesium borides can be practically 'invisible' for a traditional XRD analysis [6].

The study of MgB_2-based materials structures (bulk, wires, and thin films) by scanning electron microscopy (with quantitative X-ray microprobe analysis) and Auger spectroscopy (with quantitative analysis) showed us presence of oxygen in superconducting materials matrices (in a small amount) and presence of oxygen-enriched dispersed separate inclusions or nanolayers in the matrices. So, no pure MgO (without presence of boron) and no pure MgB_2 (without presence of oxygen) were experimentally revealed. The SEM and Auger study was performed using a JAMP–9500F which combined SEM SEI, BEI, X-ray and Auger analyzers and allow investigating of non-oxidized surfaces because the construction of the microscope gives possibility to perform etching of 1×1 mm^2 square surfaces by Ar ions directly in the exploratory chamber and even build depth profiles of distribution of present elements. In contrary to X-ray microprobe analysis which allows to estimate quantity of elements in the volume of material which is not smaller than 1-3 μm in diameter, emanated Auger electrons are collecting from a very small volume (10 nm in diameter and about two lattice parameters in depth) and thus nanoobjects can be properly analysed quantitatively by Auger spectroscopy. Besides, Auger spectroscopy is very sensitive toward light elements and simultaneous presence of B, C, O and N is not the obstacle for correct investigations.

The results of Auger spectroscopy and SEM study [6] of MgB_2-based materials with high superconducting performances: highly dense bulks [5-8], wires [9, 10] and thin films [8], would benefit from *ab-initio* modelling of boron atoms substitutions by oxygen atoms to calculate transition superconducting temperature of energetically the most preferable solid solution $MgB_{2-x}O_x$. The present paper is generalising results of the *ab-initio* modelling. The obtained data can be a basis for interpreting the unusual electronic properties of MgB_2-based materials and the development of nano-engineering of these materials with record values of superconducting characteristics.

2. Experimental

The microstructure of the materials was characterized by a JAMP–9500F. XRD patterns were obtained by X-ray powder diffraction analyses and Rietveld refinement (using "PHILIPS X'pert" and Dron-3). The critical current densities, j_c, of the bulks and thin films were estimated from data obtained by an Oxford Instruments 3001 vibrating sample magnetometer (VSM) using the Bean model [11]. Certain inconsistencies which may arise between different measurement techniques [12, 13] do not influence the result and discussion presented in this work. The values of transition temperature, T_c, the upper critical magnetic field, B_{c2}, and the irreversibility field, B_{irr}, were estimated from resistivity measurements (90% and 10% of the resistivity just above the beginning of the transition) using the four probe technique in fields from 0 to 15 T. Transition temperature, T_c, was experimentally estimated by SQUID magnetometer as well.

The list and characteristics of the materials under the discussion synthesized by the authors are shown in Table 1. The numbers assigned to the samples are used through the paper.

In order to obtain an approximate picture for the electronic structure of $MgB_{2-x}O_x$ compounds we have carried out the *ab-initio* calculations for corresponding 2×2×2 and 2×2×1 supercells of the initial MgB_2 unit cell to model various oxygen concentrations of the solid solution. The calculations were

Table 1. List and characteristics of film, wires and bulk materials.

No	Material	Initial materials	Preparation*	T_c, K
1	Wire	B (without C) and Mg – Specialty Materials Inc. (SMI)	One filament, 0.83 mm in diameter, Nb surrounding and	37.5
2	Wire	B (with C) and Mg (SMI)	monel outer sheath	31.6
3	Film	hot pressed MgB_2 target	Magnetron sputtering, 140 nm film thickness	33.5 (‖*ab*) 32.6 (‖*c*)
4	Bulk	Mg:2B (SMI)	2 GPa, 600 °C, in BN, HP	35.7
5	Bulk	Mg(chips):2B (0.66 wt. % O)	2 GPa, 1050 °C, in BN, HP	-
6	Bulk	Mg(chips):2B (0.66 wt. % O)+10% SiC	2 GPa, 1050 °C, in BN, HP	-
7	Bulk	Mg(chips):2B (1.4 wt. % O)	2 GPa, 900 °C, in BN, HP	38.2
8	Bulk	Mg(chips):2B (1.4 wt. % O)	50 MPa, 1050 °C, in C, SPS	37.5
9	Bulk	Mg(chips):2B (1.4 wt. % O)	30 MPa, 900 °C, in BN, HotP	37.1
10	Bulk	Mg(chips):2B (1.4 wt. % O)	0.1 MPa, 800 °C, in Ar flow	38.1

* HP- high pressure-high temperature conditions; HotP- hot pressed; SPS-spark plasma sintering.

based on the density-functional theory (DFT) and performed by employing the full potential linearized augmented plane waves (FP-LAPW) methods implemented in WIEN2k [14] and Elk [15] implementations) and the full-potential linear muffin-tin orbital (FP-LMTO) implemented in RSPt [16, 17]).

The exchange-correlation potential was taken into account in the local density approximation (LDA) [18] and the generalized gradient approximation (GGA) [19] of DFT. For the used full potential FP-LAPW and FP-LMTO methods any restrictions were not imposed on charge densities or potentials of studied systems that is especially important for anisotropic layered structures. The constructed supercells were based upon the experimental lattice parameters of the hexagonal MgB_2 cell, a_0=b_0=0.3086 nm; c_0=0.3524 nm. For the unit cell 1×1×1, supercells 2×2×1 and 2×2×2 the lattice parameters a and c are considered equal to a_0 and c_0, $2a_0$, and c_0, $2a_0$ and $2c_0$, respectively. In order to determine stable optimized positions of atoms and the related binding energies E_b in the $MgB_{2-x}O_x$ supercells, we applied the FP-LAPW method [14] within DFT-GGA approximation. The phonon calculations were performed within the Elk code (DFPT) for the optimized structure of $MgB_{1.75}O_{0.25}$ and the calculations of the superconducting critical temperature were conducted within Eliashberg theory using the formula [20] with the LDA [18] approach. In order to introduce oxygen in the initial MgB_2, the symmetry was reduced and the supercell 2x2x1 along the ab-plane was constructed. The scheme of MgB_2-2×2×2 supercell and the detailed descriptions of binding energy calculations as well as electronic structure variation of MgB_2 doped with oxygen are given in [21].

3. Results and discussions

Figure 1 shows the typical microstructures (after etching of their surfaces by Ar ions in the chamber of research microscope) of MgB_2 thin film (figure 1a, No 3 in table 1) wire (figure 1b, No. 1, manufactured by HyperTech from Mg and B without C addition,) and bulks synthesized from Mg:2B (amorphous) mixtures (figures 1c-f): in Ar flow at 800 °C (figure 1c, No10); at 1050 °C and quasihydrostatic 2 GPa pressure (figure 1d, No 5); at 1050 °C under 50 MPa pressure by spark plasma sintering (figure 1e, f, No 8). The differences in colors correlate with variation in concentration of magnesium, boron and admixture oxygen. The big darken gray areas in figure 1e connected with presence of MgB_4 phase (the pressure 50 MPa was not enough to suppress fully Mg evaporation during synthesis at 1050 °C). The materials superconducting characteristics (j_c, B_{c2} and B_{irr}, T_c) and XRD are presented in figure 2. The variation of stoichiometry on nanolevel has been observed (as difference in colors and microprobe quantitative analysis are shown) and the boundaries between areas with different concentration of elements (figure 1) can be responsible for pinning and high critical current density in MgB_2-based materials (figure 2). The data for the best samples synthesized under

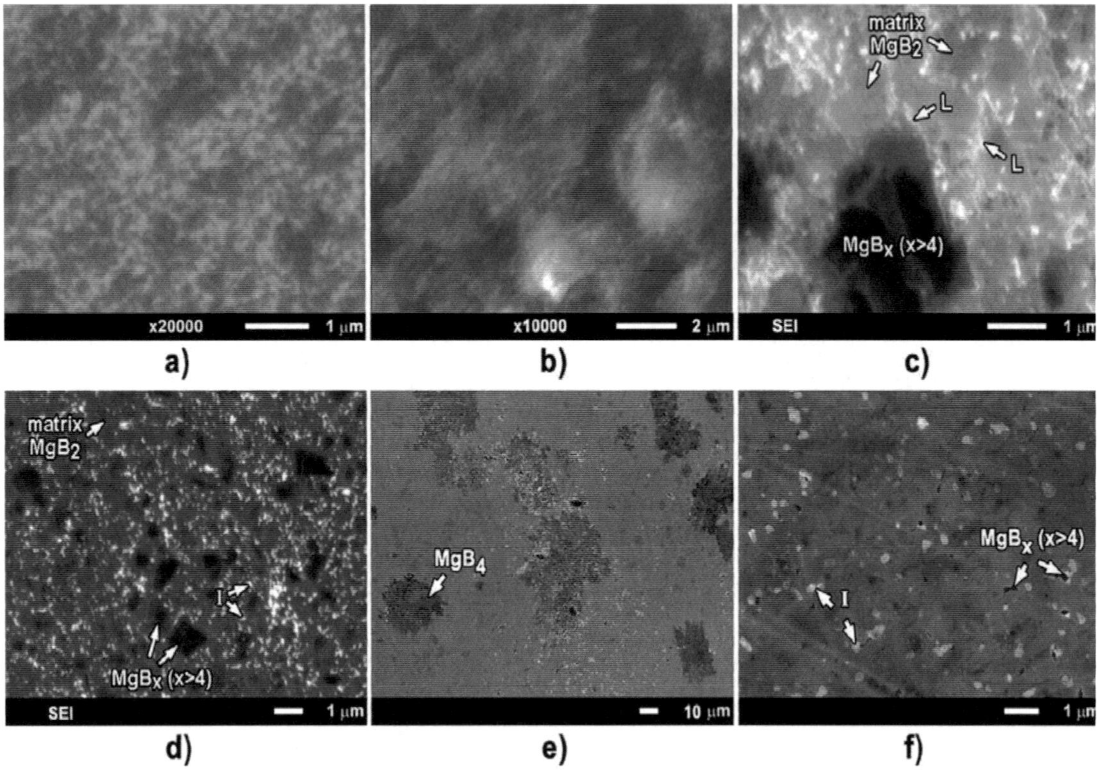

Figure 1. Typical microstructures of MgB_2 film No3 (a), wire No1 (b), and bulks: No 10 (c), No5 (d), No8 (e, f); numbering is according Table 1. All the materials contain B, Mg, admixture O and some small amount of admixture C; the brighter colors at the images correspond to lower concentration of boron and higher concentration of oxygen. By "I" and "L" the O- enriched inclusions and nanolayers, respectively, are marked.

each conditions (high pressure, HP, hot pressure, HotP, spark plasma sintering, SPS, or in Ar flow) are presented here.

Table 2 presents the results of Rietveld refinement of two samples: (1) prepared in protective from oxygenation atmosphere "with purified B" demonstrating T_c= 38.9 K [3] (figure 2g) and (2) synthesized under high pressure, No5 (table 1, figures. 2c, h). The temperature dependences of B_{c2} and B_{irr} for the material prepared "with purified B" are given in [3] (at 20 K its B_{c2} =8.8 T and B_{irr}=7 T). The results of Rietveld refinement (table 2) show $MgB_{1.8}O_{0.2}$ and $MgB_{1.74}O_{0.26}$ stoichiometries for the main phases present in the analyzed materials. The analysis of ten XRD patterns of MgB_2 materials with high superconducting performances showed $MgB_{1.8\text{-}1.68}O_{0.2\text{-}0.32}$ stoichiometries of their main phases. The XRD analysis of the diffraction pattern (figure 2h, table 2) showed the presence of 3 phases: the main phase MgB_2 and two impurity phases – BN and MgO. The Rietveld refinement revealed the presence of 5.6% on BN, 4.8% of MgO and the rest – 89.6% of MgB_2. In the case of the fully ordered MgB_2 phase with Mg solely occupying 1*a* (0 0 0) site and B occupying 2*d* (1/3 2/3 1/2) site the residuals between the calculated and experimental powder patterns were R_I = 2.96%, R_p = 9.77%, and R_{wp} = 14.0%. All attempts to refine the isotropic displacement parameters (B_{iso}) led to systematically high values for Mg atoms and very low (even negative) for B atoms. For this reason, two other models of the crystal structure were tested. The second model was refined with vacancies in the crystallographic site of Mg atoms. This led to somewhat lower residual R_I = 2.20% with 93.6% occupancy of the 1*a* site by Mg atoms. The third model was based on the statistical mixture of B and O atoms in 2*d* crystallographic site. The refinement of the third model revealed the lowest residual among three tested models: R_I = 1.92% and the 2*d* site occupied by 87% of B atoms and 13% of O

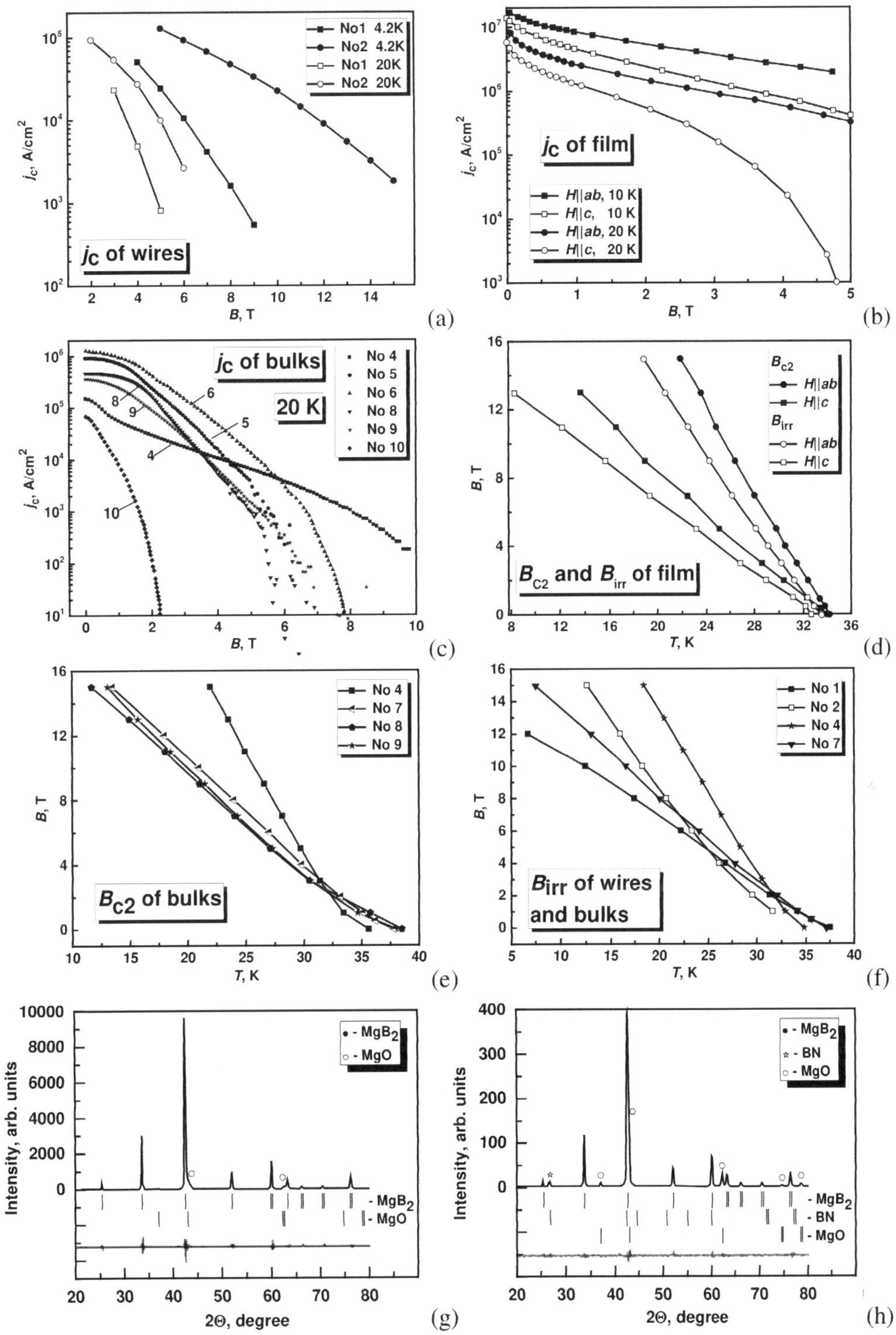

Figure 2. (a)-(c) - Dependences of critical current density, j_c, vs. magnetic field, B, for: wires (a), thin film (b) and bulks (c); (d) – temperature dependences B_{c2} and B_{irr} of thin film (Fig. 2a) at 10 and 20 K in two directions: parallel to the substrate surface ($\|ab$) and perpendicular ($\|c$); (e), (f) – temperature dependences of B_{c2} and B_{irr} of bulk and wires; (g), (h) – XRD patterns of MgB_2 bulks made in clean from oxygen conditions (the details are given in [1]) and of No5 (table 1).

Table 2. Results of Rietveld refinement.

Sample	MgB_2 (figure 2g), Ref.[3]	MgB_2 (Table 1 No 5, figure 2h)
Space group, structure type	*P6/mmm*, AlB_2	
a, *c* (nm)	0.30848(1), 0.35235(1)	0.30849(4), 0.35227(6)
Mg in 1*a*	0, 0, 0	
B_{iso} (10^2 nm^2)	0.36(5)	0.4(2)
B/O in 2*d*	1/3, 2/3, 1/2	
B_{iso} (10^2 nm^2)	1.11(7)	1.4
Occupancy B/O	1.80(2)/0.20(2)	1.74(2)/0.26(2)
R_I (%)	1.27	1.92
R_p (%)	7.60	9.65
R_{wp} (%)	11.4	13.9
Additional phases	MgO (1.9%)	BN (5.6 %), MgO (4.8 %)

giving the formula $MgB_{1.74}O_{0.26}$. All three models were tested with DFT calculations to compare their heats of formation, and the third model appeared to be the most energetically favorable. The analysis of XRD powder pattern of another bulk sample (figure 2g) showed that the sample contained only 1.9% of MgO impurity phase and the rest was MgB_2. Rietveld refinement was performed using the fully ordered model of MgB_2 and the model with B and O mixture in 2*d* site. The fully ordered model produced R_I =6.30%, while the model with B and O mixture – R_I = 1.27%. The refined oxygen and boron content in 2*d* site appeared to be very close to the first sample: 90% of B and 10% of O, which led to the formula $MgB_{1.80}O_{0.20}$.

For the DFT calculations we used the 2x2x2 supercell to model for the whole extent of dopant content (from 6 to 50 at% of substituting oxygen atoms). By means of FP-LAPW GGA calculations we optimized positions of atoms in the supercell from energetic point of view by method of step-by-step variation of the coordinates of all atoms. The criterion of optimization was taken as 0.001 Ry for energy variations. The results of the binding energies (E_b) calculations for oxygen substitutions were reported in Ref. [21] (see table 4). The analysis of the data has shown that the binding energy for almost all calculated $MgB_{2-x}O_x$ supercells is higher than that of pristine MgB_2. If we impose restriction to consider the high symmetric cells only, the monotonous increase of binding energy can be noticed, and for $MgB_{1.5}O_{0.5}$ it reaches E_b=-1.05984 Ry which is about 0.064 Ry higher than that of MgB_2.

The most energetically beneficial ordering was found to occur for the nearest (paired) oxygen atoms in the (001) type plains. The analysis of these data has shown that in the case of substitution of boron positions for oxygen the binding energy is essentially affected by the superlattice symmetry, the type of ordering of oxygen atoms and the amount of substituting oxygen atoms. In the case of 2×2×2 supercells, boron planes with and without substituting oxygen atoms alternate sequentially. The binding energy can decrease down to E_b=-1.13246 Ry or to E_b=-1.15268 Ry, i.e. below than that for MgB_2, and the substituting oxygen atoms located in the nearest positions form zigzag chains [21]. Thus the most preferable for the considered supercells is configuration when in the each boron plane (with partly oxygen substitution) the substituting oxygen atoms form zigzag chains and when the each subsequent boron plane contains only boron atoms at near pristine positions. As the calculations have shown [21], in the cases of the most preferable (from the energetic point of view) location of oxygen, when its amount is low, the shift of boron and magnesium atoms from the pristine positions (as compared with the case of ideal MgB_2 without substitutions) was not higher than 0.15 Å. It should be mentioned that results of high resolution TEM studies [22] indicate that oxygen atoms can be present regularly in every second boron plane of MgB_2. Thus in the practically undisturbed boron planes (from the point of view of binding energy and electron density distribution) there is still a possibility of high T_c superconductivity. Despite the fact that in the case of MgB_1O_1 chemical composition the formation of zigzag oxygen chains leads to even lower binding energy, E_b=-1.15712 Ry, and the amount of oxygen is so high that zigzag chains can be formed in each boron plane, such distribution in our opinion will be destructive for superconductivity, because each boron plane would be essentially distorted and the large deviations of Mg and B atoms from their pristine positions would take place. We suppose that such distribution of oxygen (zigzag chains in each boron plane) can be prevailed in the observed Mg-B-O nanolayers "L" or nanoinclusions "I" (figure 1c, d, f). This allows us to explain the tendency towards segregation of impurity O in MgB_2 structure during synthesis or sintering, and formation of Mg-B-O inclusions or nanolayers (with MgO type of structure), which effect pinning. So, seems the synthesis in the Mg-B-O system allows the realization of several phases with AlB_2 type crystal structure. Several (at least two) superconducting phases can be formed during crystallization with the structure of the same material due to the difference in melting temperatures of phases with different O content and the existing tendency for oxygen segregation. This makes it reasonable to perform analysis of such characteristic of MgB_2 as the two-band nature of the superconducting gap [1, 2] and 1.5 superconductivity (disorder of the vortex lattice) [23] in a new way.

In the case of $MgB_{2-x}O_x$ the lowest density of electronic states (DOS) about 0.46 states/eV/f.u. was found for $MgB_{1.75}O_{0.25}$ if the oxygen atoms are in neighboring positions [8]. If the concentration of oxygen is higher than in $MgB_{1.5}O_{0.5}$ the MgB_2 structure is destroyed. *Ab-initio* modelling of enthalpy of formation of boron substitution by oxygen in MgB_2 as well shows that this is energetically favourable up to the composition $MgB_{1.75}O_{0.25}$. . Enthalpy of MgB_2 formation is ΔH_f =-150.6 meV/atom and that of $MgB_{1.75}O_{0.25}$ is ΔH_f =-191.4 meV/atom. The formation of vacancies at the Mg site of both MgB_2 and substituted $MgB_{1.75}O_{0.25}$ was modelled as well, but has shown that such processes are energetically disadvantageous (ΔH_f of $Mg_{0.875}B_2$ and $Mg_{0.75}B_{1.75}O_{0.25}$ are equal to -45.5 and -93.5 meV/atom, respectively) [24].

For MgB_2 and $MgB_{1.75}O_{0.25}$ the calculations of the superconductive critical temperature (T_c) were performed within the framework of Density Functional Perturbation Theory (DFPT). First the calculations were performed on a low k-point mesh (8×8×8) to get the phonon dynamical matrices. After that the calculations of the McMillan-Allen-Dynes critical temperature T_c were performed on a high k-point mesh (16x16x16). Interesting, that the obtained T_c for the MgB_2 is 21.13 K, while for the $MgB_{1.75}O_{0.25}$ it is just a couple of degrees higher 23.3 K. It should be noted that it was hard to achieve the convergence of the phonon calculations for $MgB_{1.75}O_{0.25}$, which appeared to be unstable for the constructed supercells.

4. Conclusions

The experimental study and *ab-initio* modeling witnessed about the existence of superconductivity in *P6/mmm*, AlB_2 structural type $MgB_{1.75}O_{0.25}$. The level of superconducting characteristics (T_c, j_c in 0-10 T fields, B_{irr} and B_{c2}) of the materials with stoichiometry of the main phase $MgB_{1.8-1.68}O_{0.2-0.32}$ is the highest reported in literature. DFT calculations showed that B for O substitution in MgB_2 is energetically favorable up to $MgB_{1.75}O_{0.25}$. Other variants of substitutions and vacancy formations lead to less stable structure, which is in agreement with the results of Rietveld refinement. The calculated critical temperature T_c for $MgB_{1.75}O_{0.25}$ (23.3 K) is just a couple of degrees higher than for pure MgB_2 (21.13 K). The observed experimental T_c of the bulk MgB_2-based materials was 35.7-38.2 K. The advantages from location of substituted oxygen atoms in the nearby positions in the same boron plane allow explanation of the tendency towards segregation of impurity O in MgB_2 structure and formation of oxygen enriched areas which can affect pinning and critical current density increase.

5. References

[1] Bouquet F, Fisher R A, Phillips N E, Hinks D G and Jorgensen D 2001 *Phys. Rev. Lett.* **87** 047001
[2] Szabó P, Samuely P, Kačmarčík J, Klein T, Marcus J, Fruchart D, Miraglia S, Marcenat C and Jansen A G M 2001 *Physical review letters* **87**(13) 137005
[3] Jiang J, Senkowicz B J, Larbalestier D C and Hellstrom E E 2006 *Supercond. Sci. Technol. Rapid communication* **19**, L33
[4] Zhou S, Zhang Y, Pan A V, Dou S X, Chung K C, Kim Y K and Yoo J M 2009 *Supercond. Sci.*

Technol. **22**, 045018

[5] Prikhna T 2016 Structure and properties of bulk MgB_2 *Superconducting Wires: Basics and Applications* ed. R Fluekiger (World Scientific Series in Applications of Superconductivity and Related Phenomena 668 p) chapter 3a pp. 131-157

[6] Prikhna T A, Eisterer M, Weber H W, Gawalek W, Kovylaev V V, Karpets M V, Basyuk T V and Moshchil V E 2014 *Supercond. Sci. Technol.* **27**(4) 044013–044017

[7] Prikhna T A, Gawalek W, Savchuk Ya M, Sergienko N V, Moshchil V E, Wendt M, Habisreuther T, Dub S N, Melnikov V S, Kozyrev A V, Schmidt Ch, Dellith J, Litzkendorf D, Nagorny P A, Dittrich U, Sverdun V B, Kovalev L K, Penkin V T, Goldacker W, Rozenberg O A and Noudem J 2008 *Journal of Physics: Conference Series* **97**, 012022

[8] Prikhna T A, Romaka V V, Shapovalov A, Eisterer M, Sokolovsky V L, Weber H W, Grechnev G E, Boutko V G, Gusev A A, Kozyrev A V, Goldacker W, Podhurska V, Sverdun V.B, Habisreuther T, Schmidt Ch, Kovylaev V V, Shaternik V E, Karpets M V and Shaternik A V 2017 *IEEE Transactions on Applied Superconductivity* **27**, Issue: 4, 6200705 doi**:** 10.1109/TASC.2016.2638201

[9] Prikhna T, Eisterer M, Rindfleisch M, Ponomaryov S S, Tomsic M, Romaka V V, Moshchil V, Kozyrev A, Karpets M and Shaternik A 2019 *Journal of Superconductivity and Novel Magnetism* **31**(4) doi.org/10.1007/s10948-019-5062-z

[10] Prikhna T A, Eisterer M, Rindfleisch M, Kozyrev A V, Shaternik A V, Shaternik V E, Tomsic M, Moshchil V E, Karpets M V, Sverdun V B, Ponomaryov S S, Romaka V V and Seidel P P 2019 *IEEE Transactions on Applied Superconductivity* **29**(5) 1-5. doi: 10.1109/TASC.2019.2907662

[11] Bean C B 1964 *Rev. Mod. Phys.* **36** 31-36

[12] Pan A V, Golovchanskiy I A and Fedoseev S A, *Europhys.* 2013 *Lett.* **103**, 17006

[13] Golovchanskiy I A, Pan A V, Shcherbakova O V, and Fedoseev S A *J.* 2013 *Appl. Phys.* **114**, 163910

[14] http://www.wien2k.at/

[15] http://elk.sourceforge.net/

[16] Wills J M, Alouani M, Andersson P, Delin A, Eriksson O and Grechnev A 2010 Full-Potential Electronic Structure Method" *Energy and Force Calculations with Density Functional and Dynamical Mean Field Theory, Springer Series in Solid-State Sciences, Springer Verlag, Berlin* **167**

[17] http://fplmto-rspt.org/

[18] Barth von U and Hedin L 1972 *J. Phys. C* **5** 1629

[19] Perdew J P, Burke K and Ernzerhof M 1996 *Phys. Rev. Lett.***77**(18) 3865-3568

[20] Allen P B and Dynes R C 1975 *Phys. Rev. B.***12**(5) 905 -922

[21] Prikhna T A, Shapovalov A P, Grechnev G E, Boutko V G, Gusev A A, Kozyrev A V, Belogolovskiy M A, Moshchil V E, Sverdun V B 2016 *Low Temperature Physics* **42** 380-394 doi: 10.1063/1.4952985

[22] Klie R F, Idrobo J C, Browning N D, Regan K A, Rogado N S, and Cava R 2001 *J Appl. Phys. Lett.* **79**, 1837

[23] Moshchalkov V, Menghini M, Nishio T,. Chen Q H, Silhanek A V, Dao V H, Chibotaru L F, Zhigadlo N D and Karpinski J 2009 *Phys. Rev. Lett.* **102**, 117001

[24] Prikhna T, Romaka V, Eisterer M, Shapovalov A, Kozyrev A, Grechnev G, Boutko V, Goldacker W, Habisreuther T, Vakaliuk O, Halbedel B 2017 *IOP Conf. Series: Materials Science and Engineering* 012023. doi:10.1088/1757-899

Acknowledgments

The investigations was performed in the frame of project III-3-18 (0776) supported by the National Academy of Sciences of Ukraine.

Development of an in-situ analysis instrument for microstructure of materials with low temperature

Chen Liubiao[1], Zhao Yuanheng[1, 2], Liu Xuming[1, 2], Xi Xiaotong[1, 2], Wang Jue[1, 2], Zhou Yuan[1, 2], Wang Junjie[1, 2*]

[1] CAS Key Laboratory of Cryogenics, Technical Institute of Physics and Chemistry, Beijing 100190, China;
[2] University of Chinese Academy of Sciences, Beijing 100049, China

Corresponding Author: wangjunjie@mail.ipc.ac.cn

Abstract. The development of instruments capable of dynamically observing the microstructure at different temperatures is of great significance for the study of materials. The liquid nitrogen and a self-developed high-frequency pulse tube cryocooler were used as the cold source to develop the cryogenic system of scanning electron microscope (SEM) in this paper. The vibration and temperature control problems and solutions involved in using these two cold sources as SEM cryogenic systems were described and discussed in detail. It was found that it is necessary to fully achieve the heat balance at a certain temperature and adjust the image displacement to overcome the adverse effects caused by the thermal expansion and contraction of the low temperature components, and the cryocooler needs to use an intermittent operation mode to avoid distortion of the material microstructure image caused by vibrations of the cryocooler. Whether it is for liquid nitrogen or cryocooler, a thermal switch between the cold source and the thermal bridge is helpful for temperature control of the sample and reduced heat leakage of the system.

Introduction

The development of instruments capable of dynamically observing the microstructure at different temperatures is of great significance for the study of materials [1, 2]. The cold sources of cryogenic scientific instruments mainly include cryogens and cryocoolers. For cryogens, the most used is liquid nitrogen, but its deficiency is that the operating temperature is relatively high (77 K, by reducing the gas pressure, the minimum can be reduced to 63 K). In order to obtain a lower cooling temperature, liquid helium (about 4 K) is required, but the cost will be significantly increased. For cryocoolers, the most widely used ones are GM cryocoolers or GM pulse tube cryocoolers, which have the advantage of achieving very low cooling temperatures (less than 3K) and a large cooling capacity (larger than 1.5W/4.2K), but the large size/weight (>100 kg) and power consumption (7-8 kW) of such cryocoolers limits their application in some applications; There are also reports of using high-frequency cryocooler with smaller size/weight (10-20 kg) and lower power consumption (<0.5 kW) as cold sources for cryogenic scientific instruments [2, 3]. However, such cryocoolers have very small cooling capacity in the same temperature zone, and special attention must be paid to efficient heat transfer and effective heat insulation during the development process to minimize system heat leakage [4, 5].

In this paper, based on a commercial scanning electron microscope (SEM, model SU1510, produced by Hitachi), liquid nitrogen and self-developed high-frequency pulse tube cryocoolers are used as cold sources to develop in-situ observation instruments for material microstructure. The reason for using

IOP Conf. Series: Materials Science and Engineering **756** (2020) 012021 doi:10.1088/1757-899X/756/1/012021

liquid nitrogen as a cold source before using the cryocooler is to investigate the effect of the introduction of the external cooling system on the SEM system, and explore some general methods of vibration and temperature control.

SEM cryogenic system cooled by liquid nitrogen

The schematic and photograph of the SEM system using liquid nitrogen as a cold source are shown in Figures 1 and 2, respectively. In fact, the original solution was to fix the liquid nitrogen system on another support shelf, but found that it was necessary to introduce a special vibration reduction device to the support shelf: if without additional vibration reduction device, the image may be distorted due to the ground vibration when the magnification of the observation is high. After adopting the scheme shown in Figures 1 and 2, that is, the liquid nitrogen system is integrated with the SEM's sample chamber, and the vibration is reduced by the existing vibration reducing device, the original imaging quality of the SEM can be maintained. However, the test results show that during the cooling process, the volume of the sample and other components connected to it (such as the sample holder and the thermal bridge) change due to thermal expansion and contraction, which will cause image shift and thus affect the image quality. Therefore, in order to effectively observe the microstructure of the material, it is necessary to reach the thermal equilibrium at a certain temperature, and then observe the microscopic morphology of the next temperature. Through the stepwise operation of selecting multiple temperature points in the temperature zone to be tested, the dynamic observation of the material's microstructure at different temperatures can be realized. During the test, it is also necessary to continuously adjust the XYZ displacement adjusting device to overcome the positional deviation of the material due to thermal expansion and contraction (In fact, it will only shift in one direction), thereby realizing the observation of a certain fixed position of the material.

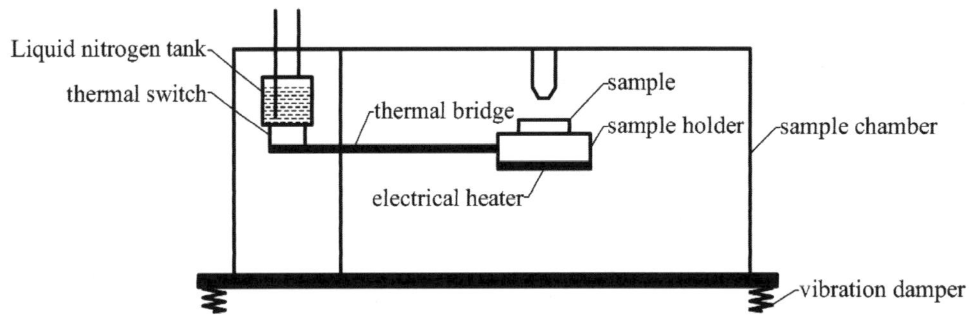

Figure 1. Schematic of SEM cryogenic system (cooled by liquid nitrogen).

Figure 2. SEM cryogenic system photo (cooled by liquid nitrogen).

For temperature control, in order to be able to control the sample to any temperature between 77 K (the internal pressure of the liquid nitrogen tank is equal to the atmospheric pressure) and room temperature, a thermal switch is required between the liquid nitrogen tank and the thermal bridge. The thermal switch is a gas gap structure that controls the thermal resistance between the liquid nitrogen tank and the thermal bridge to a suitable range by controlling the amount of helium inside. Then the PID was used to control the electric heating of the sample holder to achieve precise control of the sample temperature (Test results show that temperature control accuracy can reach ±0.1 K). Figure 3 is a photomicrograph of the sample holder without sample (oxygen-free copper) at 80 K by SEM. It should be noted that another purpose of controlling the thermal resistance between the liquid nitrogen tank and the thermal bridge through the thermal switch is to reduce the heat leakage between the two while heating the sample holder.

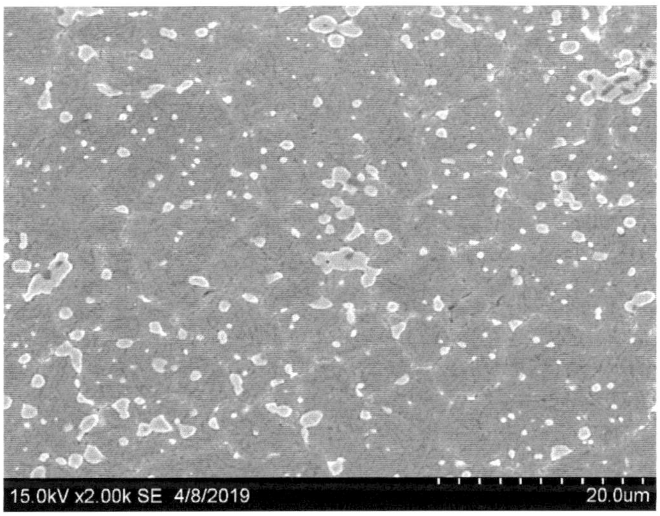

Figure 3. A photomicrograph of the sample holder (oxygen-free copper) without sample at 80 K.

SEM cryogenic system based on high frequency pulse tube cryocooler

Based on the development of above liquid nitrogen cooling system, further development was carried out using a cryocooler as a cold source. A self-developed high-frequency pulse tube cryocooler was adopted, which can achieve a no-load temperature of less than 5.7 K and provide a cooling power of 80 mW at 8 K or 400 mW at 15 K with an electric power input of 400 W [6]. Figure 4 and Figure 5 are the schematic and cooling capacity curve of the developed high-frequency pulse tube cryocooler, respectively. A prominent feature of pulse tube cryocooler is that there are no moving parts at the low temperature part, which has significant advantages in terms of low vibration and operational reliability. Therefore, the vibration of the pulse tube cryocooler mainly comes from the compressor. In order to reduce vibration, the developed pulse tube cryocooler in this paper uses an oppose-dual linear compressor to reduce vibration by the relative movement of two opposed pistons. In addition, a flexible tube connection is used between the compressor and the cold finger to further eliminate the vibration comes from the compressor. Through testing, it has been found that this vibration reduction scheme can indeed be useful under an optical microscope (tens or hundreds of magnification observations). But for SEM (magnification observations of thousands or even tens of thousands), the cryocooler's vibration still has a significant impact on the imaging of the SEM, especially when the operating temperature of the cryocooler is very low (the lower the temperature, the greater the electrical power required, resulting in greater piston movement of the compressor), image distortion will occur and it will be completely unobservable. Therefore, some new strategies to reduce vibration are needed. Figures 6 and 7 are schematic and three-dimensional design views of the SEM cryogenic system, respectively. Based on this system, we conducted research on vibration reduction measures, which will be discussed in detail below.

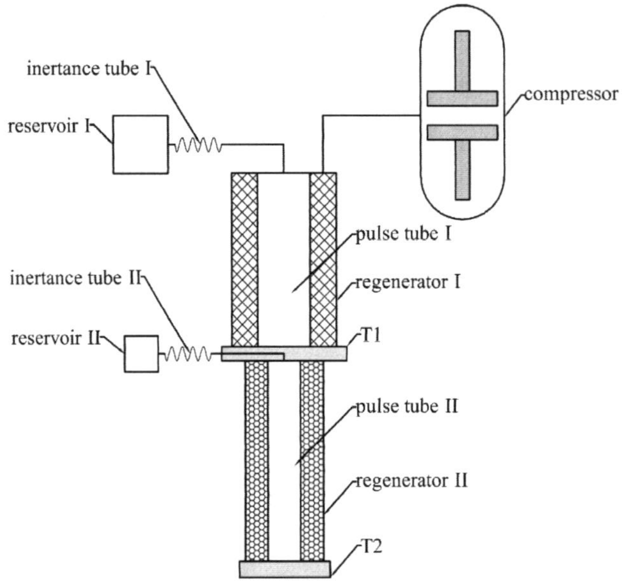

Figure 4. Schematic of the SEM cold source-high frequency pulse tube cryocooler.

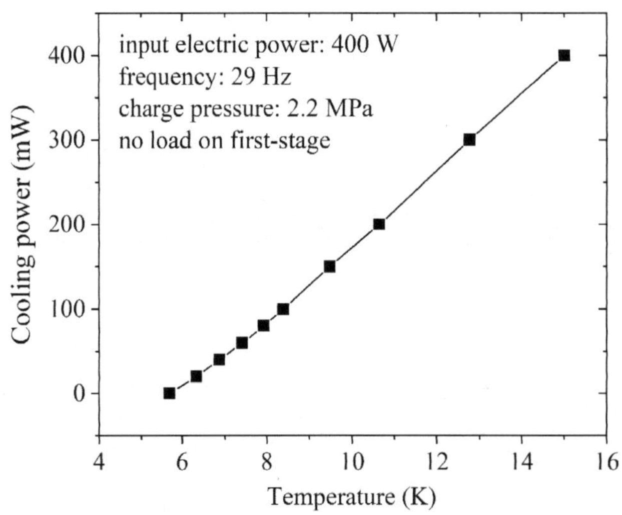

Figure 5. Cooling power of the SEM cold source-high frequency pulse tube cryocooler.

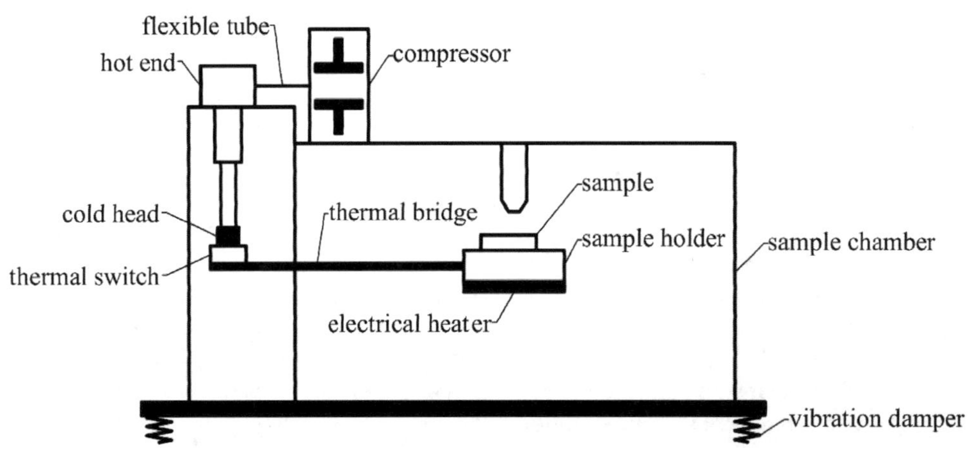

Figure 6. Schematic of SEM cryogenic system (cooled by high frequency pulse tube cryocooler).

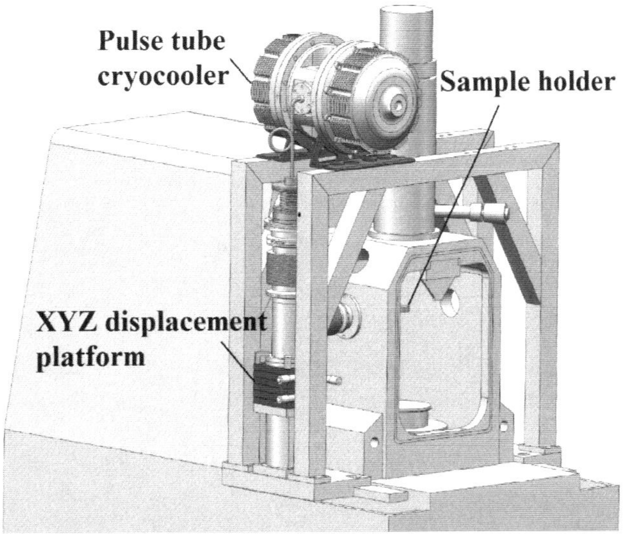

Figure 7. Three-dimensional design views of SEM cryogenic system (cooled by high frequency pulse tube cryocooler).

In order to solve the vibration problem of the cryocooler, we adopted an intermittent working mode: When the sample temperature is higher than the set temperature, start the cryocooler to cool down (high vibration state). When the sample temperature is significantly lower than the set temperature, turn off the cryocooler (no vibration), and maintaining the temperature of the sample at the set temperature by controlling the heating amount of the sample holder by PID (test results show that temperature control accuracy can also reach ±0.1 K), thus, the observation of the microscopic morphology of the sample at this temperature is completed. In fact, during the temperature control, the thermal bridge with a lower temperature than the sample is equivalent to a cold source, which is used to meet the cooling demand of the sample. In addition, the heat bridge has a large heat capacity, so that its temperature rises very slowly, which can provide sufficient time for temperature control of the sample.

It can also be seen from Figure 6 that a thermal switch is also installed between the thermal bridge and the cold head of the cryocooler. However, it is slightly different from the thermal switch function of the liquid nitrogen cooling system, it usually has only two working states (on or off): When the cryocooler is in the start-up state, the heat switch is filled with helium to enhance the heat transfer between the cold head of the cryocooler and the heat bridge; When the cryocooler is in the off state, the helium inside the thermal switch is extracted to reduce the heat transfer between the cold head of the cryocooler and the thermal bridge.

In general, after using the above-mentioned vibration reduction and temperature control methods, the SEM cryogenic system cooled by the cryocooler can work normally, and the microstructure of the material at different temperatures can be observed. However, this cooling system needs further optimization work, and in order to better observe and analyse the materials, it also needs to add some new functions, for example, on the one hand, a deformation excitation mechanism can be added to achieve deformation excitation such as stretching and compression of a material sample at different temperatures; on the other hand, a multi-stage high frequency pulse tube cryocooler capable of achieving a lower temperature (to reach liquid helium temperature range) will be used to further broaden the temperature observation range of the sample [7, 8].

Conclusions

In order to dynamically observe the microstructure of materials at different temperatures, the liquid nitrogen and self-developed high-frequency pulse tube cryocooler were used as the cold source to develop the cryogenic system of SEM. The test results show that coupling the cryogenic system to the SEM sample cavity and sharing the existing vibration reduction device of the sample cavity does not affect the imaging quality of the SEM. However, the sample and other components connected to it will expand and contract due to temperature changes, which will cause the image to shift and thus affect the image quality. Therefore, it is necessary to reach the thermal equilibrium at a certain temperature, and then observe the microscopic morphology of the next temperature in the process of cooling. Meanwhile, it is also necessary to continuously adjust the XYZ displacement adjusting device to overcome the positional deviation of the material. Whether it is for liquid nitrogen or cryocooler, a thermal switch between the cold source and the thermal bridge is helpful for temperature control of the sample and reduced heat leakage of the system. When the SEM is at a high observation magnification, the cryocooler needs to use an intermittent operation mode to avoid distortion of the material microstructure image caused by the cryocooler's vibration.

References

[1] Liubiao C, Qiang Z, Xiaoshuang Z, Yuan Z and Junjie W. 2015 *Physics Procedia* **67** 354-359.
[2] Lei Y, Qin X, Wan F, Liu P, Chen L and Wang J. 2017 *Fusion Engineering and Design* **125** 603-607.
[3] L B Chen, S X Liu, K X Gu, Y Zhou and J J Wang. *2015 IOP Conference Series: Materials Science and Engineering* **102(1)** 012011.
[4] Chen L, Jin H, Wang J, Zhou Y, Zhu W and Zhou Q. 2013 *Cryogenics* **54** 54-58.
[5] Chen L, Zhou Q, Jin H, Zhu W, Wang J and Zhou Y. 2013 *Cryogenics* **57** 195-199.
[6] Wu X L, Chen L B, Liu X M, Wang J, Zhou Y and Wang J J. 2019 *Cryogenics* **101** 7-11.
[7] Liubiao C, Xianlin W, Xuming L, Changzhao P, Yuan Z and Junjie W. 2018 *International Journal of Refrigeration* **88** 204-210
[8] L Chen, X Wu, J Wang, X Liu, C Pan, H Jin, W Cui, Y Zhou and J Wang. 2018 *Cryogenics* **94** 103-109

Acknowledgments

This research is supported by National Key R&D Program of China (No. 2018YFB0504603), the National Natural Science Foundation of China (No. 51706233, 51427806, U1831203), Strategic Pilot Projects in Space Science of China (No. XDA15010400), and supported by Key Research Program of Frontier Sciences, Chinese Academy of Sciences, Grant No. QYZDY-SSW- JSC028.

Quantum effects in graphitic materials: Colossal magnetoresistance, Andreev reflections, ferromagnetism, and granular superconductivity

N Gheorghiu[1], CR Ebbing[2], BT Pierce[3], and TJ Haugan[3]

[1]*UES Inc., Dayton, OH 45432*

[2] *University of Dayton Research Institute, Dayton, OH 45469*

[3]*The Air Force Research Laboratory, Wright-Patterson Air Force Base, OH 45433*

Nadina.Gheorghiu@yahoo.com

Abstract. Unlike the more common local conductance spectroscopy, nonlocal conductance can differentiate between nontopological zero-energy modes localized around inhomogeneities, and true Majorana edge modes in the topological phase. In particular, negative nonlocal conductance is dominated by the crossed Andreev reflection. Fundamentally, the effect reflects the system's topology. In graphene, the Andreev reflection and the inter-band Klein tunneling couple electron-like and hole-like states through the action of either a superconducting pair potential or an electrostatic potential. We are here probing quantum phenomena in modified graphitic samples. Four-point contact transport measurements at cryogenic to room temperatures were conducted using a Quantum Design Physical Property Measurement System. The observed negative nonlocal differential conductance G_{diff} probes the Andreev reflection at the walls of the superconducting grains coupled by Josephson effect through the semiconducting matrix. In addition, G_{diff} shows the butterfly shape that is characteristic to resistive random-access memory devices. In a magnetic field, the Andreev reflection counters the effect of the otherwise lowered conduction. At low temperatures, the magnetoresistance shows irreversible yet strong giant oscillations that are known to be quantum in nature. In addition, we have found evidence for seemingly granular superconductivity. Thus, graphitic materials show potential for quantum electronics applications, including rectification and topological states.

1. Introduction

While carbon(C)-based materials are known for many practical properties, such as light-weight and high strength, their inner nature is governed by the laws of quantum physics manifested as magnetism [1], conveniently used for spintronics [2], or as unconventional superconductivity [3,4] that we have previously explored [5].

In this paper we present evidence for several quantum phenomena occurring in graphitic materials: nonlinear electronic transport resulting in negative differential resistance, Landau-level quantization, colossal magnetoresistance, ferromagnetism, and Andreev reflections at the boundaries between SC grains and the surrounding semiconducting matrix.

2. Experimental details

The materials are PAN-based (polyacrylonitrile $(CH_2\text{-}CH\text{-}CN)_n$) T300 C fibers (Cytec), highly oriented pyrolytic graphite (particle size 10 μm), and graphite foil (Graphtek). Selected samples were oxygen(O)-ion implanted (Cutting Edge Ions) at 2D concentrations 7.07×10^{12}, 5.66×10^{15}, and 2.24×10^{16} ions/cm^2, respectively. For an ion energy 70 keV, the implantation depth was 125 nm, i.e., ~ 1% implantation volume. The four-wire Van Der Pauw technique was used for resistivity measurements. Scanning electron microscopy pictures and the elemental analysis for fiber samples are shown in Fig. 1.

The quality of the electrical contacts was optically checked using an Olympus BX51 microscope. The current-to-voltage gap ratio was close to the required factor of four [6]. Magneto-transport and magnetization measurements were carried out in the 1.9 K - 300 K temperature range and for magnetic fields of induction B up to 9 T using a Physical Properties Measurement System (PPMS) model 6500 (Quantum Design).

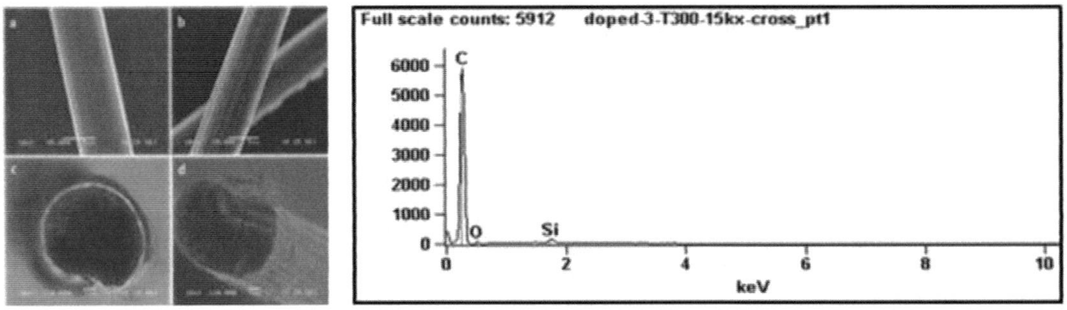

Figure 1 SEM pictures showing the effect of O-ion implantation (2.24×10^{16} ions/cm^2) on the C fiber (b, d) vs. to the raw C fiber (a, c), both along and across the fibers (upper images and lower images, respectively). Also shown are the C and O peaks on the EDS scan. The Si peak is due to the placement of the C fiber on a Si substrate during the implantation procedure. An accelerating voltage of 10.0 kV and a magnification factor ×15000 were used.

3. Results and Discussion

3.1. *Electrical resistivity and nonlocal differential conductance without magnetic field*

The temperature-dependent resistivity $\rho(T)$ for two C fibers O-implanted at 2.24×10^{16} ions/cm^2 is shown in Fig. 2. At $T \sim 250$ K, there is either an insulator-metal-insulator transition or a metal-insulator-metal transition for a sourced current $I = 20$ μA or $I = 1$ μA, respectively. These transitions reflect the T-dependence balance of charge carrier densities and their mobilities. For the larger I, ρ has a metallic behaviour below $T \sim 25$ K [7]. For the smaller I, ρ for one of the fibers becomes 'negative' for temperatures below ~150 K. The observation of negative resistance (i.e., nonlocal negative voltage) is a signature for highly viscous flow of electrons or, equivalently, zero electrical resistance in the SC state [8]. Moreover, $T \sim 150$ K is the BCS mean-field critical temperature for Cooper-pair instability in graphene [9]. Pseudo-gap behaviour below the onset temperature $T \sim 150$ K is possible.

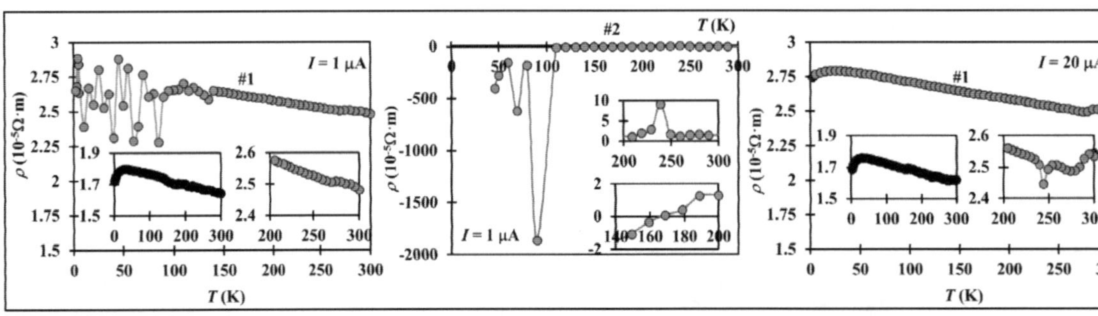

Figure 2 Temperature-dependent electrical resistivity $\rho(T)$ for two C fibers that have been O-implanted at 2.24×10^{16} ions/cm^2 (in blue) and raw fiber (in black), respectively.

Below 150 K, giant fluctuations in the nonlinear differential conductance $G_{diff} = \mathrm{d}I/\mathrm{d}V$ and the Chua's butterfly are observed at $T = 130$ K (Fig. 3). For $V < 0$, $G_{diff} > 300$ mS or $R_{diff} < 4\ \Omega$ or $\rho_{diff} < 8\ \mu\Omega\cdot$cm. The Chua nonlinear phenomenon can be also seen as a subcritical Hopf bifurcation in which new attractors are being born at the pitchfork point [10]. The inset for $V > 0$ shows that G_{diff} has the shape of the nonlinear $I(V)$ in the Chua circuit. I.e., the charge oscillates as a third-order polynomial of the magnetic flux as in a cubic memristor during a supercritical Andronov–Hopf bifurcation [11]. The butterfly wing span is $\Delta V \cong 0.5$ mV, corresponding to the critical temperature for twisted bilayer graphene, $T_c \cong 1.8$ K. Interestingly, the minimum G_{diff} (< 0) corresponding to $V \cong 60$ mV would give a BCS critical temperature $T_c = 2\Delta/3.53k_B \cong 250$ K, exactly at the insulator-metal transition. The subcritical and supercritical bifurcations meet at the tricritical point and the magnetic field of the local spins acts as a chemical potential. The oxide layer at the surface of the C fiber might be the location of both hard and soft SC fluctuations, i.e., the system might be a multigap SC where the larger gap SC phase takes over the smaller gap SC phases. The evolution of the attractor – which is a manifold of equilibrium states – is actually the evolution of the amplitude (Higgs mode) and phase for the SC order parameter.

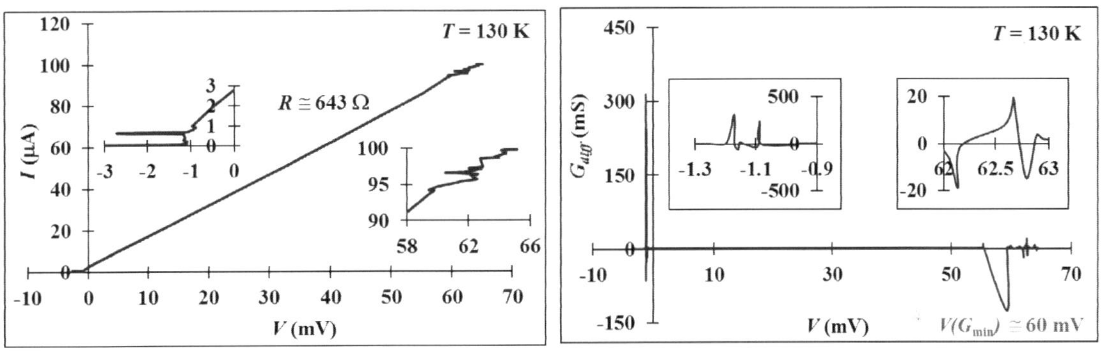

Figure 3 $I(V)$ and the corresponding $G_{diff}(V)$ data for O-implanted (at 2.24×10^{16} ions/cm^2) C fiber.

3.2. *Transport measurements in magnetic field*

3.2.1 Landau-level quantization, the universal conductance, and butterfly shapes

As known, graphite has antiferromagnetic (AFM) correlations between unlike sublattices (ABAB...) and FM correlations between like sublattices (AAA... or BBB...). A magnetic field can turn a FM into an AFM at the tricritical point, where the FM and AFM merge into the system paramagnetic (PM) phase as the temperature is increased. Under a static magnetic field, the differential resistance $R_{diff} = \mathrm{d}V/\mathrm{d}I$ (not the voltage-current characteristic $V(I)$) shows the butterfly shape (Figure 4). Significantly, Landau-level quantization results in a G_{diff} peak exactly at the universal conductance $G_0 = 2e^2/h \cong 77\ \mu$S, two peaks at $G_0/4$ (one for $V = 0$), and a minimum due to the Zeeman spin-splitting at $\cong -0.9G_0/2$. Anomalous Josephson effect induced by spin-orbit interaction and Zeeman effect in semiconductor channel between two SC grains is also possible. In a semiconductor nanowire, a $G_0/2$ conductance peak in high enough magnetic field is a signature of ballistic 1D transport, where backscattering is reduced and spin degeneracy is lifted. An even higher magnetic field (normal to the wire) can lead to the depopulation of sub-bands and $G_0/4$ peaks are observed.

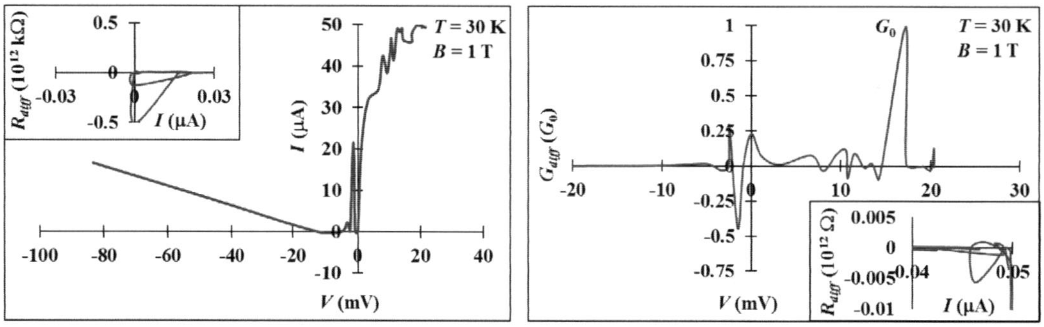

Figure 4 Nonlinear current-voltage characteristic *I(V)* and the corresponding $G_{diff}(V)$ data for O-implanted (at 2.24 × 10^{16} ions/cm^2) C fiber in transverse magnetic field.

3.2.2 Colossal magnetoresistance

Low-temperature magnetoresistance $MR(\%) = 100 \times [R(B) - R(B=0]/R(B=0)$ for a C fiber O-implanted at 2.24 × 10^{16} ions/cm^2 is shown in Figure 5. While MR takes colossal values for both small (50 nA) and large (50 μA) currents, the electronic charge transport is significantly less viscous for the former ($MR < 0$) than for the the latter (($MR > 0$). Viscosity generates vorticity, arguable playing the same role as the zero-electrical resistance does for SC [8].

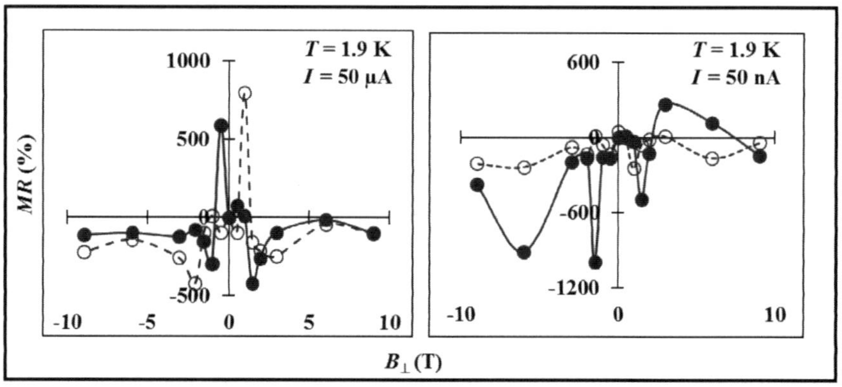

Figure 5 Magnetoresistance data for a C fiber O-implanted at 2.24 × 10^{16} ions/cm^2 at T = 1.9 K. Legend: open/closed symbols for decreasing/increasing the magnetic field with the sample thermally stabilized.

3.2.3 Andreev oscillations

We have also observed Andreev oscillations in the magnetoresistance of highly oriented pyrolytic graphite (HOPG) cylindrical samples of radius $R = 3$ mm and height 1 mm (Figure 6). While this kind of quantum resonances that are attributed to Andreev reflections between SC and semiconducting regions have been observed before [3,12], their period was significantly larger (0.1 T to 0.4 T). Here, the period for Andreev oscillations is $\Delta B = 2$ mT. We can estimate the area for the SQUID-type oscillations of two SC pathways separated by a normal/insulating region as $S = \Phi_0/\Delta B \cong 1.0 \times 10^{12}\ \mathrm{m}^2$, where $\Phi_0 \cong 2.07 \times 10^{-15}\ \mathrm{T}\cdot\mathrm{m}^2$ is the magnetic flux quanta. Given the sample's symmetry, this would give a radius $R \cong 560$ nm for the circular paths.

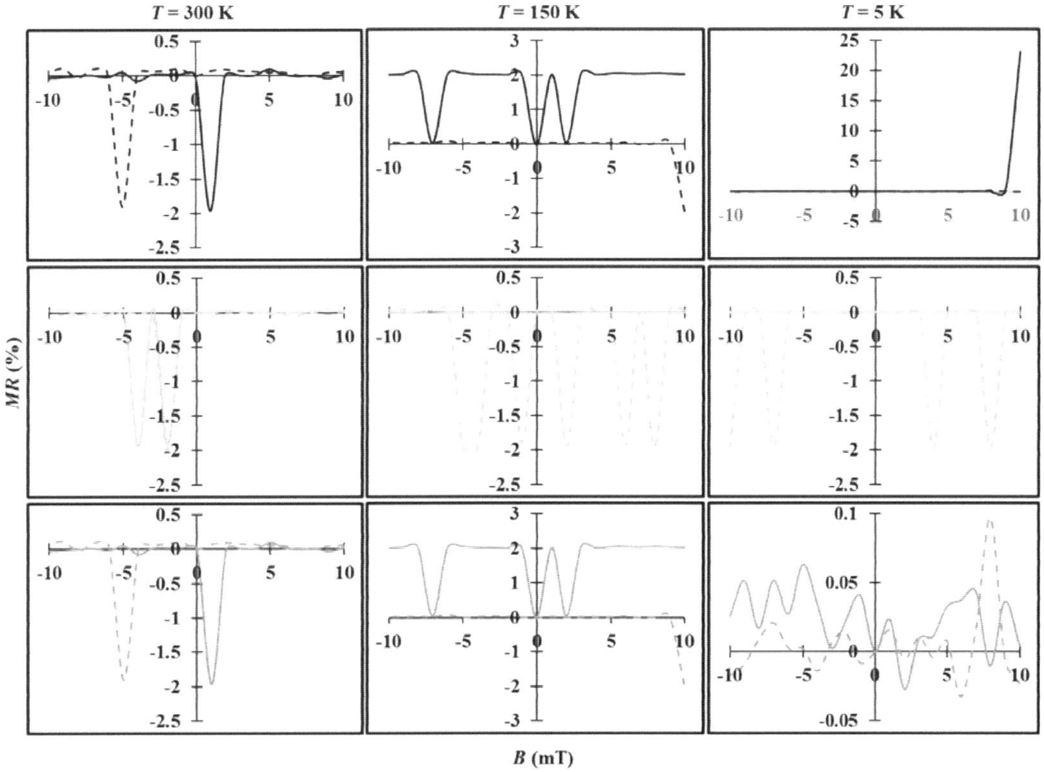

Figure 6 Andreev oscillations in the magnetoresistance of highly oriented pyrolytic graphite cylindrical samples O-implanted at 7.07 × 10^{12} ions/cm^2 (light blue), 5.66 × 10^{15} ions/cm^2 (blue), and raw sample (in black) at temperatures 5 K, 150 K, and 300 K, respectively.

3.3 Magnetization measurements

3.3.1 Zero-field cooled (ZFC) and field-cooled (FC) measurements

Temperature-dependent ZFC-FC magnetization data for O-implanted graphite foil samples having dimensions 2 mm × 2 mm × 1 mm are shown in Figure 7. While fluctuations are observed up to $T \sim$ 150 K, the irreversibility temperature is $T_{irrev} \cong 50 - 60$ K. Our results are close to the BCS mean-field calculations that found the critical temperature T_c for graphite at 60 K [13] and for graphene at 150 K [9]. An unusual feature is the upward turn of the magnetization at low temperatures that was interpreted as a "reentry" phenomena [14]. The magnetization in the SC phase has three sources: (a) the diamagnetic shielding moment, (b) the trapped flux and (c) a paramagnetic contribution (positive magnetic susceptibility χ_m) from the normal matrix. The $T_{irrev} \sim 50$ K here obtained is the same as the estimate of T_c for semiconducting structures [16], where the excitonic mechanism was proposed [15].

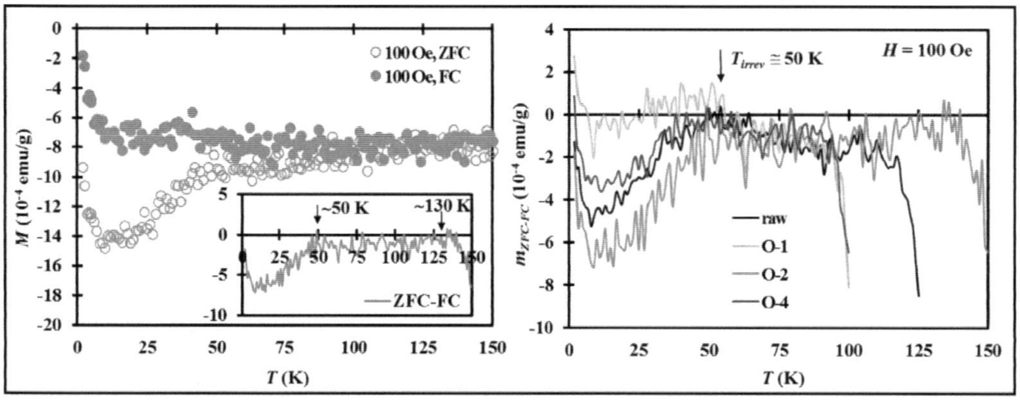

Figure 7 Zero-field cooled (ZFC) and field-cooled (FC) temperature dependent magnetization data for graphite foil samples O-implanted at 2D concentrations 7.07 × 10^{12} ions/cm^2 (O-1), 5.66 × 10^{15} ions/cm^2 (O-2), and 2.24 x 10^{16} ions/cm^2 (O-3). Data for a raw sample is also shown.

3.3.2 Magnetization loops

Magnetic-field dependent magnetization loops at T = 300 K for the described before graphite foil samples are shown in Figure 8. Except for the prolonged tail, these high-temperature magnetization loops are clearly similar to the ones found for hard SCs. In addition, these loops also show the kink characteristic to granular SCs such as the ones found in [16]. Low-field loops (Figure 9a) show the shielding effect due to the pseudo-Meissner effect [17]. The $M(H)$ loops for HOPG and graphite samples are similar (Figure 9b). The magnetic coherence length $\xi_m = \left(\frac{\Phi_0}{2\pi B_{c2}}\right)^{1/2}$ is in the 100 – 200 nm range.

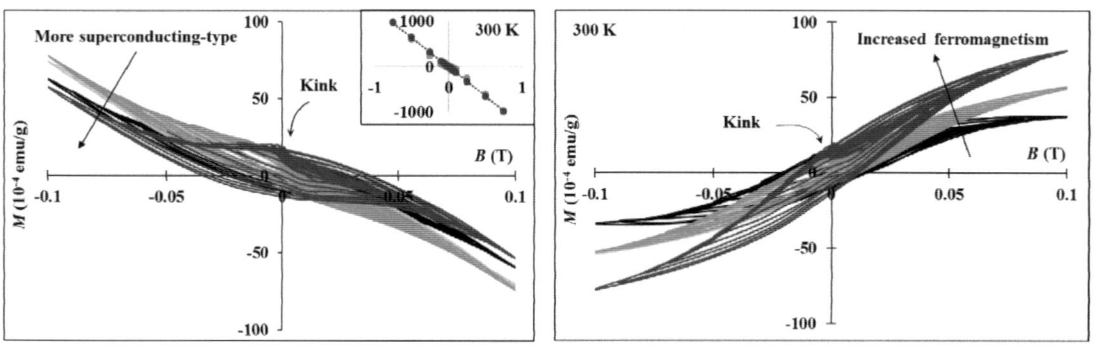

Figure 8 Nonlinear and irreversible $M(B)$ loops at T = 300 K for O-implanted graphite foil samples: raw (in black) and O-implanted (in blue, description in Figure 7) before (left) and after (right) the subtraction of their own diamagnetic background (inset).

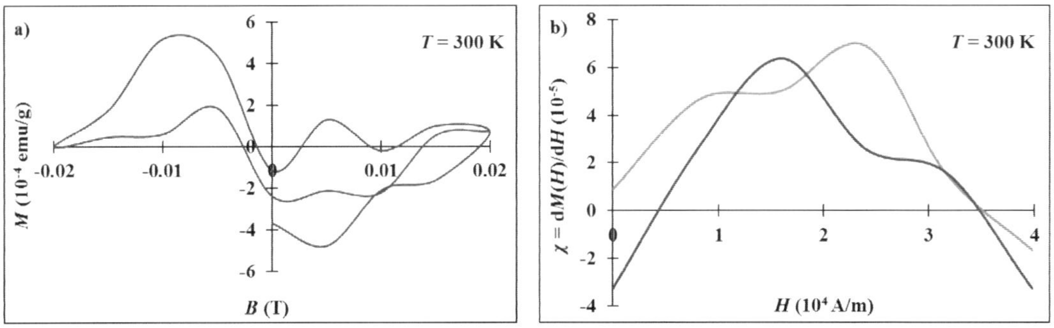

Figure 9 a) Superconducting-type magnetization loop for a graphite foil sample (O-ion implantation dose 5.64 x 10^{15} ions/cm^2). b) Field-dependence of the magnetic susceptibility χ for an O-implanted sample at 2.24 x 10^{16} ions/cm^2 (in dark blue) and for an O-implanted (at 5.64 x 10^{15} ions/cm^2) graphite foil sample (in blue).

4. Conclusions
Graphitic materials are extremely rich in quantum phenomena. In this work, we have presented experimental evidence for Andreev reflections, colossal magnetoresistance, ferromagnetism and granular superconducting-like behaviour. Future work will explore in more detail these and other related phenomena.

5. References
[1] D. Spemann, P. Esquinazi, R. Höhne, A. Setzer, M. Diaconu, H.Schmidt, T. Butz 2005 Magnetic carbon: A new application for ion microbeams *Nuclear Instruments and Methods in Physics Research B* **231(1-4)** 433
[2] T.L. Makarova *et al.* 2015 Edge state magnetism in zigzag-interfaced graphene via spin susceptibility measurements *Scientific Reports* **5** 13382
[3] P. Esquinazi *et al.* 2008 Indications for intrinsic superconductivity in highly oriented pyrolytic graphite *Phys. Rev. B* **78** 134516
[4] Y. Cao *et al.* 2018 Correlated insulator behaviour at half-filling magic angle graphene superlattices, *Nature* **80** 56
[5] B.T. Pierce, J.L. Burke, L.B. Brunke, T.J. Bullard, D.C. Vier, and T.J. Haugan 2013 Search for Superconductivity in Doped Amorphous Carbon Thin Films *IEEE Transactions on Applied Superconductivity* **23(3)** 7000205
[6] J.W. Ekin 2006 Experimental Techniques for Low-temperature measurements *Oxford University Press*
[7] I.L. Spain, K.J. Volin, H.A. Goldberg, and I.L. Kalnin 1983 Unusual electrical resistivity behavior in carbon fibers *Solid State Commun.* **45(9)** 817
[8] L. Levitov and G. Falkovich 2016 Electron viscosity, current vortices and negative nonlocal resistance in graphene *Nat. Phys.* **12** 672
[9] D.V. Kveshchenko 2009 Massive Dirac fermions in single-layer graphene *J. Phys.: Condes. Matter* **21(7)** 075303
[10] N.V. Stankevich, N.V. Kuznetsov, G.A. Leonov, and L.O. Chua 2017 Scenario of the Birth of Hidden Attractors in the Chua Circuit *International Journal of Bifurcation and Chaos* **27(12)** 1730038
[11] I.A. Korneev, and V.V. Semenov 2017 Andronov–Hopf bifurcation with and without parameter in a cubic memristor oscillator with a line of equilibria *Chaos* **27** 081104
[12] J. Ridderbos,et al. 2019 Multiple Andreev reflections and Shapiro steps in a Ge-Si nanowire Josephson junction *Phys. Rev. Materials* **3** 084803
[13] N. García and P. Esquinazi 2009 Mean Field Superconductivity Approach in Two Dimensions *J*

Supercond Nov Magn **22** 439
[14] Y. Yeshurun, I. Felner, and H. Sompolinski 1987 Magnetic properties of a high-T_c superconductor $YBa_2Cu_3O_7$: Reentrylike features, paramagnetism, and glassy behavior *Phys. Rev. B* **36(1)** 840
[15] E.D. Cherotchenko, T. Espinosa-Ortega, A.V. Nalitov, I.A. Shelykh, A.V. Kavokin 2016 Superconductivity in semiconductor structures: The excitonic mechanism *Superlattices and Microstructures* **90** 170
[16] S. Saha and B.K. Das 1993 A realistic two-component critical state model: successful derivation of low-field M-H loops of granular superconductors *Supercond. Sci. Technol.* **6** 685
[17] M.D. Sumption, E. Lee, E.W. Collings, X.L. Wang, and S.X. Dou 2002 Suppression of AC (hysteretic) loss of magnetic shielding of MgB_2/Fe superconductors: The pseudo-Meissner effect *Advances in Cryogenic Engineering: Proceedings of the International Cryogenic Materials Conference* **48A and B** *AIP Conference Proceedings* **614** 824

Acknowledgments
This work was supported by The Air Force Office of Scientific Research (AFOSR) for the LRIR #14RQ08COR & LRIR #18RQCOR100 and the Aerospace Systems Directorate (AFRL/RQ). We acknowledge J.W. Lawson for the SEM scans, J.P. Murphy for the cryogenics, and Dr. T.J. Bullard for suggestions. Dr. G.Y. Panasyuk is acknowledged for his experienced view on condensed matter physics topics and for continuous support and inspiration to this paper's first author.

The Effect of Reinforcement Substrate Alloy Selection on Mechanical Properties of REBCO Coated Conductors

K. J. Radcliff[1], R. P. Walsh[1], D. C. Larbalestier[1] and Seungyong Hahn[1, 2]

[1]National High Magnetic Field Laboratory, 2031 E. Paul Dirac Drive, Tallahassee, FL 32310, US

[2]Department of Electrical and Computer Engineering, Seoul National University, Gwanak-ro 1, Gwanak-gu, Seoul, Korea 08826

Abstract. Rare earth Barium Copper Oxide (REBCO) coated conductors are promising candidates for high field (>25 T) user magnets. However, as the demand for higher fields increase, so does the potential to overstrain the conductors being used. Coated conductor substrates, such as 310s stainless steel and the super-alloy Hastelloy C276, serve as the backbone for mechanical strength in these conductors. Both substrate alloys share similar properties when optimally processed into strips prior to manufacturing of the REBCO coated conductor. We find that with subsequent REBCO manufacturing processes the strength of the substrate changes, the magnitude of which depends on whether Hastelloy C276 or 310s stainless steel is used. In this study, we investigate the stress-strain variability found in coated conductors and how the manufacturing process affects the mechanical properties. The manufacturing step of concern is the short time that the substrate is exposed to high temperature (700 to 800 C) during the REBCO deposition process. To better relate manufacturing processes and mechanical properties, we subjected bare substrates to different heat treatments at 700, 750, and 800 C for 15 minutes each. With post heat-treatment room-temperature tensile tests, we found that the 310s stainless steel substrate was sensitive to the variations of time and temperature, exhibiting yield strength reductions of 20 to 50 % depending on the heat treatment. By contrast, Hastelloy C276 did not weaken and initially showed strengthening effects with exposure to the lower temperature heat treatments. Coated conductor manufactures may prefer 310s stainless steel as their substrate due to cost and availability, however, moving to Hastelloy C276 will offer better mechanical robustness and reproducibility of mechanical properties within their coated conductor.

1. Introduction

REBCO coated conductors are increasingly becoming more popular for next generation high field magnets. With their high current density and relatively high operating temperature allowing a wide range of magnet applications. However, these tape conductors are quite sensitive to the effects of strain during manufacturing and in use operations. Mechanical properties of REBCO coated conductors show variable strength depending on the type of substrate used to reinforce the conductor. Evidence shows this variability could come from the vapour deposition process during manufacturing of the coated conductor. The two types of commonly used substrates are Hastelloy C276 and 310s stainless steel are used to perform this systematic study on the effects of deposition temperature during the manufacturing process.

2. Material Properties

310s stainless steel is a high nickel austenitic steel alloy that allows increased resistance to strain induced martensitic transformations. Hastelloy C276 is a nickel- tungsten superalloy with high strength and non-magnetic properties [1]. Both alloys are used in high temperature applications for their strength allowing them to be strong candidates to handle annealing temperatures during the vapour deposition process. Table 1 shows the chemical composition of both 310s stainless steel and Hastelloy C276.

Table 1: Chemical composition in wt % of 310s stainless steel and Hastelloy C276.

	Ni	Mo	Cr	Fe	W	Mn	Co	C	V	P	S	Si
Hastelloy C276	Balance	15-17	14.5-16.5	4.0-7.0	3-4.5	1.0	2.5	0.01	0.35	0.4	0.03	0.08
310s Stainless Steel	19-22	-	24-26	Balance	-	2.0	-	0.25	-	0.045	0.030	1.50

The 310s stainless steel samples were 100 μm thick and four mm. The Hastelloy samples were 30 and 50 μm thick and 12 mm wide.

3. Procedure

Four different conditions of both substrates were tested. Two samples for each substrate were tested in the as received condition, after 700 °C, 750 °C, and 800 °C heat treatment. To simulate the temperature excursions during the vapour deposition process, three heat treatment conditions were done on 100 μm 310s stainless steel and on 50 and 30 μm Hastelloy C276. Figure 1 shows the temperature profiles for each heat treatment; each heat treatment took about three hours to reach desired temperature was held at temperature for a total of 15 minutes inside an argon environment and then was air cooled. The temperatures were chosen due to a previous report on IBAD temperature during YBCO manufacturing being approximately 750 °C [2].

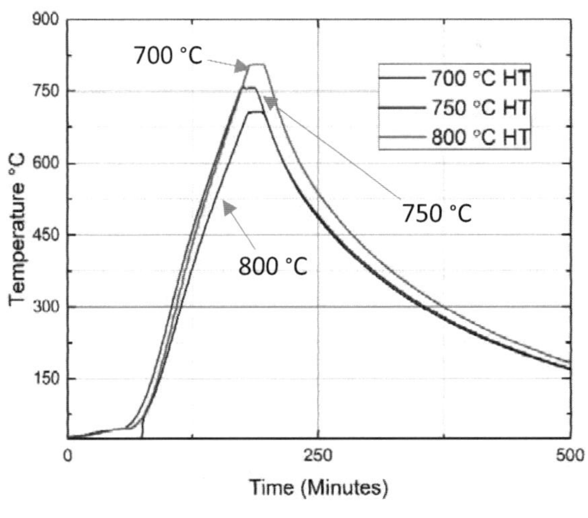

Figure 1: The annealing temperature profiles on bare 310s stainless steel and Hastelloy C276 substrates [9].

Room temperature tensile test are performed according to the ASTM E8 standard [2]. Tests are conducted in displacement control at a rate = 0.5 mm/min (for an elastic strain rate = $1.67*10^{-4}$ Ɛ s^{-1}) using a clip-on extensometer. An unload cycle is performed during the tensile test at 0.15% strain for modulus determination and strain is measured with a 50 mm gage length clip-on extensometer.

4. Results and Discussion

The tensile test results are shown in Table 2. The initial results of the as received condition of the Hastelloy C276 and 310s stainless steel seen in figure 2(A) shows that each substrate starts in the cold-worked condition. The 310s stainless steel has a yield strength of about 1237 MPa which corresponds to about 40 % cold worked [5] while the as-received yield strength of Hastelloy C276 is approximately 1464 MPa, independent of Hastelloy C276 thickness. Figures 2(A) – (D) show the effect of heat-treatment temperature on the stress-strain relationship of the Hastelloy C276 and 310s stainless steel substrates. Intuition suggest that relatively no change should occur to the substrates at after the 700 °C heat treatment since both materials are used for high temperature applications. Figure 2(B) suggest that is not necessarily the case for the 310s stainless steel. The yield strength of the 310s stainless steel dropped 22 percent to 965 MPa from the as-received condition.

Table 2: Tensile test results of bare 310s stainless steel and Hastelloy C276 substrates tested at 293 K

Material	Condition	Test Temp	Young's Modulus	0.2 % Yield Strength
		K	GPa	MPa
310s Stainless Steel	AR	293	174	1237
	700 °C	293	163	965
	750 °C	293	164	619
	800 °C	293	179	527

Material	Condition	Test Temp	Young's Modulus	0.2 % Yield Strength
		K	GPa	MPa
Hastelloy C276	AR	293	181	1464
	700 °C	293	209	1524
	750 °C	293	209	1477
	800 °C	293	202	1224

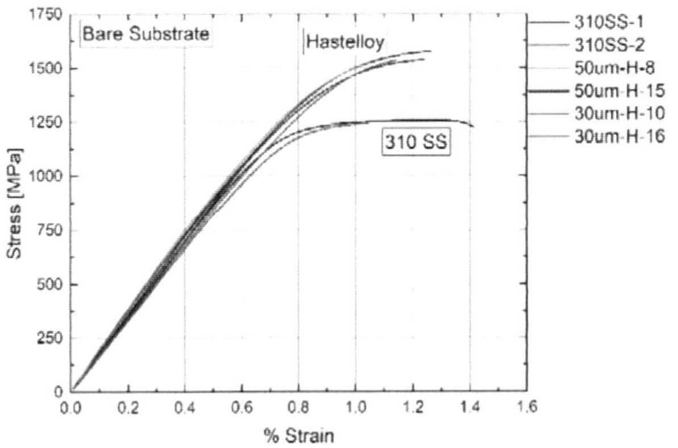

Figure 2(A) : As received stress-strain relationship of Bare Hastelloy C276 and 310s stainless steel at 293 K [9].

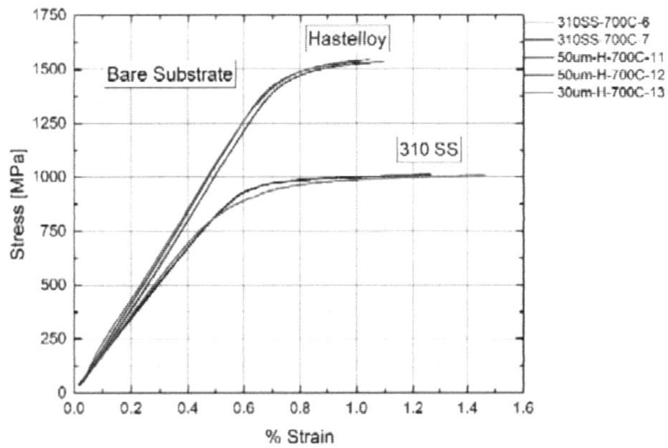

Figure 2(B): 700 C heat treatment effect on stress-strain relationship of Bare Hastelloy C276 and 310s stainless steel at 293 K [9].

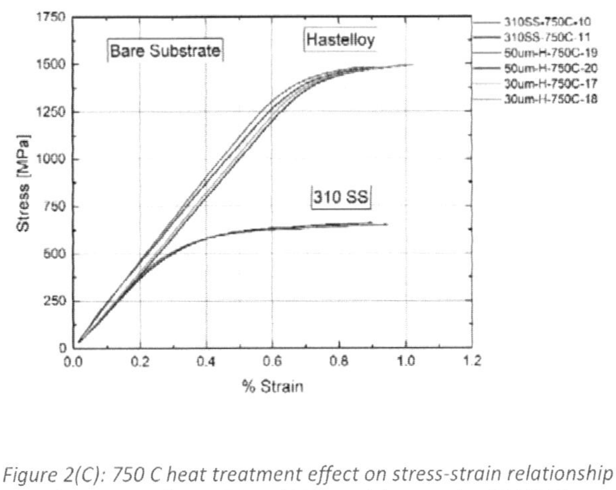

Figure 2(C): 750 C heat treatment effect on stress-strain relationship of Bare Hastelloy C276 and 310s stainless steel at 293 K [9].

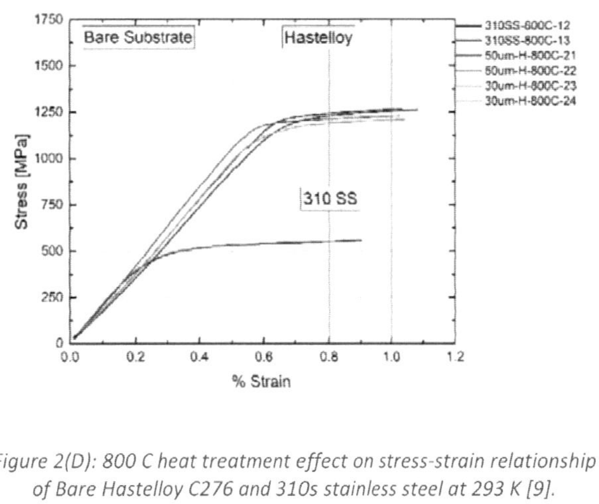

Figure 2(D): 800 C heat treatment effect on stress-strain relationship of Bare Hastelloy C276 and 310s stainless steel at 293 K [9].

The mechanical properties of Hastelloy C276 after the 750 °C heat treatment showed no apparent difference when compared to the 700 C heat treatment. The 310s stainless steel however, showed another distinct drop in yield strength. The yield strength of the 310s stainless steel dropped to 619 MPa, about half the strength of the as-received condition. Figure 2(D) shows the effects of heating the substrates at 800 °C. Surprisingly, Hastelloy C276 was affected by this annealing temperature causing a 240 MPa drop in yield from the as received condition. 310s stainless steel yield strength dropped to 510 MPa. One potential reason the Hastelloy C276 experience some weakening is possibly due to the 800 °C heat treatment is close the maximum precipitation temperature of 871 C [5]. Slight variation can also be seen in figure 2 (C-D) for Hastelloy C276 yield strength but with similar elastic modulus. This scatter in data has been previously reported in Clickner et. al [5] between batches of Hastelloy C276 substrate.

Figure 3 shown below, gives an average trend of the samples tested in each of the four conditions. It is immediately apparent that the 310s stainless steel is unmistakably affected by temperature excursions during the REBCO manufacturing process. Hastelloy C276 on the other hand is weakly effected by these temperatures in terms of yield strength. Interestingly, Hastelloy C276 had a strengthening effect on the elastic modulus from the 700 and 750 °C heat treatment. It can also been seen that the 800 °C annealing on the 310s stainless steel gave it a slight curvature to the modulus. Barth et. al. [7] showed a similar rounding or curvature effect in REBCO coated conductors that utilized 310s stainless steel conductors.

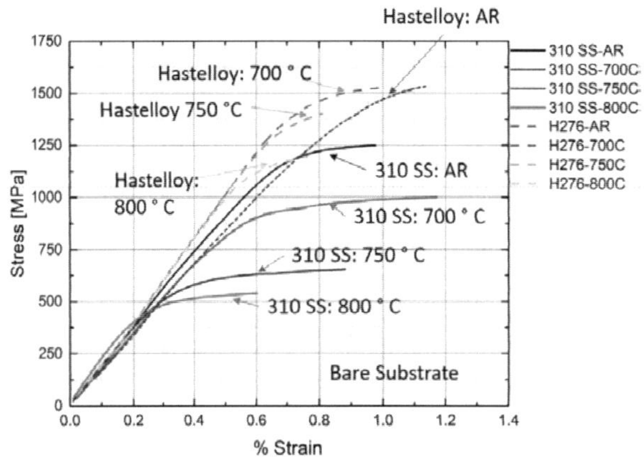

Figure 3: Average stress-strain trend of as received and heat treated 310s stainless steel and Hastelloy C276 bare substrate tested at 293 K [9].

The yield strength vs heat-treat temperature can be seen in figure 4 for both substrates. The trend of strengthening of the Hastelloy C276 and weakening of the 310s stainless steel becomes more apparent. When the 310s stainless steel is compared to the Hastelloy C276, the difference of yield strength is striking. Over the range of temperatures, Hastelloy C276 outcompetes with the 310s stainless steel. The trend of which the 310s stainless steel weakens is similar to previously reported effects in the ASM handbook of stainless steels [5] and in Shi. L et al [8].

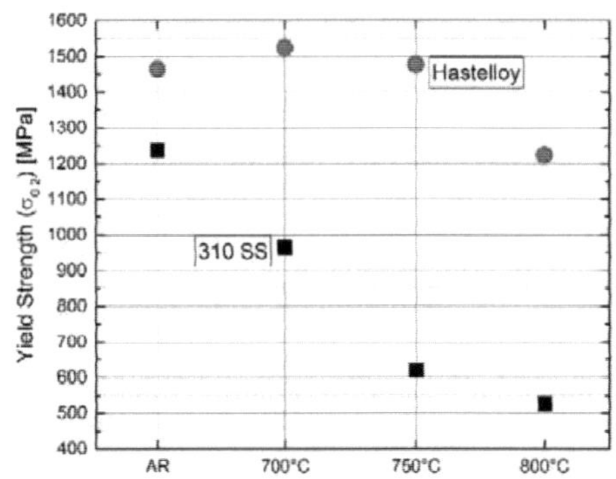

Figure 4: Yield strength vs heat treatment on 310s stainless steel (black square) and Hastelloy C276 (red circle) [9].

5. Conclusion

Simulating the temperatures that could be seen during the manufacturing of REBCO coated conductors shows how two common types of substrates are affected. 310s stainless steel being a high nickel and chromium alloy is no match in terms of retaining strength to Hastelloy C276. Hastelloy C276 suffered no drastic weakening effect, even though after the 800 °C heat treatment it was approximately the same yield strength of the 310s stainless steel in its initial cold-worked state. Substrates used for REBCO coated conductors should display no changes in mechanical properties after manufacturing to ensure strain tolerance of the coated conductor is ideal [10].

6. References

[1] T. Nagaishi, Y. Shingai, M. Konishi, T. Taneda, H. Ota, G. Honda, T. Kato, K. Ohmatsu, Development of REBCO coated conductors on textured metallic substrates, Physica C: Superconductivity, Volume 469, Issues 15–20, 2009, Pages 1311-1315,

[2] S. Gnanarajan and N. Savvides,\Dual ion beam assisted magnetron deposition of biaxially textured ysz and ybco/ysz thin _lms," Surface and Coatings Technology, vol. 305, pp. 116 { 122, 2016.

[3] ASTM International. *E8/E8M-16a Standard Test Methods for Tension Testing of Metallic Materials.* West Conshohocken, PA, 2016. Web. 17 Jul 2019. <https://doi.org/10.1520/E0008_E0008M-16A>

[4] J. R. Davis, "Low-Temperature Properties." ASM Specialty Handbook: Stainless Steels, 1994, pp. 495-504

[5] R. B. Leonard, \Thermal stability of hastelloy alloy c-276," CORROSION, vol. 25, no. 5, pp. 222{232, 1969. [Online]. Available: https://doi.org/10.5006/0010-9312-25.5.222

[6] C.C. Clickner, J.W. Ekin, N. Cheggour, C.L.H. Thieme, Y. Qiao, Y.-Y. Xie, A. Goyal, Mechanical properties of pure Ni and Ni-alloy substrate materials for Y–Ba–Cu–O coated superconductors, Cryogenics, Volume 46, Issue 6, 2006, Pages 432-438.

[7] Barth, C., Mondonico, G., and Senatore, C., Electro-mechanical properties of REBCO coated conductors from various industrial manufacturers at 77 K, self-field and 4.2, 19 T, IOP publishing, Volume 28.

[8] L. Shi, D.O. Northwood, The mechanical behaviour of an AISI type 310 stainless steel,Acta Metallurgica et Materialia, Volume 43, Issue 2, 1995, pp. 453-460,

[9] K. Radcliff, 2017, Mechanical Properties of SuperPower and SuNAM REBCO Coated Conductor, Master Thesis, Florida State University, Tallahassee.

[10] R. P. Walsh, D. McRae, W. D. Markiewicz, J. Lu and V. J. Toplosky, "The 77-K Stress and Strain Dependence of the Critical Current of YBCO Coated Conductors and Lap Joints," in *IEEE Transactions on Applied Superconductivity*, vol. 22, no. 1, pp. 8400406-8400406, Feb. 2012, Art no. 8400406.

Acknowledgments

This work was supported by the National High Magnetic Field Laboratory (which is supported by the National Science Foundation under NSF/DMR-1157490/1664779), the State of Florida. A part of work by S. Hahn was supported by the National Research Foundation of Korea as a part of Mid-Career Research Program (No. 2018R1A2B3009249).

Improvement of in-field performance for EuBCO with heavily doped BHO coated conductors by PLD method with high temperature deposition and low temperature annealing

A Ibi[1], T Machi[1], K Nakaoka[1], M Sato[1], T Izumi[1], J Nishimura[2], M Miura[2], D Yokoe[3], T Kato[3], T Kato[3], T Hirayama[3]

[1]National Institute of Advanced Industrial Science and Technology (AIST), 1-2-1 Namiki, Tsukuba, Ibaraki 305-8564, Japan
[2]Seikei University, 3-3-1 Kichijoji-Kitamachi, Musashino-shi, Tokyo 180-8633, Japan
[3]Japan Fine Ceramics Center (JFCC), 2-4-1 Mutsuno, Atsuta-ku, Nagoya, 456-8587, Japan

akira-ibi@aist.go.jp

Abstract. Recently, we have found that $BaHfO_3$ (BHO)-doped $EuBa_2Cu_3O_{7-X}$ (EuBCO) coated conductors by the combination of the IBAD and PLD methods show high critical current (I_c) even in an applied magnetic field. However, for the wide application of $BaMO_3$ (BMO, M: metal)-doped $REBa_2Cu_3O_{7-X}$ (REBCO) coated conductors to industrial and commercial applications, much higher in-field performance is required. It is known that the critical temperature (T_c) of BMO-doped REBCO layers, especially by the PLD method, decreases with the increase in the amount of doped BMO apparently due to the strain of the REBCO induced by BMO doping. Therefore, it is difficult to improve the critical current density (J_c) in the applied magnetic field of BMO doped REBCO coated conductors only by increasing the quantity of BMO especially at high temperatures such as 77 K. To solve this problem, we tried to optimize the deposition conditions, especially the deposition temperature and O_2 annealing processes for heavily BHO doped-EuBCO layers fabricated by the PLD method. As a result, the combination of high temperature deposition and low temperature O_2 annealing was effective in obtaining high T_c and high in-field performance of heavily BMO-doped REBCO coated conductors. The T_c of 10 mol% BHO-doped EuBCO coated conductors was 93.9 K (setting a deposition temperature of 1150 ° C and O_2 annealing temperature of 280 ° C) which is nearly the same as that for non-doped EuBCO coated conductors. On the other hand, over-doping is preferred for high in-field J_c. Therefore, a high J_c under a magnetic field was obtained in the BMO-doped REBCO layer annealed at a low temperature. The $J_{c\ (min.)}$ of 5 mol% BHO-doped EuBCO coated conductors was 0.62 MA/cm^2 at 77 K and 3 T (setting deposition temperature of 1150 ° C and O_2 annealing temperature of 250 ° C). Using these results, we confirm the successful fabrication of heavily BHO-doped EuBCO coated conductors showing high in-field performance by the PLD method.

1. Introduction

Long artificial pinning centers (APCs)-doped $REBa_2Cu_3O_{7-X}$ (REBCO, RE: rare earth element) coated conductors are being developed for industrial and commercial applications. Recently, APCs-doped and non-doped REBCO coated conductors having length ranging from several hundred meters to several kilometers are being commercialized by many companies around the world. We have been also investigated $BaMO_3$ (BMO, M: metal) doped and non-doped long REBCO coated conductors by the combination of the ion-beam assisted deposition (IBAD) [1, 2] and pulsed laser deposition (PLD) methods [3-16]. In recent years, it has been found that $BaHfO_3$ (BHO)-doped $EuBa_2Cu_3O_{7-X}$ (EuBCO) coated conductors by the IBAD/PLD method is effective for obtaining high in-field performance. The BHO-doped EuBCO coated conductor shows a high critical current (I_c) of 144 A/cm-W at 77 K in 3 T and 411 A/cm-W at 65 K in 3 T [17]. Especially, we optimized the growth conditions of BHO-doped EuBCO layers by which the mode of EuBCO layer growth changed from the vapor-solid (VS) mode to the vapor-liquid-solid (VLS) mode in order to deposit BHO-doped EuBCO layers having high superconducting characteristics [18]. Furthermore, we can fabricate 100 - 200 m long BHO-doped EuBCO layers at high production rate and maintaining high in-field performance over the entire length [19]. However, the I_c values in the magnetic field at high temperature of long BMO-doped REBCO coated conductors need to be further improved for industrial and commercial applications.

As a candidate, it is thought to be effective in increasing the doping amount of BMO for further improvement of the in-field performance of BMO-doped REBCO coated conductors. However, it is known that the critical temperature (T_c) of REBCO layers decreases with the increase in the amount of BMO for the vapor deposition process including the PLD method because of the strain of the REBCO phase by BMO doping [20, 21]. Therefore, it is difficult to improve the critical current density (J_c) in the applied magnetic field of BMO-doped REBCO coated conductors simply by increasing doping amount of BMO. To prevent the reduction of T_c and J_c of the REBCO layers as the quantity of BMO increases, we tried to optimize the deposition conditions, especially the deposition temperature and O_2 annealing processes for BHO-doped EuBCO layers fabricated by the PLD method. As a result, we improved the in-field performance of heavily BHO-doped EuBCO layers by the PLD method with high temperature deposition and low temperature O_2 annealing process.

2. Experimental details

BHO-doped EuBCO layer underwent deposition by a Reel-to-Reel PLD system on 10 mm wide CeO_2 (300 - 700 nm) / $LaMnO_3$ (7 nm) / IBAD-MgO (5 nm) / Y_2O_3 (14 nm) / $Gd_2Zr_2O_7$ (56 nm) / Hastelloy C276 (100 μm) substrates (Fig. 1). The value of in-plane and out-of-plane degrees of texturing ($\Delta\phi$ and $\Delta\omega$) of the CeO_2 cap layer were approximately 1 - 3 ° and 0.5 - 1 ° degrees, respectively. The texture of the CeO_2 and BHO-doped EuBCO layers were measured by the analysis of the X-ray diffraction pattern and X-ray pole figure. The $\Delta\omega$ (full width at half maximum of the X-ray diffraction rocking curve) and $\Delta\phi$ (full width at half maximum from X-ray F-scan of the CeO_2 (220) and BHO-doped EuBCO (103) peak) were applied as the out-of-plane and in-plane texturing degrees, respectively. A 200 W industrial XeCl excimer laser with a wavelength of 308 nm and typical pulse duration (FWHM) of approximately 22 ns was used at a pulse energy in the range of 500 - 600 mJ (the size of the laser beam is approximately 420 mm^2) and a pulse repetition rate of 177 Hz. The optical system is synchronized to the laser pulse and scanning of the laser beam on a sintered a BHO-doped EuBCO target was also controlled. The laser repetition rate at 177 Hz is divided into 4 plumes which are nearly identical in size and shape (Fig. 2). The laser energy density directly above the sintered the BHO-doped EuBCO target is approximately 2 - 3 J/cm^2. To realize the VLS growth mode for BHO-doped EuBCO layers, we used the commercially available sintered target of off-stoichiometric $EuBa_2Cu_{3.2}O_X$ containing 3.5 - 10 mol% doped BHO with a diameter of 6 inches and a thickness of 5 mm for BHO-doped EuBCO layer depositions. The amount of BHO doped is 3.5, 5, 7.5 and 10 mol% correspondings to approximately 1.5 vol%, 2.1 vol%, 3.3 vol% and 4.4 vol%, respectively. Typically, the pulse repetition rate of 177 Hz and pulse energy of 500 - 600 mJ cause a deposition rate of approxiamtely12 nm/s for a BHO-doped EuBCO layer. The setting deposition temperatures were 1035 - 1150 ° C for a substrate with a transfer speed of 30 m/h. The setting deposition temperatures were determined from the surface temperature of the CeO_2 layer measured using a K-type thermocouple. The surface temperature of the CeO_2 layer was approxmately 880 ° C when the setting deposition temperature was 1095 ° C. The oxygen pressure was maintained at 600 mTorr (80 Pa) with a flow of 10 sccm (standard cubic centimeter per minute) oxygen. The T-S (distance between the sintered BHO-doped EuBCO target and the substrate) was set at 90 - 100 mm. The typical thickness value of a BHO-doped EuBCO layer was in the range of 0.50 – 0.55 μm for one time deposition with a substrate transfer speed of 30 m/h. The I_c value was measured by the conventional four-terminal method with a criterion of 1 μV/cm. The J_c value of the BHO-doped EuBCO coated conductors was calculated from the cross-sectional area of a BHO-doped EuBCO layer. The thickness and cross-section images of the BHO-doped EuBCO layer were observed using a scanning electron microscope (SEM) and a transmission electron microscope (TEM). Samples several centimeters in length were cut out from the ends of long BHO-doped EuBCO coated conductors to examine the angular dependence of I_c values on the applied magnetic field angle, which was changed from 0 to 180 ° to the *c*-axis (180 ° corresponds to a magnetic field parallel to the *c*-axis direction).

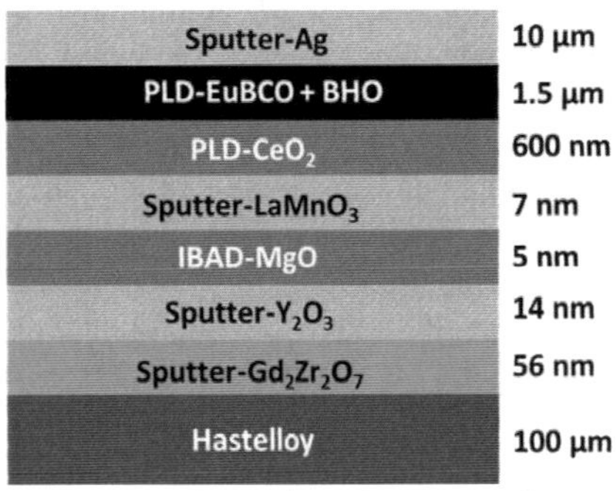

Figure 1. Schematic diagram of our BHO-doped EuBCO coated conductors by IBAD/PLD method.

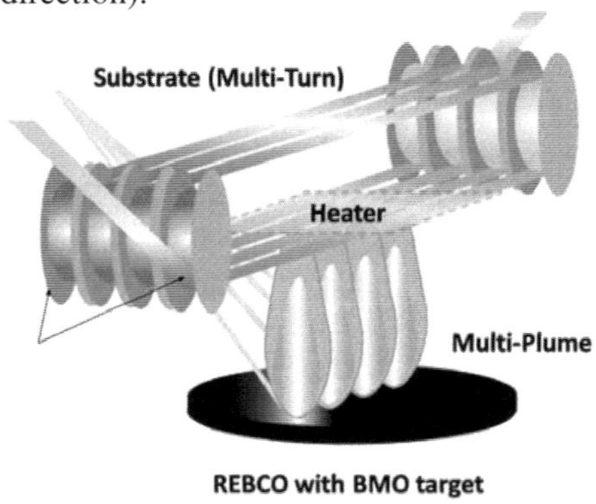

Figure 2. Schematic diagram of our multi-plume and multi-turn (MPMT) PLD method.

3. Results and discussion

We fabricated 10 mol% BHO-doped EuBCO coated conductors at several different deposition temperatures by the PLD method. Figure 3 shows the TEM cross-section images of 10 mol% BHO-doped EuBCO coated conductors that underwent (a) low temperature (1095 ° C) deposition and (b) high temperature (1135 ° C) deposition. Although both have nearly the same structure, the BHO-doped EuBCO coated conductor obtained through low temperature deposition seems to have a slight decrease in the straightness of the distribution of BHO nano-rods compared to a BHO-doped EuBCO coated conductor at high temperature deposition. The difference in the distribution of BHO nano-rods is considered to be due to the deposition temperature [22]. The EuBCO and BHO phases grow at the same interface. The BHO nano-rod is formed due to the interfacial energy between the EuBCO and BHO. If the interfacial energy is large enough comparing with the supersaturation, which is the driving force of the overall growth, the BHO phase should be continually grown. On the other hand, if the interfacial energy is relatively small, the degree of continuity and straightness of the nano-rod should deteriorate. Because the growth at a low temperature means high supersaturation, which might lead to the latter ease, the short rods are tilted in the film grown at a low temperature.

Figure 4 shows the setting deposition temperature dependence of (a) T_c and I_c at 77 K in a self-field and (b) T_c and *c*-axis length of 10 mol% BHO-doped EuBCO coated conductors. The 10 mol% BHO-doped EuBCO coated conductors showed a high T_c of 93.9 K. The T_c and *c*-axis length of 10 mol% BHO-doped EuBCO coated conductors were improved by high temperature deposition. It is considered that EuBCO phase is stretched by the strain between EuBCO phase and BHO rods as a cause of low T_c and long *c*-axis length at low temperature deposition. Figure 5 shows the schematic diagram of the reason for the improvement of T_c and *c*-axis length. The strain in REBCO phase surrounding the BMO rods is suppressed by high temperature deposition, since the plasticities of both phases may the higher at high temperatures. The decrease in T_c is attributed to be distortion of the

REBCO phase in the vicinity of the BMO phase due to the difference in lattice constant between the REBCO and BMO phases. It is thought that strain regions are reduced by high temperature deposition.

We fabricated 3.5, 5, 7.5 and 10 mol% BHO-doped EuBCO coated conductors by the PLD method with a high deposition temperature of 1150 ° C. Additionally, we tried to anneal it for oxygen doping at a low temperature of 250 ° C. Figure 6 shows the schematic diagram of our O_2 annealing profile for BHO-doped EuBCO coated conductors by the PLD method. Figure 7 shows the effect of doping on BHO dependence of J_c at 77 K in the self-field of BHO-doped EuBCO coated conductors. Critical currents were measured by the conventional four-terminal method in laser patterned surface of BHO-doped EuBCO layers with bridges 10 - 20 μm in width. J_c values of EuBCO with a doping of up to 5 mol% BHO coated conductors were improved to the same value as non-doped EuBCO coated conductor by low temperature O_2 annealing. It is expected that the EuBCO phase is in an over-doped state because the J_c value increases although T_c is decreases by low temperature O_2 annealing.

From these results, high temperature deposition and low temperature O_2 annealing are effective in improving the T_c and J_c at 77 K in the self-field for BMO-doped REBCO coated conductors by the PLD method.

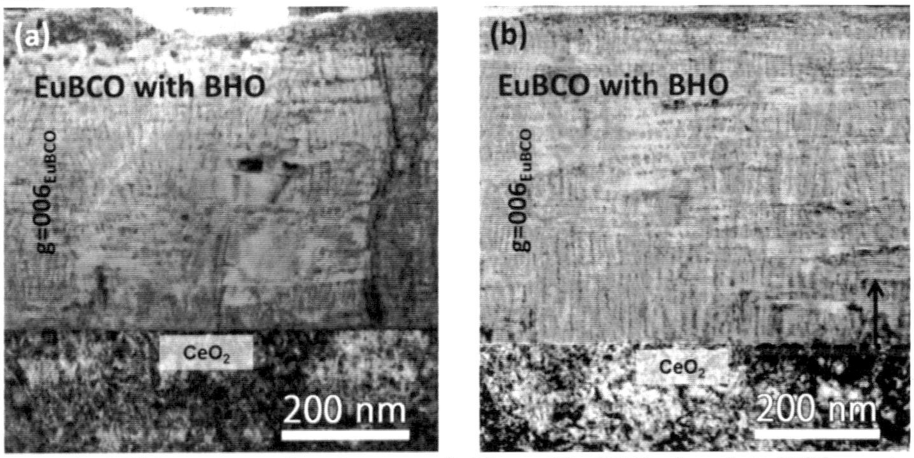

Figure 3. TEM cross-section images of 10 mol% BHO doped EuBCO coated conductors with (a) low temperature (1095 ° C) deposition and (b) high temperature (1135 ° C) deposition. Although both have nearly the same structure, BHO doped EuBCO coated conductor with low temperature deposition seems to have a slight decrease in straightness of the BHO nano-rod distribution compared to BHO-doped EuBCO coated conductor with high temperature deposition.

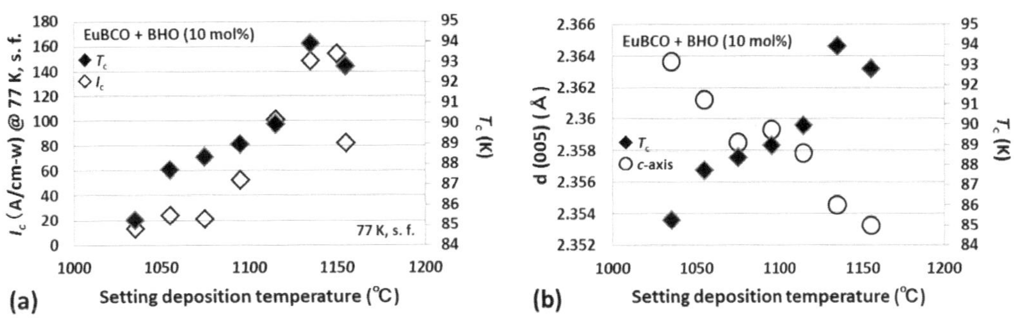

Figure 4. Setting deposition temperature dependence of (a) T_c and I_c at 77 K in the self-field and (b) T_c and c-axis length of 10 mol% BHO-doped EuBCO coated conductors. The 10 mol% BHO-doped EuBCO coated conductors showed a high T_c of 93.9 K by high temperature deposition (setting deposition temperature of 1150 ° C and O_2 annealing temperature of 280 ° C), which is almost same as a non-doped EuBCO coated conductors.

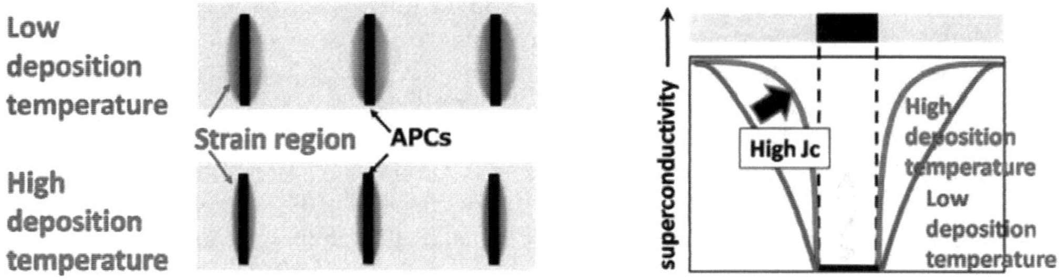

Figure 5. Schematic diagram of improved suppression of strain regions between BMO (APCs) doped REBCO phase by high temperature deposition. The cause of T_c decrease is considered to be the distortion of the REBCO phase in the vicinity of the BMO phase due to the difference in lattice constant between these phases. It is thought that strain regions are reduced by high temperature deposition.

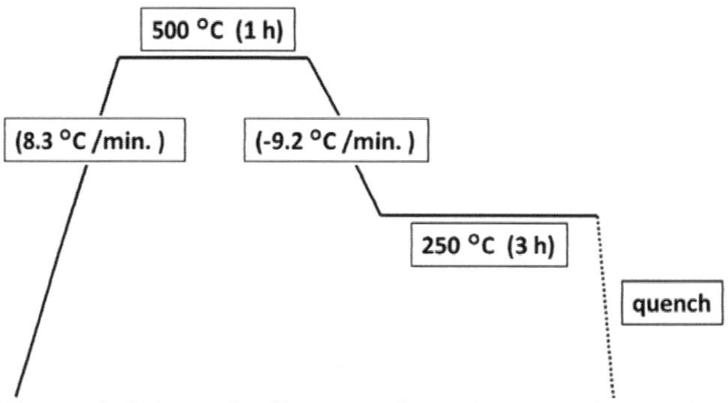

Figure 6. Schematic diagram of our O_2 annealing profile for BHO doped EuBCO coated conduictors by PLD method.

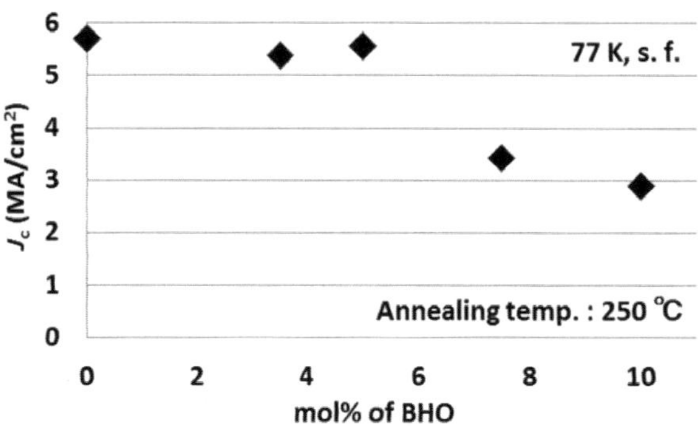

Figure 7. Plot of J_c vs. quantity of BHO at 77 K in self-field of BHO doped EuBCO coated conductors by PLD method. J_c values of EuBCO with a doping of up to 5 mol% BHO coated conductors were improved to the same value as non-doped EuBCO coated conductor by low temperature O_2 annealing.

On the other hand, we examined the effect on the improvement of high superconducting characteristics of BMO-doped REBCO coated conductors by the PLD method with low temperature

O_2 annealing only. Figure 8 shows the setting deposition temperature dependence of I_c at 77 K in the self-field of 5 mol% BHO-doped EuBCO coated conductors by the PLD method with the low temperature of O_2 annealing. It can be considered that the T_c value is improved by reducing the strain at the interface between the EuBCO and BHO phases due to high temperature deposition. Furthermore, the over-doped state is realized because the O_2 is easily diffused into the BHO-doped EuBCO phase during the annealing process thereby reducing the strain. Therefore, heavily BHO-doped EuBCO coated conductors does not obtain high superconducting characteristics by high temperature deposition and low temperature O_2 annealing alone.

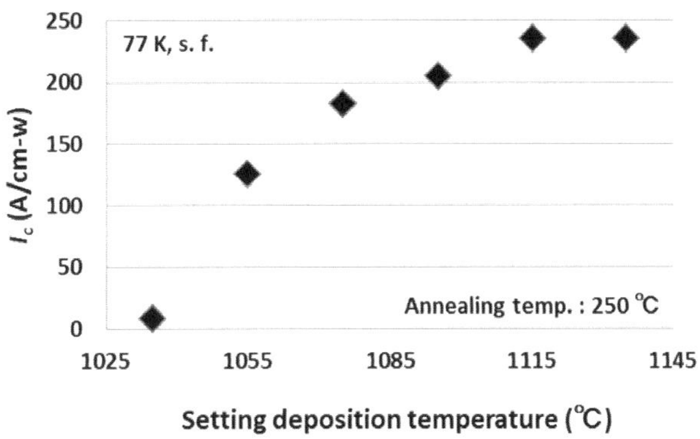

Figure 8. Plot of I_c vs. setting deposition temperature at 77 K in the self-field of 5 mol% BHO-doped EuBCO coated conductors by PLD method with low temperature annealing. Good in-field performance heavily BHO-doped EuBCO coated conductors is not obtained by high temperature deposition and low temperature annealing alone.

We fabricated 3.5, 5, 7.5 and 10 mol% BHO-doped EuBCO coated conductors by the PLD method with a high setting deposition temperature of 1150 ° C and low O_2 annealing temperature of 250 ° C. Figure 9 indicates (a) the magnetic field dependence of J_c for 3.5, 5, 7.5 and 10 mol% BHO-doped EuBCO coated conductors and (b) the magnetic field angular dependence of J_c at 77 K in 3 T for 3.5, 5, 7.5 and 10 mol% BHO-doped EuBCO coated conductors by the PLD method with laser patterned surface of BHO-doped EuBCO layers with bridges 10 - 20 μm in width. The J_c (min.) of 5 mol% BHO-doped EuBCO coated conductor was approximately 0.62 MA/cm^2 at 77 K in 3 T. Clear improvement in the in-field performance for 5 mol% BHO-doped EuBCO coated conductors can be recognized in J_c-B-θ properties too by high temperature deposition and low temperature O_2 annealing. It seems that the J_c-B-θ properties depending on the quantity of BHO and manner of distribution of BHO nano-rods in the EuBCO layer. However, the correlation and trend between the amount of BHO doping and J_c-B-θ properties have not yet been clarified. It should be discussed based on further microstructural analysis.

Although the T_c of 10 mol% BHO-doped EuBCO coated conductor was improved (from Fig. 4) by using high temperature deposition and low temperature O_2 annealing, and it is considered that the improvement of superconducting characteristics is locally achieved because the J_c at 77 K and the self-field decreases as the proportion of doped BHO increases (from Fig. 7). The in-field performance were improved since the pinning effect is stronger in the relatively rightly BHO-doped (3.5 -5 mol%) EuBCO coated conductors, but it is thought that the in-field performance also deteriorated in the EuBCO with heavily BHO-doped (7.5 - 10 mol%) EuBCO coated conductors. Therefore, to improve the J_c at self and magnetic field of more heavily BMO-doped REBCO coated conductors, it is necessary to improve both high temperature deposition and low temperature O_2 annealing process.

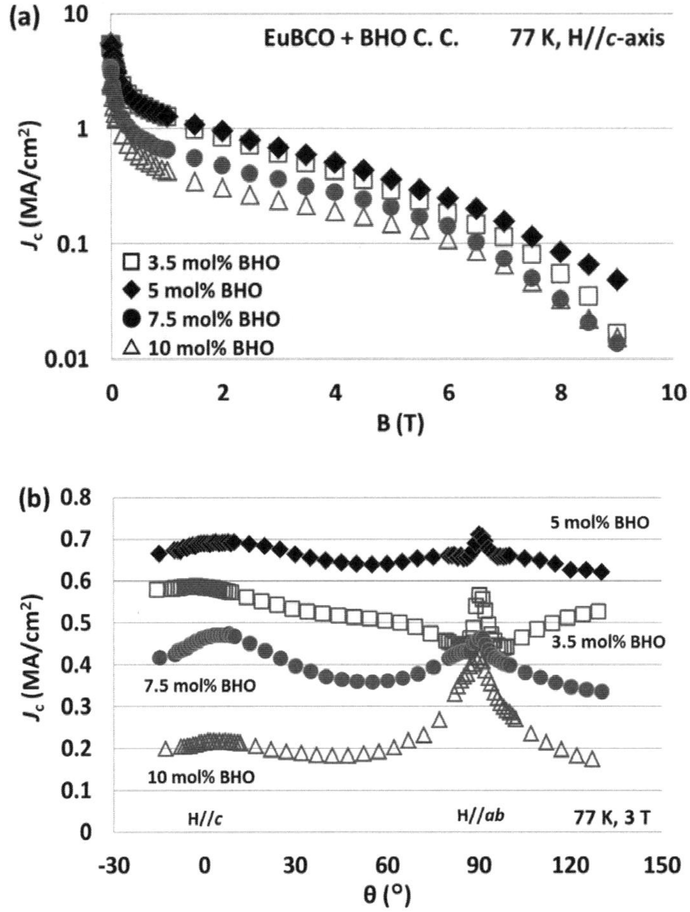

Figure 9. (a) magnetic field dependence of J_c for 3.5, 5, 7.5 and 10 mol% BHO-doped EuBCO coated conductors and (b) magnetic field angular dependence of J_c at 77 K in 3 T of 3.5, 5, 7.5 and 10 mol% BHO-doped EuBCO coated conductors by PLD method. The $J_{c\ (min.)}$ of 5 mol% BHO-doped EuBCO coated conductor was about 0.62 MA/cm^2 at 77 K in 3 T.

4. Conclusions

We tried to optimize the fabricating conditions in the PLD method, especially the deposition temperature and O_2 annealing processes to reduce the reduction of T_c and J_c of REBCO layers with simply increasing the amount of BMO because of the strain of the REBCO by doping of BMO. We fabricated 3.5, 5, 7.5 and 10 mol% BHO-doped EuBCO coated conductors by the PLD method with high setting deposition temperature of 1150 ° C and low O_2 annealing temperature of 250 ° C. Conventionally, 3.5 mol% BHO-doped EuBCO coated conductors showed the highest in-field performance in our laboratory. However, by using the high temperature deposition and low temperature annealing processes, the T_c of 10 mol% BHO-doped EuBCO was improved to almost the same value of 93.9 K as pure-EuBCO (non BHO-doped EuBCO) coated conductors and the 5 mol% BHO-doped EuBCO coated conductor had the highest J_c (min.) at 77 K in 3 T of approxmately 0.62 MA/cm^2. The cause of T_c decrease is considered to be the distortion of the REBCO phase in the vicinity of BMO phase due to the difference in lattice constant between the REBCO and BMO phases, resulting is a situation in which oxygen finds it difficult to enter the REBCO phase. It is thought that these strain regions are reduced by high temperature deposition and REBCO is in an over-doped state by low temperature O_2 annealing. From these results, we improved the in-field performance of heavily BMO-doped REBCO using the PLD method with high temperature deposition and low temperature O_2 annealing.

Acknowledgments
This work is supported by the New Energy and Industrial Technology Development Organization (NEDO), Advanced Medical Services from the Japan Agency for Medical Research and development (AMED) and Ministry of Economy, Trade and Industry (METI).

We would like to thank Editage (www.editage.com) for English language editing.

5. References

[1] Iijima Y, Tanabe N, Kohno O, Ikeno Y 1992 *Appl. Phys. Lett.* 60 769
[2] Miyata S, Muroga T, Iwai H, Ibi A, Watanabe T, Yamada Y, Shiohara Y, Kato T, Hirayama T 2005 *IEEE Trans. Appl. Supercond.* 15 2671
[3] Sievers S, Mattheis F, Krebs H U, Freyhardt H C 1995 *J. Appl. Phys.* 78 5546
[4] Iwai H, Muroga T, Watanabe T, Miyata S, Ibi A, Yamada Y, Shiohara Y, Kato T, Hirayama T 2004 *J. Japan Inst. Met.* 68 718
[5] Goyal A, Kang S, Leonard K J, Martin P M, Gapud A A, Varela M, Paranthaman M, Ijaduola A D, Specht E D, Thompson J R, Christen K, Pennycook S J, List F A 2005 *Supercond. Sci. Technol.* 18 1533
[6] Matsumoto K, Horide T, Mele P, Yoshida Y, Mukaida M, Ichinose A, Horii S 2005 *Physica C* 426-431 1091
[7] Yamada Y, Takahashi K, Kobayashi H, Konishi M, Watanabe T, Ibi A, Muroga T, Miyata S 2005 *Appl. Phys. Lett.* 87 132502
[8] Ibi A, Fukushima H, Kuriki R, Miyata S, Takahashi K, Kobayashi H, Konishi M, Watanabe T, Yamada Y, Shiohara Y 2006 *Physica C* 445-448 525
[9] Takahashi K, Kobayashi H, Yamada Y, Ibi A, Fukushima H, Konishi M, Miyata S, Shiohara Y, Kato T, Hirayama T 2006 *Supercond. Sci. Technol.* 19 924
[10] Ibi A, Fukushima H, Yamada Y, Miyata S, Kuriki R, Takahashi K, Shiohara Y 2006 *Supercond. Sci. Technol.* 19 1229
[11] Ibi A, Miyata S, Kuriki R, Kobayashi H, Fukushima H, Kinoshita A, Kato T, Hitrayama T, Yamada Y, Shiohara Y 2008 *Physica C* 468 1514
[12] Chikumoto N, Lee S, Nakao K, Tanabe K 2009 *Physica C* 469 1303
[13] Lee S, Chikumoto N, Yokoyama T, Machi T, Kakao K, Tanabe K 2009 *IEEE Trans. Appl. Supercond.* 19 3192
[14] Kakimoto K, Igarashi M, Hanyu S, Sutoh Y, Takemoto T, Hayashida T, Handa Y, Nakamura N, Kikutake R, Kutami H, Iijima Y, Saito 2011 *Physica C* 471 929
[15] Tobita H, Notoh K, Higashikawa K, Inoue M, Kiss T, Kato T, Hirayama T, Yoshizumi M, Izumi T, Shiohara Y 2012 *Supercond. Sci. Technol.* 25 062002
[16] Miura M, Maiorov B, Sato M, Kanai M, Kato T, Kato T, Izumi T, Awaji S, Mele P, Kikuchi M, Matsushita T 2017 *NPG Asia Materials* 9 e447
[17] Yoshida T, Ibi A, Takahashi T, Yoshizumi M, Izumi T, Shiohara Y 2014 *Physica C: Superconductivity* 504 42
[18] Ibi A, Yoshida T, Izumi T, Shiohara Y, Yokoe D, Kato T, Hirayama T 2016 *Physics Procedia* 81 97
[19] Ibi A, Machi T, Kakaoka K, Izumi T 2018 *Journal of Physics: Conf. Series* 1054 012032
[20] Gurevich A, Pashitskii E A 1997 *Phys. Rev. B* 56 6213
[21] Cantoni C, Gao Y, Wee S H, Specht E D, Gazquez J, Meng J, Pennycook S J, Goyal A 2011 *ACSNANO* 5 4783
[22] Miura S, Yoshida Y, Ichino Y, Ichinose A, Matsumoto K, Awaji S 2014 *Journal of Physics: Conference Series* 507 022021

IOP Conf. Series: Materials Science and Engineering **756** (2020) 012025 doi:10.1088/1757-899X/756/1/012025

The effect of APC/YBCO interface on the angular range of effective pinning by one-dimensional artificial pinning centers in $YBa_2Cu_3O_{7-x}$ nanocomposite films

V Ogunjimi[1], B Gautam[1], M. A. Sebastian[2,3], T Haugan[2], J Wu[1]

[1]*Department of Physics and Astronomy, the University of Kansas, Lawrence, Kansas 66045, USA*

[2]*U.S Air Force Research Laboratory, Aerospace Systems Directorate, WPAFB, OH 45433, USA*

[3]*University of Dayton Research Institute, Dayton, OH 45469, USA*

E-mail: jwu@ku.edu; and victorogunjimi@ku.edu

Abstract. The APC/YBCO interface has been reported to directly affect the pinning efficiency of 1D APCs at *B*//*c*-axis. This raises a question on how the APC/YBCO interface affects angular range of the pinning effectiveness for a given 1D APC. In this work, two types of 1D APCs of different APC/YBCO interfaces and hence pinning efficiencies were selected to understand the correlation of the pinning efficiency at *B*//*c*-axis and the angular range of the effectiveness. Specifically, $BaZrO_3$ (BZO) and $BaHfO_3$ (BHO) 1D APCs were selected for a comparative study in the APC/YBCO nanocomposite films. The BZO and BHO 1D APCs have comparable diameters in the range of 5-6 nm. In the doping range of 2-6 vol.%, both BZO and BHO form *c*-axis aligned 1D APCs in YBCO films. However, differences are present at their interfaces with YBCO. While the BZO/YBCO interface is semi-coherent, a coherent BHO/YBCO interface has been found to be critical to the higher pinning efficiency of the BHO 1D APCs. Therefore, they provide ideal systems for investigation of the angular range of pinning effectiveness by 1D APCs. By evaluating the nanocomposites' maximum pinning force density ($F_{p,\,max}$) and its location B_{max}, normalized to that of the reference YBCO film as functions of magnetic field (*B*) orientation at temperatures of 65–77 K, a quantitative correlation between the pinning efficiency of the BZO 1D-APCs and their effective angular range was obtained. Our results indicate that all 1D APCs can provide enhanced B_{max} over certain angular ranges away from the *c*-axis. However, 1D APCs with higher pinning efficiency, such as BHO 1D APCs can have enhanced $F_{p,\,max}$ over the entire angular range of *B*-orientations at temperatures of 65-77 K with respect to that for the reference YBCO sample.

1. Introduction

Nanoscale artificial pinning centers (APCs) in $YBa_2Cu_3O_{7-x}$ (YBCO) and other high T_c superconducting films on various substrates have been reported to enhance magnetic vortex pinning and hence critical current density J_c in applied magnetic fields (*B*) [1-7]. Most APCs form through addition of secondary dopants like $BaZrO_3$ (BZO), $BaHfO_3$ (BHO) and other insulator oxides that aggregate into APCs via strain mediated self-assembly in APC/YBCO nanocomposites during growth [8]. In particular, *c*-axis aligned one-dimensional APCs (1D-APCs) have been reported to provide strong correlated pinning in YBCO nanocomposites at *B*//*c*-axis [5-7, 9-18]. Since 1D-APCs are expected to be effective within a limited angular range around *B*//*c*-axis, investigating how this angular range is affected by APC materials is crucial to achieving an optimal strong and *B*-field orientation-independent pinning landscape for practical applications.

BZO and BHO both form *c*-axis aligned 1D-APCs in YBCO nanocomposites with comparable diameters in the range of 5-6 nm. However, the BZO 1D-APC/YBCO and BHO 1D-APC/YBCO interfaces differ considerably due to the difference in the lattice mismatches and elastic properties of the materials involved, which result in differences in the microstructures and pinning efficiencies of the two types of the 1D-APCs [19, 20]. Specifically, the BZO 1D-APC/YBCO interface is semi-coherent with a high concentration of defects such as dislocations in a YBCO column of a few nm in thickness around the BZO 1D-APC [21]. While the local strain is significantly reduced in the column with respect to the theoretical values based on the BZO/YBCO lattice mismatch, oxygen deficiency has been revealed as a consequence of the defects. Therefore, a lower interface T_c in this YBCO column as compared to the global T_c of the nanocomposite can be projected, which can lead to a reduced pinning potential across the BZO 1D-APC/YBCO interface. In contrast, the BHO 1D-APC/YBCO interface can be made approximately coherent with much reduced defect concentration and hence less degraded interface T_c in the YBCO column around the BHO 1D-APC [20]. This explains the significantly higher pinning efficiency of the BHO 1D-APC as compared to its BZO counterpart at *B*//*c*-axis in terms of high pinning force density ($F_p(B) = J_c \times B$), especially the peak value ($F_{p,\,max}$) and its location (B_{max}) [20].

Considering the correlated pinning by *c*-axis aligned 1D-APCs is typically illustrated as a peak of J_c as function of the *B* orientation in the plane perpendicular to the J_c, defined using angle θ with θ=0 for *B*//*c*-axis *and* θ=90° for *B*//*ab*-plane, a question arises on how the angular range of the 1D-APC pinning correlates with the pinning efficiency of the 1D-APCs. In fact, many previous studies have investigated $J_c(\theta)$ on various nanocomposites including BZO and BHO ones [22-34]. This work adopts a slightly different approach by measuring $J_c(B)$ and $F_p(B)$ at θ angles. This allows extraction of the $F_{p,\,max}(\theta)$ and $B_{max}(\theta)$ on BZO and BHO 1D-APC/YBCO nanocomposite films. By plotting these values normalized to that of reference YBCO film, we aim to shed light on the effect of the 1D-APC pinning efficiency on the angular range the correlated pinning extends.

2. Experimental

2 vol.% and 4 vol. % singly doped BZO/YBCO and BHO/YBCO nanocomposite films (to be regarded henceforth as 2% BZO, 4% BZO, 2% BHO and 4% BHO) and a reference YBCO film were selected for this study. The samples were fabricated using pulse laser deposition (PLD) technique on (100) $SrTiO_3$ (STO) single crystal substrates under optimal PLD conditions. The fabrication details have been reported elsewhere [20, 34, 35]. Briefly, the substrate temperatures of 825 °C and 810 °C, respectively,

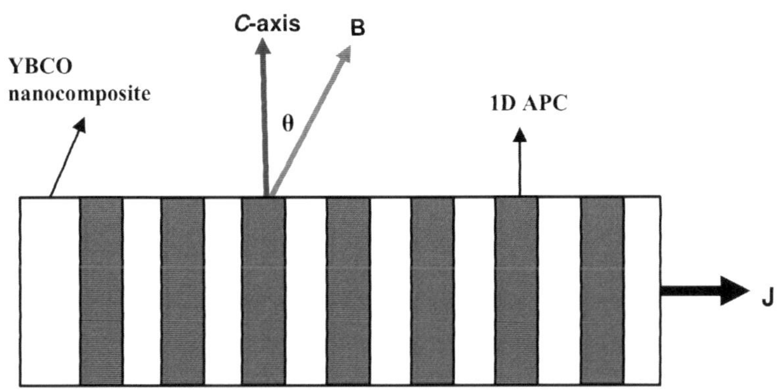

Figure 1. Schematic showing the deviation angle θ of the magnetic field, *B* away from the *c*-axis of APC/YBCO nanocomposite films and the direction of the electric current in the film.

were employed for PLD BZO and BHO 1D-APC/YBCO nanocomposite films in 300 mTorr oxygen using a Lambda Physik LPX 300 KrF excimer laser (248 nm). The laser energy density and repetition frequency were set at 1.6 J/cm^2 and 8.0 Hz respectively. The samples were in situ annealed in oxygen for 30 minutes at 500°C. The sample thickness is around 140 nm and silver contacts were sputtered on all samples after the PLD growth. The crystallinity and microstructure of these samples were reported in our previous works [9, 20, 36, 37]. The samples are all *c*-axis oriented epitaxial films. In all nanocomposite samples, *c*-axis aligned 1D-APCs have been confirmed with BZO 1D-APC diameter in range of 5.2-5.7 nm and BHO 1D-APC diameter around 5.0 nm. The T_c values are 89.27 K, 87.48 K, 89.36 K and 87.40 K for 2% BZO, 4% BZO, 2% BHO and 4% BHO respectively. Two microbridges of length 500 μm and widths of 20 μm and 40 μm were patterned on each sample using photolithography. The samples were wired to pins on a sample stage with 50 μm in diameter platinum wires. This stage rotation was done in the plane perpendicular to the $J_c(B)$ as depicted schematically in Figure 1 at different angles such that the applied B field orientation angle with respect to the *c*-axis, θ, can be varied from $\theta \sim 0°$ to $\theta \sim 90°$. J_c was determined using the 1 μVcm^{-1} criterion Specifically for this work, $J_c(B)$ measurements were carried out in a Physical Property Measurement Systems (Evercool II, Quantum Design) at 65 and 77 K with B up to 9 T and at several selected θ angles between 0° ($B//c$-axis) and 90° ($B//ab$-plane). $F_p(B) = J_c \times B$ curves were then derived from the measured $J_c(B)$ curves at these θ angles for analyses.

3. Results

Figure 2 depicts the F_p (B) curves taken on the reference YBCO (Figures 2a-b), 2% BZO (Figures 2c-d) and 4% BZO SD (Figures 3e-f) nanocomposite samples at different θ angles of 0°, 22°, 37°, 45°, 67° and 90°. These θ angles were selected based on a preliminary test to ensure the difference between the F_p (B) curves at these angles are well distinguishable. It should be noted that the B_{max} values may be above the limit of experimental apparatus of 9.0 T on some F_p (B) curves in this work, such as the 4% BHO/YBCO sample at 65 K. In this case, $B_{\max}$ is assigned the value of 9.0 T as the lower limit. In the reference YBCO sample, the intrinsic pinning by the *ab*-planes provides the strongest pinning at $B//ab$ (θ = 90°) as expected at both 77K and 65 K. This is illustrated in the highest F_p (B) at θ = 90° and decreasing F_p (B) with decreasing θ angles. The 77 K behavior shows the 90° pinning force clearly over performing the other orientations. At 65 K, this trend persists although the difference in F_p (B) curves at different θ angles is narrowed in the B-field range of 0-9 T. In the 2% BZO sample (Figures 2c-d), the addition of the *c*-axis aligned 1D-APCs is demonstrated in the enhanced F_p (B) at θ=0° and other smaller θ angles as compared to that in the reference YBCO, although the F_p (B) at θ =90° remains similar to that of YBCO. Quantitatively, the pinning efficiency of the BZO 1D-APCs differs at different temperatures. At 77 K, the enhancement in F_p is moderate while at 65 K, greater $F_{p,\,max}$ enhancement is seen at 0° and 22°. In contrast, as the BZO doping is increased to 4% (Figure 2e-f), the F_p (B) at these two field orientations outperform that at θ =90° and other angles. At 77 K, the $F_{p,\,max}$ at θ =0° (circle) and 22° (square) ~ 13.3 and ~ 10.8 GN/m^3 respectively compared to only about ~ 6 GN/m^3 or lower for other angles. In addition, due to stronger correlated pinning in this sample its $F_{p,\,max}$ at θ =90° underperforms those at these two orientations. The results indicate that in terms of pinning enhancement relative to the reference YBCO sample, the 4% BZO sample exhibits the best performance. Specifically, for θ < 67°, the pinning enhancement by the BZO 1D APCs in the 2% BZO sample (relative to YBCO) is only illustrated in a shift of the B_{max} to ~ 3.0 T from ~ 2.0 T in the YBCO reference sample. The performance of 4% BZO over 2% BZO may be attributed to the enhanced strain field overlap in the 4% BZO sample. Enhanced strain field overlap results in the formation of larger number of defects density in the 4% BZO. These defects reduce the strain in the BZO/YBCO interface thereby minimizing the formation of oxygen vacancies in the interface.

Figures 3 compares the $F_p(B)$ curves measured at θ angles of 0°, 45° and 90° for 2% BZO 1D-APC/YBCO (solid) and BHO 1D-APC/YBCO (open) nanocomposites at 77 K and 65 K (Figures 3a-3b), and for 4% counterparts at the same temperatures (Figures 3c-3d), respectively. For both 2% and 4% APC doping cases, the $F_{p,\,max}$ at $\theta = 0^o$ of the BHO samples are significantly higher than that of the

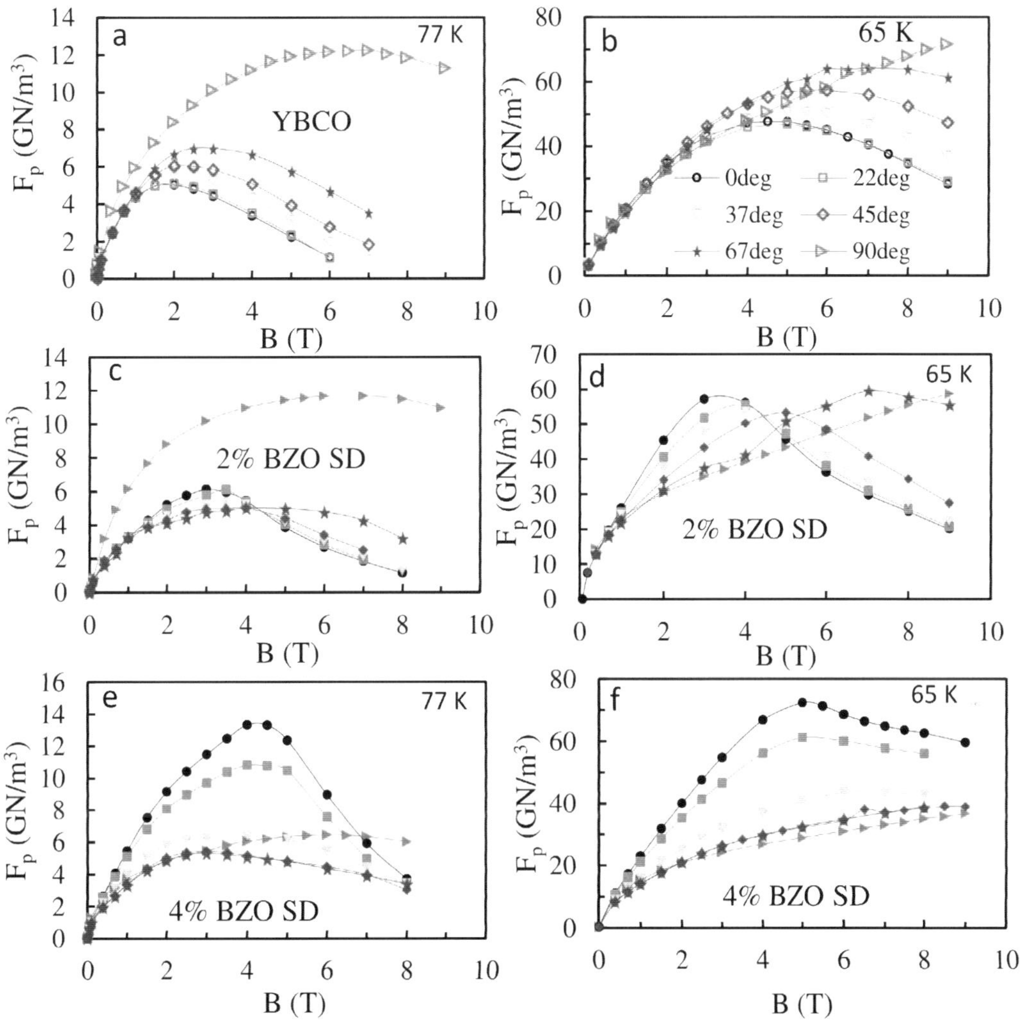

Figure 2. F_p (B) curves for each θ measured at 77 K (solid symbols) and 65 K (open symbols) on 2% BZO SD (a and b), 4% BZO SD (c and d), and undoped YBCO (e and f).

BZO samples. For instance, the $F_{p,\,max}$ values of the 2% BHO sample are 12.9 GN/m^3 and 76.9 GN/m^3 at 77 K and 65 K respectively. These values are significantly higher than the $F_{p,\,max}$ values of 6.1 GN/m^3 at 77 K and 57.1 GN/m^3 at 65 K for the 2% BZO sample. The superior performance in the BHO sample is even more remarkable when the 4% samples are compared at 65 K. At this lower temperature, the $F_{p,\,max}$ value of 182.3 GN/m^3 at $\theta = 0^o$ for the 4% BHO sample is about two and half times the value of its BZO counterpart. This strong correlated pinning by the BHO 1D-APCs may be ascribed to its coherent BHO/YBCO interface [20]. The lower (relative to BZO) lattice mismatch of the BHO with YBCO may ensure less interfacial dislocations and lower oxygen vacancy density around the BHO/YBCO interface. The difference in $F_p(B)$ magnitude of the samples is less dramatic at 45° and 77 K. Nevertheless, the BHO 1D-APCs are still more effective at this orientation and temperature as shown in Figures 3a and 3c. At 65 K, the $F_p(B)$ curve of 4% BHO at $\theta = 45^o$ is consistently higher at all fields than that of 4% BZO (Figure 3d). Interestingly, the $F_{p,\,max} \sim 94.9$ GN/m^3 in the 4% BHO is not only more than double of

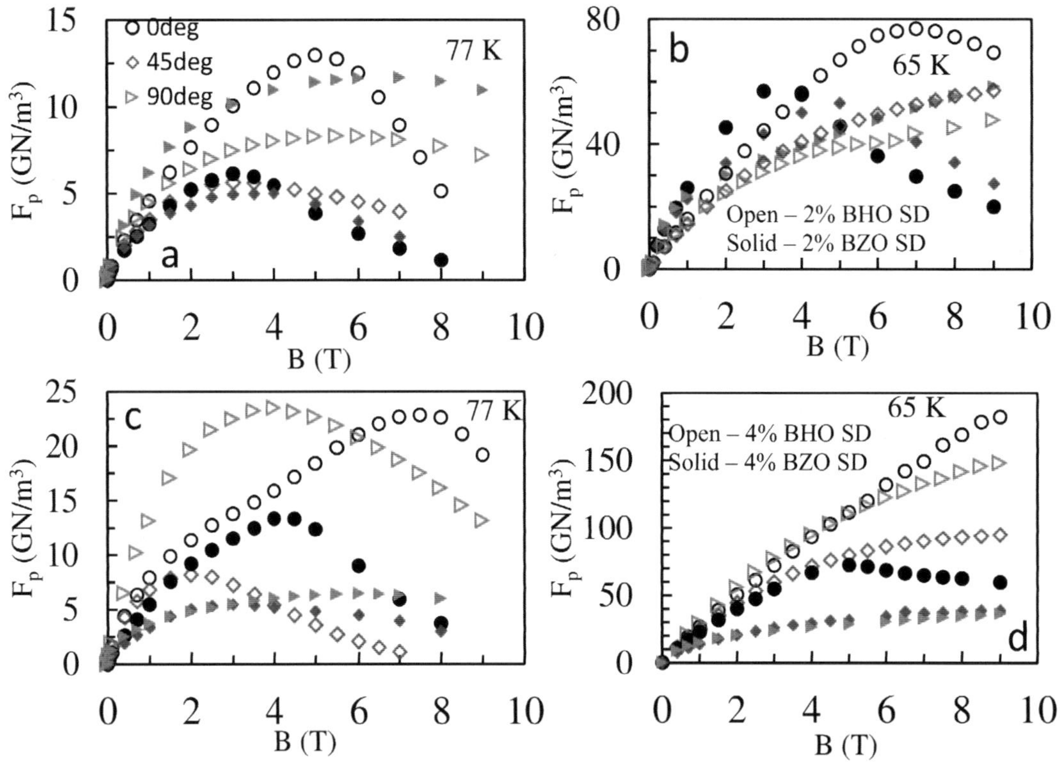

Figure 3. Comparison of the F_p *(B)* curves of) (a and b) 2% BHO SD (open) and 2% BZO SD (solid); and (c and d) 4% BHO SD (open) and 4% BZO SD (solid) at 77 K and 65 K.

that of the 4% BZO film, it is also higher than the BZO sample's $F_{p,\ max}$ at $\theta = 0^o$. In fact, while the $F_p(B)$ curve of the 4% BHO sample at $\theta = 45^o$ is comparable to that of the 4% BZO's below 5.0 T, beyond this field, it continues increasing with *B*-field, in contrast to the decreasing trend for the BZO sample at higher fields. In other words, at the same doping level, the BHO 1D-APCs are stronger pins at 45° than the BZO 1D-APCs at *B*//*c*. This behaviour points to a relatively wider angular range of the pinning enhancement in the BHO sample. The $F_p(B)$ trend of the nanocomposite samples is bifurcated at $\theta = 90^o$. The $F_{p,\ max}$ values in 2% BZO of 11.7 (77 K) and 58. 7 GN/m^3 (65 K) are slightly higher than that measured in the 2% BHO film. However, the 4% BHO values of 23.5 (77 K) and 148.5 GN/m^3(65 K) are about quadrupled the results from the 4% BZO films. It is also important to note that the $F_{p,\ max}$ values of 12.3 (77 K) and 71.7 GN/m^3 (65 K) at $\theta = 90^o$ measured on the reference YBCO (Figures 2e – 2f) are significantly smaller than those of the 4% BHO film [20].

The angular range of the effective pinning by the BZO and BHO 1D-APCs is quantified by the $F_{p,\ max}$ (Figures 4a-4b) and B_{max} (Figures 4c–4d) values of the nanocomposite films normalized to that of the reference YBCO film as functions of θ at 77 K (solid symbols) and 65 K (open symbols). At both temperatures, enhanced $F_{p,\ max}$ is observed for all four nanocomposite samples at lower angles up to $\theta \sim 40^o$ except the 4% BHO sample that shows enhancement in the entire angular range up to $\theta \sim 90^o$. The amplitude of the $F_{p,\ max}$ enhancement is the highest at $\theta = 0^o$ and decreases monotonically with increasing θ. Differing from the trend that the normalized $F_{p,\ max}$ is smaller than 1 (or worse than that of YBCO) at $\theta > 40^o$ for the other three samples, the normalized $F_{p,\ max}$ for the 4% BHO nanocomposite exhibits a second peak at $\theta = 90^o$, which is indicative of the benefit of high pinning efficiency associated to the coherent BHO/YBCO interface. At 77 K, the data show the amplitude of the $F_{p,\ max}$ enhancement of 2% BZO going below the reference YBCO value at ~ 37° while those of the 4% BZO and 2% BHO fall below 1 at ~ 37°. At 65 K the point at which the amplitude crosses the reference line has contracted to

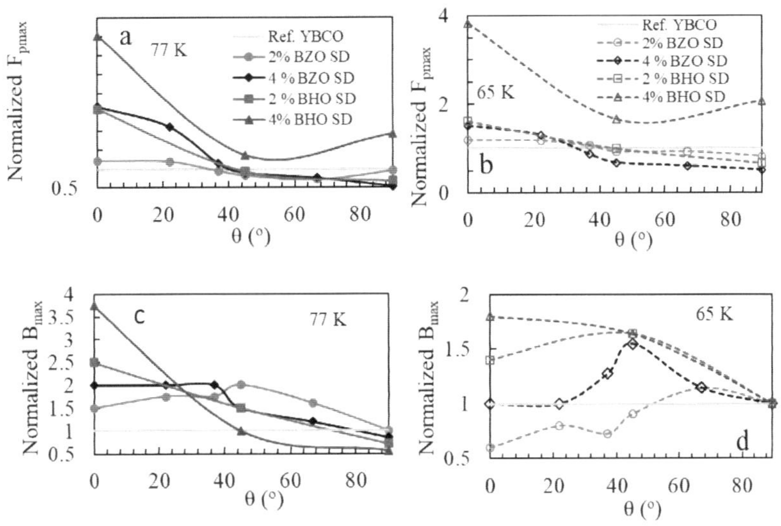

Figure 4. The ratio of $F_{p,\,max}$ (a and b) and B_{max} (c and d) to reference YBCO values vs θ at 77 K (solid symbols) and 65 K (open symbols) for the samples- 2% BZO SD (circle), 4 % BZO SD (diamond), 2% BHO SD (square) and 4 % BHO SD (triangle) with Ref. YBCO (horizontal line).

less than ~ 37° in the 4% BZO and remains roughly same for the 2% samples. The highest normalized $F_{p,\,max}$ up to 4.53 is observed in the 4% BHO at 77 K followed by ~ 2.6 in the 4% BZO and 2% BHO sample at the same temperature. This result suggests that in terms of correlated pinning and the angular range of pinning, the 2% BHO is comparable to the 4% BZO. Doubling the BHO concentration to 4% almost doubles the correlated pinning and extends the angular range to ~ 90°. This might be connected to the coherent BHO/YBCO interface. Perhaps, the coherent interface provides such a strong correlated pinning that some of pinning benefits at *B*//*c* are extended to other orientations. The normalized B_{max} data show that at 77 K, all samples (except 4% BHO) show enhancement up to about 67°. At 65 K the trends of the amplitude of the normalized B_{max} of the BHO films differ from (and are significantly higher than) those of the BZO samples across the angular window. In the 2% BHO film, the amplitude increases from ~ 1.4 at 0° until it reaches a maximum at 45° and decreases monotonically until it crosses the reference value at 90°. The 4% BHO also goes below the YBCO value at ~ 90° except its maximum is at 0° and it decreases monotonically with increasing θ. However, while both BZO films show a smaller normalized B_{max} and narrower B_{max} enhancement range, only the 2% BZO records negative B_{max} enhancement (between 0° and 45°). The amplitude of the B_{max} enhancement of the 2% BZO reaches a peak at about 67° then decreases to YBCO level at 90°. The amplitude of the 4% BZO stays at 1 until ~ 22° then increases and attains a maximum at ~ 45° before falling monotonically.

4. Conclusion

In summation, F_p (*B*) curves were studied at selected *B*-orientations on 2% and 4% BZO and BHO doped 1D-APC/YBCO nanocomposite films at 65 K and 77 K. To quantify the *B*-orientation range of the enhanced pinning by the 1D-APCs, the $F_{p.max.}$ and B_{max} normalized to that of the reference YBCO film were plotted as functions of θ. Several important observations have been made in this comparative study on BZO and BHO 1D-APCs of different interfaces with YBCO. It has been found that the pinning benefit of the *c*-axis aligned 1D-APCs extends beyond θ=0° (*B*//*c*-axis) as shown in both the normalized $F_{p.max,}$ and B_{max} exceeding 1. Quantitatively, the enhancement and angular range vary with APC concentrations and APC/YBCO interface quality. The enhanced B_{max} up to ~ 67° was observed in all the nanocomposite films with the BHO samples exhibiting remarkable B_{max} enhancement especially at 65 K. In particular, the $F_{p.max.}$ enhancement angular range correlates well with the pinning efficiency of the 1D-APCs and the highest $F_{p.max,}$ enhancement over the entire θ range has been observed in the 4% BHO 1D-APCs/YBCO nanocomposite sample. This result has therefore revealed interesting correlation between the pinning efficiency of the 1D-APCs, which is dictated by the quality of the APC/YBCO interface and the *H*-orientation angular range, which is important to further improvement of pinning landscape.

IOP Conf. Series: Materials Science and Engineering **756** (2020) 012025 doi:10.1088/1757-899X/756/1/012025

References

[1] Larbalestier, D et al., *High-Tc superconducting materials for electric power applications.* Nature, 2001. **414**(6861): p. 368-377.

[2] Malozemoff, A P *High Tc for the power grid.* Nature Materials, 2007. **6**(9): p. 617-619.

[3] Luongo, C A et al., *Next Generation More-Electric Aircraft: A Potential Application for HTS Superconductors.* IEEE Transactions on Applied Superconductivity, 2009. **19**(3): p. 1055-1068.

[4] Holtz, R.L., et al., *High Temperature Superconductors for Naval Power Applications.* Materials Science and Technology: 2006 NRL Review, 2006: p. 1-3.

[5] Obradors, X., et al., *Growth, nanostructure and vortex pinning in superconducting YBa2Cu3O7 thin films based on trifluoroacetate solutions.* Superconductor Science and Technology, 2012. **25**(12): p. 123001.

[6] Matsumoto, K. and P. Mele, *Artificial pinning center technology to enhance vortex pinning in YBCO coated conductors.* Superconductor Science and Technology, 2009. **23**(1): p. 014001.

[7] Macmanus-Driscoll, J.L., et al., *Strongly enhanced current densities in superconducting coated conductors of YBa2Cu3O7-x+BaZrO3.* Nature Materials, 2004. **3**(7): p. 439-443.

[8] Wu, J. and J. Shi, *Interactive modeling-synthesis-characterization approach towards controllable in situ self-assembly of artificial pinning centers in RE-123 films.* Superconductor Science & Technology, 2017. **30**(10): p. 103002.

[9] Baca, F., et al., *Control of BaZrO 3 nanorod alignment in YBa 2 Cu 3 O 7− x thin films by microstructural modulation.* Applied Physics Letters, 2009. **94**(10): p. 102512.

[10] Baca, F.J., et al., *Interactive Growth Effects of Rare-Earth Nanoparticles on Nanorod Formation in YBa2Cu3Ox Thin Films.* Advanced Functional Materials, 2013. **23**(38): p. 4826-4831.

[11] Emergo, R., et al., *The effect of thickness and substrate tilt on the BZO splay and superconducting properties of YBa2Cu3O7− δ films.* Superconductor Science and Technology, 2010. **23**(11): p. 115010.

[12] Feldmann, D., et al., *Improved flux pinning in YBa2Cu3O7 with nanorods of the double perovskite Ba2YNbO6.* Superconductor Science and Technology, 2010. **23**(9): p. 095004.

[13] Goyal, A., et al., *Irradiation-free, columnar defects comprised of self-assembled nanodots and nanorods resulting in strongly enhanced flux-pinning in YBa2Cu3O7− δ films.* Superconductor Science and Technology, 2005. **18**(11): p. 1533.

[14] Harrington, S., et al., *Self-assembled, rare earth tantalate pyrochlore nanoparticles for superior flux pinning in YBa2Cu3O7− δ films.* Superconductor Science and Technology, 2008. **22**(2): p. 022001.

[15] Mele, P., et al., *Ultra-high flux pinning properties of BaMO3-doped YBa2Cu3O7− x thin films (M= Zr, Sn).* Superconductor Science and Technology, 2008. **21**(3): p. 032002.

[16] Varanasi, C., et al., *Thick YBa 2 Cu 3 O 7− x+ BaSnO 3 films with enhanced critical current density at high magnetic fields.* Applied Physics Letters, 2008. **93**(9): p. 092501.

[17] Wang, X., et al., *Eliminating thickness dependence of critical current density in YBa 2 Cu 3 O 7− x films with aligned BaZrO 3 nanorods.* Journal of Applied Physics, 2010. **108**(11): p. 113911.

[18] Wee, S.H., et al., *Formation of self-Assembled, double-perovskite, Ba2YNbO6 nanocolumns and their contribution to flux-pinning and Jc in Nb-Doped YBa2Cu3O7-δ Films.* Applied physics express, 2010. **3**(2): p. 023101.

[19] Gautam, B., et al., *Transformational dynamics of BZO and BHO nanorods imposed by Y_2O_3 nanoparticles for improved isotropic pinning in $YBa_2Cu_3O_{7-\delta}$ thin films.* AIP Advances, 2017. **7**(7): p. 075308.

[20] Gautam, B., et al., *Probing the effect of interface on vortex pinning efficiency of one-dimensional BaZrO3 and BaHfO3 artificial pinning centers in YBa2Cu3O7-x thin films.* Applied Physics Letters, 2018. **113**(21): p. 212602.

[21] Cantoni, C., et al., *Strain-driven oxygen deficiency in self-assembled, nanostructured, composite oxide films.* Acs Nano, 2011. **5**(6): p. 4783-4789.
[22] Awaji, S., et al., *High-performance irreversibility field and flux pinning force density in BaHfO3-doped GdBa2Cu3Oy tape prepared by pulsed laser deposition.* Applied Physics Express, 2015. **8**(2): p. 023101.
[23] Cayado, P., et al., *Large critical current densities and pinning forces in CSD-grown superconducting GdBa2Cu3O7− x-BaHfO3 nanocomposite films.* Superconductor Science and Technology, 2017. **30**(9): p. 094007.
[24] Civale, L., et al., *Angular-dependent vortex pinning mechanisms in YBa 2 Cu 3 O 7 coated conductors and thin films.* Applied Physics Letters, 2004. **84**(12): p. 2121-2123.
[25] Engel, S., et al., *Enhanced flux pinning in Y Ba 2 Cu 3 O 7 layers by the formation of nanosized Ba Hf O 3 precipitates using the chemical deposition method.* Applied physics letters, 2007. **90**(10): p. 102505.
[26] Foltyn, S.R., et al., *The role of interfacial defects in enhancing the critical current density of YBa2Cu3O7− δ coatings.* Superconductor Science and Technology, 2009. **22**(12):p.125002.
[27] Gautam, B., et al., *Microscopic adaptation of BaHfO3 and Y2O3 artificial pinning centers for strong and isotropic pinning landscape in YBa2Cu3O7–x thin films.* Superconductor Science and Technology, 2018. **31**(2): p. 025008.
[28] MacManus-Driscoll, J., et al., *Strongly enhanced current densities in superconducting coated conductors of YBa 2 Cu 3 O 7–x+ BaZrO 3.* Nature materials, 2004. **3**(7): p. 439.
[29] Maiorov, B., et al., *Synergetic combination of different types of defect to optimize pinning landscape using BaZrO 3-doped YBa 2 Cu 3 O 7.* Nature materials, 2009. **8**(5): p. 398.
[30] Maiorov, B., et al., *Influence of naturally grown nanoparticles at the buffer layer in the flux pinning in YBa2Cu3O7 coated conductors.* Superconductor Science and Technology, 2006. **19**(9): p. 891.
[31] Miura, S., et al., *Characteristics of high-performance BaHfO3-doped SmBa2Cu3Oy superconducting films fabricated with a seed layer and low-temperature growth.* Superconductor Science and Technology, 2015. **28**(6): p. 065013.
[32] Tobita, H., et al., *Fabrication of BaHfO3 doped Gd1Ba2Cu3O7− δ coated conductors with the high Ic of 85 A/cm-w under 3 T at liquid nitrogen temperature (77 K).* Superconductor Science and Technology, 2012. **25**(6): p. 062002.
[33] Tsuruta, A., et al., *Enhancement of critical current density in the force-free sate of BaHfO3-doped multilayered SmBa2Cu3Oy film.* Japanese Journal of Applied Physics, 2014. **53**(7): p. 078003.
[34] Chen, S., et al., *Generating mixed morphology BaZrO3 artificial pinning centers for strong and isotropic pinning in BaZrO3–Y2O3 double-doped YBCO thin films.* Superconductor Science and Technology, 2017. **30**(12): p. 125011.
[35] Sebastian, M.A.P., et al., *Study of the flux pinning landscape of YBCO thin films with single and mixed phase additions BaMO3+ Z: M= Hf, Sn, Zr and Z= Y2O3, Y211.* IEEE Transactions on Applied Superconductivity, 2017. **27**(4): p. 1-5.
[36] Wu, J. and J. Shi, *Interactive modeling-synthesis-characterization approach towards controllable in situ self-assembly of artificial pinning centers in RE-123 films.* Superconductor Science and Technology, 2017. **30**(10): p. 103002.
[37] Wu, J.Z., et al., *The effect of lattice strain on the diameter of BaZrO3 nanorods in epitaxial YBa2Cu3O7− δ films.* Superconductor Science and Technology, 2014. **27**(4): p. 044010.

Acknowledgments

This research was supported in part by NSF contracts Nos: NSF-DMR-1337737 and NSF-DMR-1508494, the AFRL Aerospace Systems Directorate, the Air Force Office of Scientific Research (AFOSR), and the U.S. National Science Foundation (DMR-1565822) for TEM characterization

IOP Conf. Series: Materials Science and Engineering **756** (2020) 012026 doi:10.1088/1757-899X/756/1/012026

Flux Pinning Enhancements of $YBa_2Cu_3O_{7-x}$ with Nanosize Magnetic Additions

M A Sebastian[1], N A Pierce[2], I Maartense[1], G. Kozlowski[2], T J Haugan[3]

[1] University of Dayton Research Institute, Dayton, OH 45469 USA
[2] Wright State University, Dayton, OH 45435 USA
[3] U.S. Air Force Research laboratory, Aerospace Systems Directorate, WPAFB, OH 45433 USA
Author to whom correspondence should be addressed: sebastianm1@udayton.edu
Acknowledgements: Support from the Air Force Office of Scientific Research (AFOSR) and LRIR #18RQCOR100, and the Aerospace Systems Directorate (AFRL/RQ)

Abstract. Different methods of flux pinning are being tested world-wide to enhance critical currents (I_cs) of high temperature superconductor $YBa_2Cu_3O_{7-x}$ (YBCO) coated conductors exposed to high magnetic fields. Magnetic materials are interesting to consider as flux pinning additions because of their potential for very strong pinning strength. To our knowledge, there have been limited demonstrations of magnetic pinning additions to YBCO conductors. This paper describes the study of different M magnetic phase additions to YBCO including M = $BaFe_{12}O_{19}$, $La_{0.67}Ca_{0.33}MnO_3$, and other oxide phases. Nanosize additions were incorporated by depositing multilayer $(M/YBCO)_N$ films to minimize degradation of T_c, and testing volume % additions of M phase from 0.5 % to 5%. Results indicate that T_c onsets are depressed with magnetic additions, however in some cases interestingly without degrading the transition width. With optimization of magnetic additions a 50% increase of critical current density has been obtained, for low magnetic fields of < 10,000 Oe at 65K to 77K. Microstructural and superconducting properties are summarized, including SEM analysis.

1. Introduction

Increased current density for $YBaCuO_{7-x}$ (YBCO) thin films can be accomplished by adding nano-size insulating phases [1]. Research has focused on addition of many single and multi-phases as inclusions of nanoparticles and nanorods such as: Y_2BaCuO_5 (Y211), Y_2O_3, Ba_2YTaO_6 (BYTO), $BaSnO_3$ (BSO), Ba_2YNbO_6 (BYNO), $BaHfO_3$ (BHO), and $BaZrO_3$ (BZO) [2]–[10]. These nanosized inclusions (additions) can be classified as one, two, and three dimensional pinning centers [11]. Resulting defects pin the normal core of the vortices of the type II superconductor. However, core pinning is limited by the coherence length of high temperature superconductors, which is usually on the order of ξ = 25Å. The condensation energy of the Cooper pairs within the core, U, is equated to the maximum pinning energy.

$$U = (H_c^2/8\pi)\pi\xi^2 = (\Phi_0/8\pi\lambda_L)^2 \quad (1)$$

H_c = critical field
Φ = flux quantum

λ_L = London penetration depth

In contrast, with magnetic pinning, the goal is theoretically to pin the magnetic flux of the vortex, instead of just the core. In this case, the pinning is limited by the magnetic penetration depth λ, which is typically 1500 Å. The magnetic pinning strength is related to the penetration depth according to equation [2].

$$\text{Magnetic pinning strength} = MH(\pi\lambda^2) \sim \Phi_0 M \tag{2}$$

This results in a much larger columnar volume of pinning for magnetic pinning versus the core pinning [12]–[15].This larger columnar volume results in the magnetic pinning strength being 100-1000 times the core pinning strength (figure 1). When an external magnetic field is applied perpendicular to the film plane of a YBCO film containing magnetic dopants, an S-shaped hysteresis loop versus a square shaped loop, indicates the presence of domain wall movement and magnetic pinning in the film. With perpendicular anisotropy, the magnetic pinning scales with the saturation magnetization. With in-plane anisotropy, magnetic pinning due to vortex and domain wall interaction occurs below the coercive field [14].

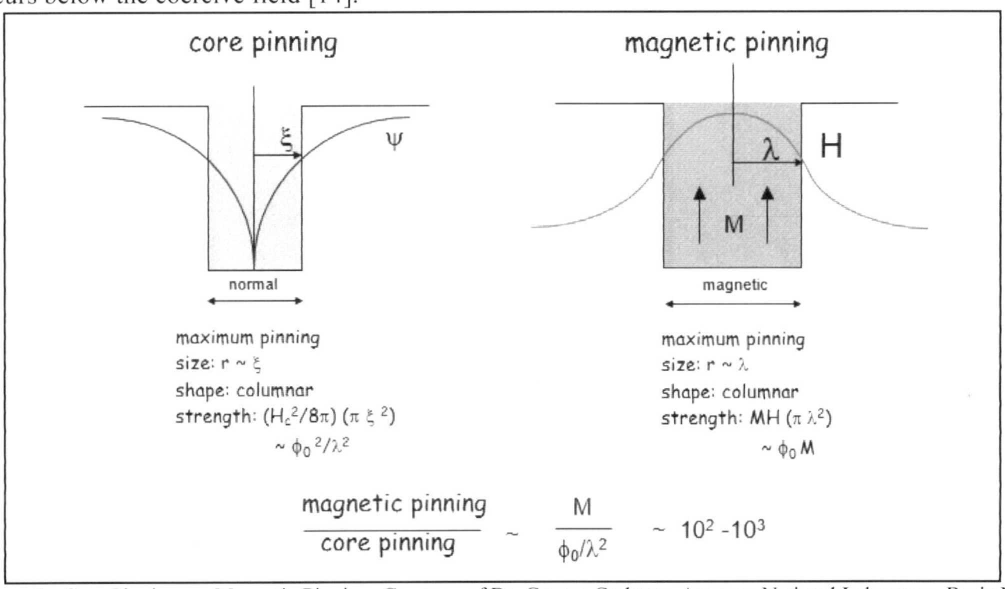

Figure 1. Core Pinning vs. Magnetic Pinning. Courtesy of Dr. George Crabtree, Argonne National Laboratory, Basic Needs for Superconductivity Workshop May 8-11, 2006

Several research groups have harnessed this additional pinning strength by ferromagnetic, antiferromagnetic, and ferrimagnetic additions to YBCO films. Chen Feng Tsai *et al.* compared and contrasted the addition of antiferromagnetic Fe_2O_3 as a buffer layer for a YBCO film, as a multilayer composite with YBCO, and as a cap layer on top of YBCO film with and without the addition of CeO_2. Their research found that the magnetic nanoparticles provided pinning in addition to the conventional defect pinning [16]. The majority of other research involved addition of soft and hard ferromagnets and ferrimagnets. Garcia-Santiago *et al.* produced YBCO films with hard ferromagnetic $BaFe_{12}O_{19}$ (BFO) bilayer via pulsed laser deposition (PLD), utilizing a yttria-stabilized-zirconia (YSZ) buffer layer to decrease lattice mismatch and chemical reactivity. When compared to a YBCO film, the irreversibility line (IL) of the bilayer YBCO-BFO film shifted upward and above the IL of the YBCO film. The results explained by Bulaevskii's theoretical approach are that "the effective magnetic field B in the YBCO layer is a sum of the applied magnetic field H and the demagnetizing field $4\pi M$ in the BFO layer: $B = H + 4\pi M$. The magnetic domain walls are pinned if the applied field is lower than the coercive field, which is of the order of 5kOe for the BFO layer." The stripe magnetic domain structure pins the vortices until the BFO film saturates at 10 kOe. The pinning enhancement of a factor of two was estimated by comparing the ratio of the irreversibility fields for the YBCO-BFO film versus the YBCO film [12]. S. Wimbush *et al.* utilized mixed targets of 1-5 at. % iron oxides with YBCO to produce films by PLD. Results obtained showed that ferromagnetism and superconductivity were both present in the films at 77 K, and that the additions increased current density [17]. ($Y_3Fe_5O_{12}$ (YIG) is ferromagnetic, and has been studied by several research groups. M. Gasmi *et al.* varied the concentration of YIG from 1-20 wt. % in YBCO powder pellets that were sintered. Magnetometer results showed that for an applied field H less than a threshold value, H^*, superconducting domains dominate. The analysis showed that H^* decreased with increasing YIG concentration. This result is indicative of the YIG nanoparticles increasing the number of ferromagnetic domains in comparison with the YBCO superconducting domains. H^* also was found to decrease with temperature. When the applied field is greater than H^*, the superconductor spin domains align with the applied field, forming a ferromagnetic phase along with the ferromagnetic domains on the surface. When the field is below H^*, the spins in the superconducting domains reform Cooper pairs [18]. Q. Jia *et al.* also produced YBCO films with polycrystalline YIG substrates using YSZ and CeO_2 buffer layers [19]. A. Bobyl *et al.* investigated a structure of YIG film on microstrip transducers and YBCO film, where the YBCO film acts as a shield for magnetostatic wave propagation [20]. Colossal magnetoresistance (CMR) compounds of $La_{2/3}Ca_{1/3}MnO_3$ (LCMO) and $La_{0.7}Sr_{0.3}MnO_3$ (LSMO) have also been studied with YBCO films. J. Chen *et al.,* sputtered 8 layers of YBCO/LCMO. It was found that a decrease in the thickness of the LCMO layer, resulted in a decrease of the magnetoresistance of the multilayer film. At low temperatures, the multilayers were superconducting, while they were ferromagnetic at high temperatures [21]. H. Habermeier *et al.* used PLD to deposit bilayers of LCMO and bilayers of $SrRuO_3$ with YBCO on $SrTiO_3$ (STO) substrates. Magnetizing perpendicular to the film surface was found to increase current density. Current density was also found to increase with increasing field after zero field cooling, in contrast to the usual decrease in current density with increasing field seen in YBCO films [14]. J. Albrecht studied the effect of decoupling YBCO and LCMO with a STO space layer. It was found that the critical current density is affected by the magnetization of the LCMO layer, with significant magnetic pinning contributions shown. Substrate choice had an impact on induced strain, which affected the temperature dependence of the magnetic pinning contribution [22]. J. Huang researched LSMO + YBCO bilayers with CeO_2, and found increased magnetic pinning at lower temperatures [23]. A. Meledin also investigated YBCO thin films with 6 mol. % $MnFe_2O_4$ (MFO) produced by chemical solution deposition, which resulted in 3D inclusions and lamellas within the YBCO film. These nanoinclusions resulted from reactivity and decomposition of the MFO with YBCO, and serve as artificial pinning centers [24].

Several issues need to be considered when incorporating magnetic nanoparticles into YBCO thin films, such as whether to utilize a hard or soft magnet phase, crystal epitaxy, particle orientation, magnetic domain size, chemical reactivity, operation temperature, oxides necessary for the formation of YBCO, and feasibility. The choice of substrate determines whether the epitaxial films are grown in tension or compression [22].

Table 1. Magnetic Properties

		σ_s (Am^2/kg) (emu/g) @ 298K	σ_s (Am^2/kg) (emu/g) @ 77K	M_0 (emu/cm^3) @ 273K	T_{curie} (K)
$La_{0.67}Ca_{0.33}MnO_3$	ferromagnetic	3.5 μ_B[g] 7 emu/g[k]	93[j]	113 (@100K)[i]	240[a]
$SrRuO_3$	ferromagnetic	1.5 μ_B[g]	20[b]		160[b]
$Y_3Fe_3O_{12}$	ferrimagnetic	25[c]		75[h]	560[d]
$BaFe_{12}O_{19}$	ferromagnetic	75[c]			723[e]
Fe	ferromagnetic	197[c]		1740[f]	1043[f]

[a][25][b][26][c][27][d][28][e][29][f][30][g][14] [h][32][i][36][j][37][k][38]

Based on the magnetization properties and lattice mismatch found in Tables I and II, it was chosen to compare multilayer films of YBCO with the following additions: $BaFe_{12}O_{19}$, LCMO, $SrRuO_3$, and $Y_3Fe_5O_{12}$.

Table 2. Magnetic Compounds Lattice Parameters and Mismatch with respect to YBCO
Lattice Parameters of Epitaxy are $\bar{a} = (a,b)_{avg}$, $a'' = a/2$, $a^* = a/\sqrt{2}$, $\bar{a}^* = \bar{a}/\sqrt{2}$, $c''' = c/3$, $c'''' = c/6$

Material	Lattice Type	Lattice Parameters (nm)	Lattice Parameter of Epitaxy (nm)	a-b in Plane Lattice Mismatch to Y123 a (%)	Lattice Mismatch c-direction to Y123 (%)
YBCO[a]	Orthorhombic	a = 0.3825 b = 0.3886 c = 1.166	$\bar{a}$ = 0.3855 c''' = 0.3887		
$La_{0.67}Ca_{0.33}MnO_3$[b]	Orthorhombic	a = 0.5457 b = 0.5451 c = 0.7708	$\bar{a}^*$ = 0.3857 c'' = 0.3854	+0.04	-0.8
$SrRuO_3$[c]	Pseudo-cubic	a = 0.393		+1.95	+1.1
$Y_3Fe_5O_{12}$[d]	Cubic	a = 1.2376	a''' = 0.41253	+7.0	+6.1
$BaFe_{12}O_{19}$[e]	Hexagonal	a =0.58945 c = 2.3215	a^* = 0.41680 c'''' = 0.3869	+8.1	-0.05

[a]ICDD 00-040-0159 [b][25] [c][31][d][32][e]ICDD 00-039-1433

2. Experimental Procedures

2.1. Target Preparation

The YBCO target was produced via solid state processing of dried commercial Nexans YBCO powder. The target was pressed utilizing a Carver die press with a 1.25 inch diameter die, and then sintered at 850°C for 60 hours and 920°C for 156 hours, achieving a final percentage of theoretical density of 93.4 %. $BaFe_{12}O_{19}$, $La_{0.67}Ca_{0.33}MnO_3$, $Y_3Fe_5O_{12}$, and $SrRuO_3$ were commercial targets purchased from SCI Engineered Materials Inc., STMC LLC JR Gaines, Kurt Lesker, and SCI respectively.

2.2. Thin Film Production via Pulsed Laser Deposition (PLD)

Multilayer composite films were produced via PLD with a KrF excimer laser (λ = 248nm) using the above targets and $LaAlO_3$ (LAO) and $SrTiO_3$ (STO) substrates. The $BaFe_{12}O_{19}$-YBCO composite films consisted of a layer of YBCO followed by a layer of BFO, repeating 28 times. PLD conditions were a heater block temperature of 790°C and a 300 mTorr O_2 atmosphere, with an energy of 650 mJ and pulse rate of 4 Hz. Pulses were varied and attenuating mirrors utilized to vary the thickness of the BFO layer, resulting in volume percent additions of 0.6, 1.4 and 2.7 vol. %. The LCMO-YBCO composite films were produced under the same conditions, with the exception of using a pulse rate of 2 Hz for the LCMO layers. Volume percent of LCMO additions were 0.8, 1.2, and 2.4%. The $SrRuO_3$ (SRO)-YBCO composite films repeated layers of YBCO and SRO 38 times under similar PLD conditions with the exception of a temperature of 800°C and a repetition rate of 2 Hz. with pulses and attenuating mirrors also varying the volume percent additions for SRO from 1.2, 2.4, and 4.7 vol.%. The $Y_3Fe_5O_{12}$ (YIG)-YBCO composite films were similar to the SRO composites, with the exception of a temperature of 790°C. Volume percent additions varied from 3.6, 7.0, and 11.1 vol. %. All films were annealed for 30 minutes at 500°C and an oxygen atmosphere. All nominal vol.%'s were based on PLD calibration deposition rates.

2.3. Film Characterization

Film thickness was determined by etching the thin films with nitric acid and measuring the stair step thickness with a profilometer. Magnetic current density (J_{cm}) was measured with a Quantum Design Physical Properties Measurement System (PPMS) with a vibrating sample magnetometer (VSM) probe. The VSM data was attained for conditions of 77 K, 65 K, and 40 K with an applied field varied from 0-9 T for *H* // c-axis of the films. The measured magnetic data was used to determine the current densities by utilizing a simplified form or the Bean Critical State Model: $J_{cm} = 30\Delta M/da^3$ (A/cm^2), where ΔM is the change in magnetization, d is the film thickness, and a is the side length dimension of the square films [33]–[35]. Superconducting transition temperature (T_c) was attained via AC susceptibility measurements, with the amplitude varied from 0.025 Oe to 2.2 Oe, at a frequency of 4 Hz. Transport current density (J_{ct}) was measured on microbridges 0.05 cm wide and 0.3 cm long, by a 4-point contact method with a 1μv/cm criteria. A FEI Sirion Scanning Electron Microscope (SEM) with ultra-high resolution at 5 kV and a spot size setting of 3 for nano-scale was used to obtain images of the films' surface. X-ray diffraction analysis utilized the Bruker D8 Discover diffractometer with the following operation conditions: high resolution, Co tube 1.78897 Å, Power: 40kV, 35mA; 2-Theta: 5-80°. A Tescan Lyra FIB/SEM system was used to prepare a foil of YBCO/SRO sample for TEM examination by using TALOS TEM SYSTEM FSE under the acceleration voltage of 200 kV.

3. Results and Discussion

Magnetic nanoparticle addition effects on the onset critical temperature ($T_{c\text{-onset}}$) and the self-field current density ($J_{ct\ self\text{-}field}$) can be found in figure 2. Increased vol. % of BFO dramatically decreases

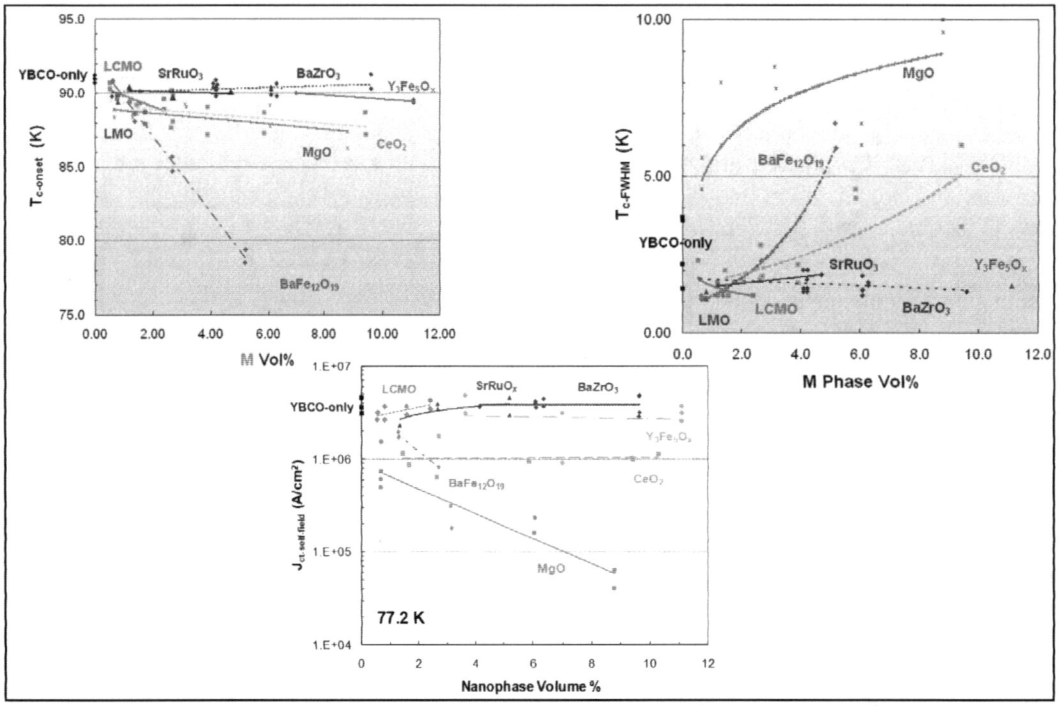

Figure 2. Comparison of $T_{c\text{-onset}}$, $T_{c\ FWHM}$, and $J_{c\ self\text{-}filed}$ results for multilayer magnetic doping of Y123 films

$T_{c\text{-onset}}$ & $J_{c\ self\text{-}field}$ and increases $T_{c\ FWHM}$, compared to LCMO, SRO, and YIG, and may be the result of reaction products and strain development due to a higher degree of lattice mismatch with YBCO. Microstructure studies involved SEM, TEM, and XRD analysis. SEM imaging in Figure 3 depicts typical island growth and insulating phase nanoparticle charging. In Figure 4, TEM depicts very nice epitaxial film growth and characteristic imaging of the YBCO film. Elemental analysis confirmed presence of the expected elements, (Ru and Cu are shown for example) and areas of the film that are copper rich. TEM was also in very good agreement with measured film thickness of 236 nm.

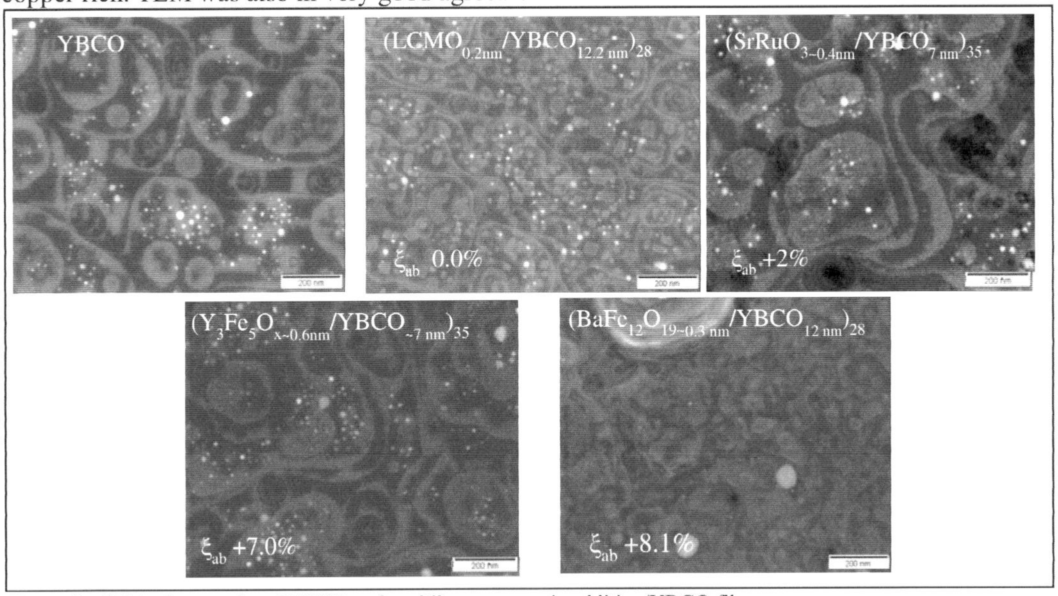

Figure 3. SEM micrographs of YBCO and multilayer magnetic addition/YBCO films.

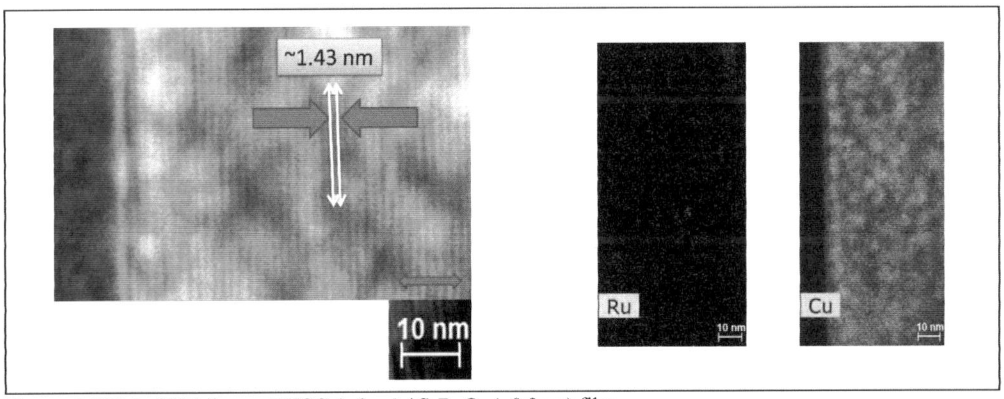

Figure 4. TEM of 38 bilayers YBCO (~6nm) / $SrRuO_3$ (~0.3nm) film

Initial XRD analysis also indicates epitaxial growth of YBCO film identified by *(00l)* and substrate peaks. Magnetic addition peaks were minimal due to the low volume percent of the films analyzed (1.4-3.6%). Magnetic current densities, (J_{cm}), were attained at 77, 65, and 40K with the field ramped from 1 to 9T for each of the dopant systems, varying the volume percent. LCMO-YBCO 28 layer bi-layer films' thickness averaged 347nm, with YBCO layers ~ 11.0 nm/ layer. From Figure 5(a) at 77K, 1.2 and 2.4vol.% LCMO films attained higher J_{cm} than an undoped YBCO film at fields < 1000 Oe and at fields <7000 Oe for 65K. At 40K, all three vol.% LCMO (0.8, 1.2, and 2.4) films attained J_{cm} higher than YBCO. BFO-YBCO 28 layer bi-layer films' thickness averaged 300 nm, with YBCO layers ~ 11.0 nm/layer. From Figure 5(b) at 77K, the 1.4 vol.% BFO film attained higher J_{cm} for fields < 5000Oe. BGO films of 1.4% and 0.6% performed better than YBCO films at 65K and 40K. Due to its poor results at 77K, 2.7 vol.% was not measured at lower temperature. YIG-YBCO 38 layer bi-layer films' thickness averaged 254 nm, with YBCO ~6nm/layer. From Figure 5(c), at 77K, the multilayer films J_{cm} is higher than YBCO for fields > 10,000Oe, whereas all vol. % YIG outperform YBCO at 65K and 40K. SRO-YBCO 38 layer bi-layer films' thickness averaged 240 nm, with YBCO layer ~ 6nm/layer. From Figure 5(d), all vol.% SRO (1.2, 2.4, and 4.7%) outperformed YBCO film at 65, and 40K.

4. Conclusion

A comparison of the best vol.% film from each multilayer system studied shows that 3.6% YIG and 2.4 vol.% SRO outperform LCMO & BFO at fields greater than 10,000 Oe for all temperatures studied (77, 65, and 40K) (Figure 5(e)). The YIG and SRO films have very good T_c's of 90K, with $T_{c\ FWHM}$ of 1.6 and 1.8K, indicating good YBCO film structure. From magnetic hysteresis loops, H_{sat} is 3000 Oe for the YIG multilayer film and 2500 Oe for the SRO multilayer film. The "S" shaped hysteresis loop is indicative of domain wall movement and magnetic pinning in the films (Figure 6). The YIG (3.6%) and SRO (2.4%) multilayer films showed an increase of 25% performance of J_{cm}/J_c(YBCO) for H< 0.5T for 40-77K, and an increase of 35% and 70% respectively for J_{cm}/J_c(YBCO) for H>1T for 40-77K. This research shows the impact of combining magnetic and defect pinning in YBCO superconducting films.

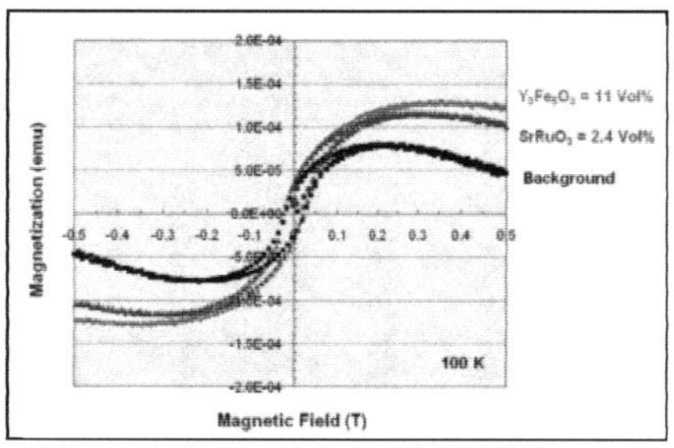

Figure 6. Hysteresis Loops for SRO & YIG/YBCO multilayer films

Acknowledgments

We wish to acknowledge Dr. Michael Sumption and Dr. Michael Susner for the magnetic characterization data acquisition at Ohio State University.

5. References

[1] Feighan J, Kursumovic A and MacManus-Driscoll J 2017 *Supercond. Sci. and Technol.* **30** 123001
[2] Haugan T, Barnes P, Wheeler R, Meisenkothen F and Sumption M 2004 *Nature* **430** pp 867–70
[3] Gapud AA, Kumar D, Viswanathan S, Cantoni C, Varela M, Abiade J, Pennycook SJ and

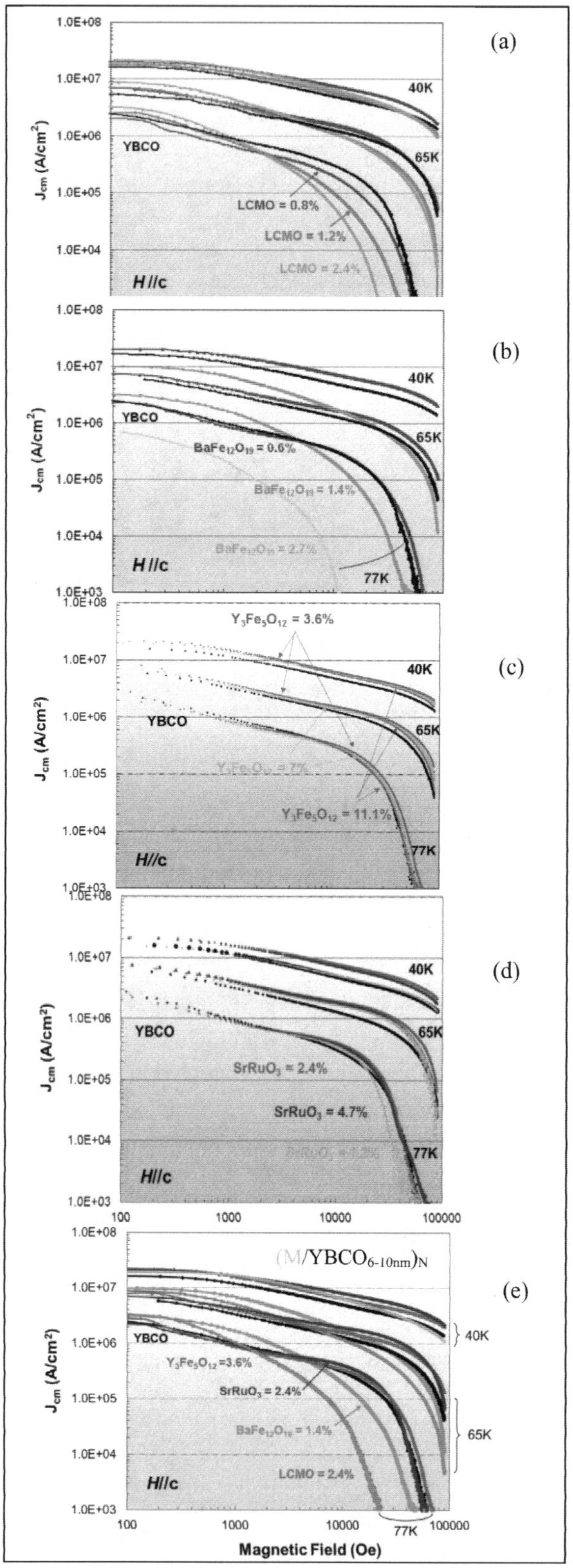

Figure 5. J_{cm} vs magnetic field for YBCO multilayer films with varying vol.% (a) LCMO, (b) BFO, (c) YIG, (d) SRO, (e) comparison of vol. % dopant optimized multilayer films.

Christen, DK 2005 *Supercond. Sci. and Technol.* **18** pp 1502–05
[4] Opherden L, *et al.* 2016 *Sci Rep* **6** 21188
[5] Coll M, Ye S, Rouco V, Palau A, Guzman R, Gazquez J, Arbiol J, Suo H, Puig T and Obradors X 2012 *Supercond. Sci. and Technol.* **26** 015001
[6] Varanasi C, Barnes P, Burke J, Brunke L, Maartense I, Haugan T, Stinzianni E, Dunn K and Haldar P 2006 *Supercond. Sci. Technol.* **19** L37–41
[7] Mele P, *et al.* 2014 *Supercond. Sci. and Technol.* **28** 024002
[8] Wu J, Shi J, Baca J, Emergo R, Ell *IEEE Trans. Appl. Supercond.* **25** 8000205

Figure 5. J_{cm} vs magnetic field for YBCO multilayer films with varying vol.% (a) LCMO, (b) BFO, (c) YIG, (d) SRO, (e) comparison of vol. % dopant optimized multilayer films.

[9] Gautam B, *et al.* 2017 *AIP Adv.* **7** 075308
[10] Sebastian M A, *et al.* 2017 *IEEE Trans. Appl. Supercond.* **27** 7500805
[11] Matsumoto K and Mele P 2010 *Supercond. Sci. Technol.* **23** 014001
[12] Garcıa-Santiago A, Sanchez F, Varela M and Tejada J 2000 *Appl. Phys. Lett.* **77** pp 2900–02
[13] Bulaevskii L, Chudnovsky E and Maley M 2000 *Appl. Phys. Lett.* **76** 2594–96
[14] Habermeier H, Albrecht J and Soltan S 2004 *Supercond. Sci. and Technol.* **17** S140–44
[15] Dew-Hughes D 1974 *Philos Mag* **30** pp 293–305
[16] Tsai C, Zhu Y, Chen L and Wang H 2011 *IEEE Trans. Appl. Supercond.* **21** pp 2749–52
[17] Wimbush S, Durrell J, Bali R, Yu R, Wang H, Harrington S and MacManus-Driscoll J 2009 *IEEE Trans. Appl. Supercond.* **19** pp 3148–51
[18] Gasmi M, Khene S, and Fillion G 2013 *J Phys Chem Solids* **74** pp 1414–18
[19] Jia Q, Arendt P, Foltyn S, Holesinger T and DePaula R 2001 *IEEE Trans. Appl. Supercond.* **11** pp 3489–92
[20] Bobyl A, Suris R, Karmanenko S, Semenov A, Melkov A, Konuhov S and Olshevski A 2002 *Physica C* **372** pp 508–10
[21] Chen J, Yang H, Wang S, Wang L and Horng H 2001 *IEEE Trans. Appl. Supercond.* **11** pp 3301–04
[22] Albrecht J, Soltan S and Habermeier H 2005 *Phys. Rev. B* **72** 092502
[23] Huang J, Fan M, Wang H, Chen L, Tsai C, Li L and Wang H 2016 *Ceram Int* **42** pp 12202–09
[24] Meledin A, Turner S, Cayado P, Mundet B, Solano E, Ricart S, Ros J, Puig T, Obradors X and Van Tendeloo G 2016 *Supercond. Sci. and Technol.* **29** 125009
[25] Lim K, Ng S, Halim S, Chen S and Wong J 2009 *Am. J. Appl. Sci.* **6** pp 1153–57
[26] Callaghan A, Moeller C and Ward R 1966 *Inorg Chem* **5** pp 1572–76
[27] Özgür Ü, Alivov Y and Morkoç H 2009 *J. Mater. Sci-Mater. El.* **20** pp 789–834
[28] Nguyet D, Duong N, Satoh T, Anh L and Hien T 2012 *J. Alloy Cmpd,* **541** pp 18-22
[29] Sardjono P and Djauhari N R 2016 *JPhys Conf Ser* **739** 012089
[30] Lee E W 1970 *Magnetism: An Introductory Survey* (New York: Dover Publications Inc.)
[31] Koster G, Klein L, Siemons W, Rijnders G, Dodge J S, Eom C, Blank H A and Beasley M R 2012 *Rev Mod Phys* **84** pp 253–98
[32] Arsad A Z and Ibrahim N B 2018 *J. Magn. Magn. Mater* **462** pp 70–7
[33] Kell J W, Haugan T J, Locke M F and Barnes P N 2005 *IEEE Trans Appl Supercond* **15** pp 3726-29
[34] Bean C P 1964 *Rev Mod Phys* **36** pp 31–9
[35] Bean C P 1962 *Phys Rev Lett* **8** pp 50–53
[36] Brück S, Treiber S, Macke S, Audehm P, Christiani G, Soltan S, Habermeier H-U, Goering E and Albrecht J 2011 *New J Phys* **13** 033023
[37] Shim I B, Bae SY, Oh YJ, Choi SY 1998 *Solid State Ionics* **108** pp 241–7
[38] Das D, Srivastava CM, Bahadur D, Nigam AK and Malik SK 2004 *J Phys-CondensMat* **16** 4089

IOP Conf. Series: Materials Science and Engineering **756** (2020) 012027 doi:10.1088/1757-899X/756/1/012027

AC Loss Analysis on Coated Conductors at different Sinusoidal Frequencies for Electric Propulsion Applications

Abhinav Kumar[1*], JV Muruga Lal Jeyan[2], Ashish Agrawal[3]

[1,2]School of Mechanical Engineering, Lovely Professional University, Punjab, India

[3]Department of Mechanical Engineering, Madanpalle Institute of Technology and Sciences, Andhra Pradesh, India

drabhinav@ieee.org*

Abstract. Superconducting magnets are one of the superior contenders in achieving the targets of electric aircraft industry successfully as they have very high power densities compared to other battery storage systems. Many aviation research agencies are looking at superconducting magnets as one of the alternate in replacing the conventional jet engines completely (electric aircrafts) or partially (hybrid aircrafts). National Aeronautics and Space Administration (NASA) and Air Force Research Laboratory (AFRL), USA has reported that high temperature superconducting magnets possesses higher specific energies (Wh/kg) and have infinite number of recharge cycles compared to other storage technologies employed for energy requirements. The other advantage of using such magnets is that there will be no hazardous disposals like batteries which lower the overall pollution. Superconducting magnets are DC operated systems however during charging or discharging transient behaviour of current results in the losses which would further ends up with heat generation. Heat generation during charging or discharging period would cause quenching of the superconductor due to sudden temperature rise.

In this work, electromagnetic analysis on superconducting magnet having capacity of 1 MJ has been performed where a 2D numerical model is developed using H-formulations in order to estimate the AC losses for a high temperature superconducting tape manufactured by SuperPower (SCS 12050) having 330 A critical current at 77 K. AC current having different load factors has been fed through the stacked tapes at 50 Hz, 60 Hz and 70 Hz frequency and AC losses have been evaluated. It has been found that at higher frequencies the AC losses are found to be larger than lower frequencies. Overcritical currents have been found in the current density distribution due to the application of E-J relationship for the homogeneous 2D numerical model.

1. Introduction

Electrical Aircrafts will be going to be the future of aviation industry as such aircrafts can deliver the expected outcomes without pollution compared to the conventional aviation transportation using jet fuels. Currently, a concept of hybrid aircrafts has been widely introduced by the aviation communities as they have to meet the pollution norms imposed by the parent governmental agencies. It includes the contribution of both conventional jet engines technology and novel electric aircraft systems where electrical power can be used during take-off, climbing and landing of the aircraft as it only contributes to 25% of the total power consumption [1]. For this to achieve, various battery operated systems have been evolved during past few years such as Lead acid, Nickel metal hydride batteries and Lithium ion batteries. However, these batteries have limited storage capacities per unit weight or surface area

(Figure 1 (a) and Figure 1 (b)) and therefore for larger power requirement, the overall weight and surface area will increase which directly affects the lift force.

To conquer such challenges, a novel concept of using high temperature superconducting magnetic energy storage (HT-SMES) systems have been introduced by various research agencies like NASA [2] and Air Force Research Laboratory, USA [3]. The agencies have reported that for fully electric passenger aircrafts, one has to acquire SMES systems than battery storage units as they have higher power densities and capacities per unit weight and do not involve any hazardous disposal like chemical batteries. SMES system can provide more specific power (kW/kg) and specific energy (Wh/kg) (Figure 1 (a)) than supercapacitors, fuel cells, Nickel Hydride and Lithium ion batteries [3]. Air Force Research Laboratory, USA [3] has reported that SMES systems can store more energy per unit weight than other storage technologies (Figure 1 (b)). After noticing the importance of this concept, it has been identified that such systems can be useful for hybrid or fully electrical aircraft industry as they can provide efficient flight with lower risk as no moving part is involved in the design of SMES.

Usually, SMES is a DC operated device that store energy in magnetic field induced due to the flow of current through the superconducting tapes. However, during the charging or discharging period, the current needs time to achieve a study state thus implies current has to go through transient phase for few milliseconds or more depending upon the inductance of the coil [4]–[7]. This transient state is responsible for the AC losses among the pancake coils [4]. It has been well understood that self-fields can also affect the critical current of the tape [8] as well as AC losses among such coils. Also, in our earlier publication, it has been found that AC losses can be controlled by increasing the substrate layer thickness however overall flux density is found to decrease [9]. In the present study, a comprehensive analysis on HT-SMES has been performed at 77 K where magnetic field and current density distribution along with AC losses (W/m) have been examined. An alternating current having magnitude 200 A has been transported through HTS tape manufactured by SuperPower [10] at a frequency of 50 Hz. COMSOL MultiPhysics software package has been used to achieve the simulations where H-formulations model has been incorporated for the analysis.

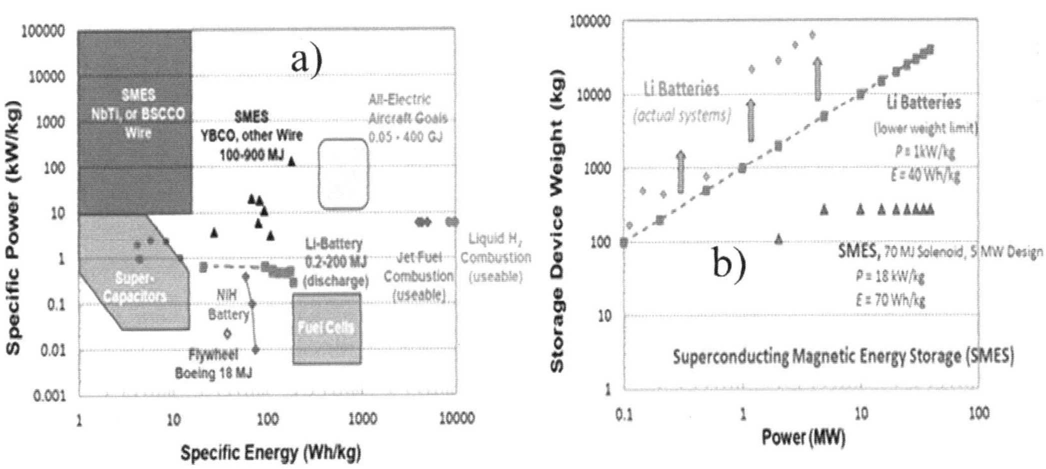

Figure 1. (a) Specific Power Vs Specific Energy, (b) Storage device weight Vs Power comparison for various Energy storage systems [3]

2. Electromagnetic Modelling of the HT-SMES

2.1. H-formulations

A similar formulation has been adopted as used by Brambilla et al. [11] and H-formulation has been selected as it is easy to implement and simple in operation. The governing equations involved in the electromagnetic modelling of the SMES are as follows [12]:

$$\nabla \times \rho \nabla \times H = -\mu \frac{\partial H}{\partial t} \quad \text{inside domain} \qquad \text{Equation 1}$$

$$H = H_{self} + H_{ext} \quad \text{on boundaries} \qquad \text{Equation 2}$$

$$\rho_{HTS} = \frac{E_c}{J_c}\left|\frac{J}{J_c}\right|^{n-1} \qquad \text{Equation 3}$$

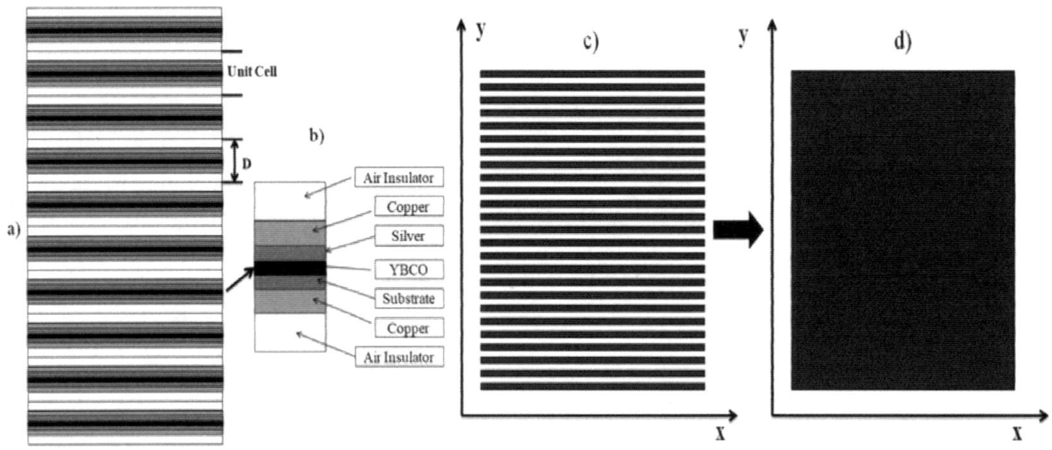

Figure 2. (a) Stacked coated conductors periodic arrangement, (b) zoomed section of unit cell, (c) stacked tapes representing turns around the coil, and (d) homogenized computational domain for the analysis.

Figure 2 shows the stacked tape periodic arrangement and homogenized domain respectively. In order to estimate magnetic field H inside the domain Equation 1 has been used along with the boundary condition imposed as Equation 2. Values for J, E and B can be calculated from H by making use of Ampere's Law $J = \nabla \times H$ and the following constitutive relations $E = \rho J$ and $B = \mu H$ respectively. The resistivity of the HTS tape can be modelled by incorporating Power Law (Equation 3).
Instantaneous AC losses (W/m) can be estimated using following relation (Equation 4):

$$\xi = \int_{Inside\,domain} E.J\, dxdy \qquad \text{Equation 4}$$

A Kim like model (Equation 5) for the J_c (***B***) dependence of HTS tape at 77K was incorporated where B_0=0.04265T, Jc (77K,SF)=2.75×10^{10} A/m^2, k=0.29515, α=0.7 and B_{II}, $B_{\perp}$ are, respectively, the parallel and perpendicular components of magnetic field density with respect to tape's surface [12].

$$J_c(B) = \frac{J_c(77K,SF)}{\left(1+\frac{\sqrt{k^2|B_{\parallel}|^2+|B_{\perp}|^2}}{B_0}\right)^{\alpha}} \qquad \text{Equation 5}$$

3. Electromagnetic Design of Superconducting Coil and AC Losses

Electromagnetic analysis for 1 MJ SMES is done using COMSOL MultiPhysics software where 2D model has been used to evaluate AC losses occur in HTS tape due to variable frequencies (50 Hz, 60 Hz and 70 Hz) and operating currents (200 A, 225 A, 250 A, 275 A and 300 A). The design parameters of 1 MJ SMES are available in Table 1. 2D analysis is done where 60 turns around a single

IOP Conf. Series: Materials Science and Engineering **756** (2020) 012027 doi:10.1088/1757-899X/756/1/012027

pancake coil has been considered and through homogenized approach, the problem is solved using H-formulations.

4. Results and Discussions

4.1. Losses due to frequency change

AC losses have been calculated for the 2D numerical model using H-formulations. Figure 3 shows the instantaneous losses and Figure 4 shows the average AC losses (W/m) for the stacked superconducting tapes. It can be noticed that at higher frequencies, the losses get increased (Figure 3 (a)) and the losses variation is somewhat linear with frequency.

4.2. Losses due to current

AC losses have been computed using Eq. 4 where Figure 3 (b) shows the Instantaneous AC losses for different operating currents. It can be observed that the AC losses are found to higher at higher currents. It is also worth noting that the rate of increase in AC loss is found to higher at higher operating currents (275 A to 300 A) than at lower currents (200 A to 225 A). Figure 4 (b) shows the average loss variations with operating currents and the variation has been found exponential after fitting the data whose constants are tabulated in Table 2 along with standard errors, where I_{op} is the operating current. This exponential function can be used to estimate the AC losses among the operating current varying from 200 A to 300 A. Average AC loss of 368 W/m is found for 300 A operating current which is near to the critical value (330 A) and 80 W/m is for 200 A. This implies that AC losses are very sensitive to operating currents and found to increase with faster rates as operating current approaches to critical value.

Table 1. Design parameters for the HT-SMES

Description	Value
Stored Energy	1MJ
Bore Diameter	780mm
Coil Inductance	1.43 H
Cu cover height (each side)	20e-6m
Air gap and insulator height	2e-4m
Ag cover height	4e-6m
Substrate height	55e-6m
HTS layer height	1e-6m
Tape height	1e-4m
Tape width	12e-3m
Number of tapes	60
Coil height	0.018m
Resistivity of Air	1 m*V/A
Resistivity of Ag	2.7e-9 m*V/A
Resistivity of Cu	1.97e-9 m*V/A
Resistivity of substrate	1.25e-9 m*V/A
Frequency of transport current	50Hz

Table 2. Curve Fitting Parameters

Average AC Loss= exp(a+ b*I_{op}+ c*I_{op}^2)		
Constants	**Values**	**Standard Error**
a	1.55471	0.7187
b	0.01387	0.00551
c	2.10E-06	1.05E-05

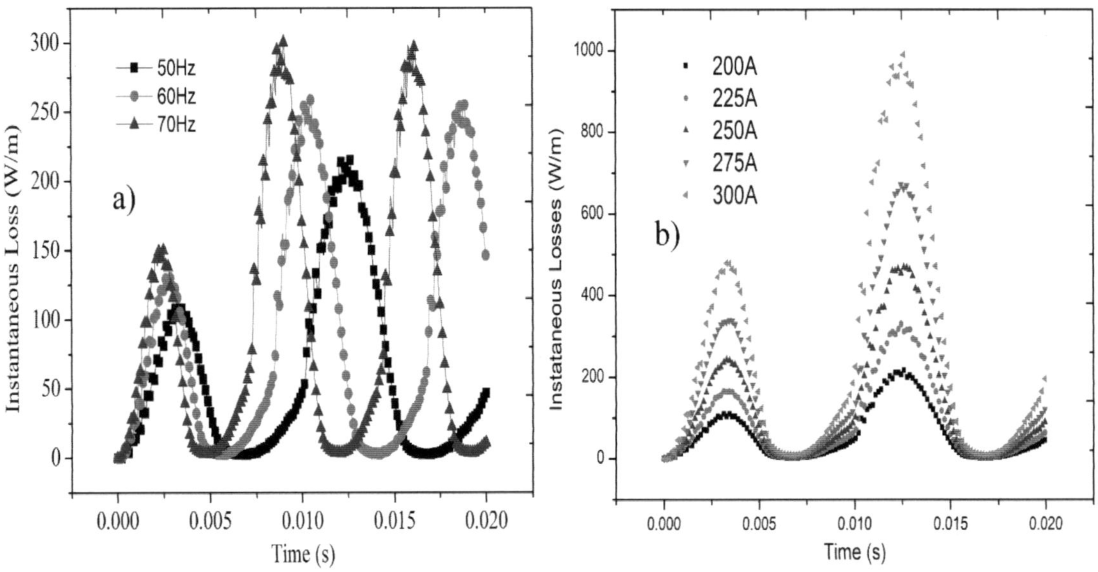

Figure 3. Instantaneous Losses variation with (a) frequency and (b) current

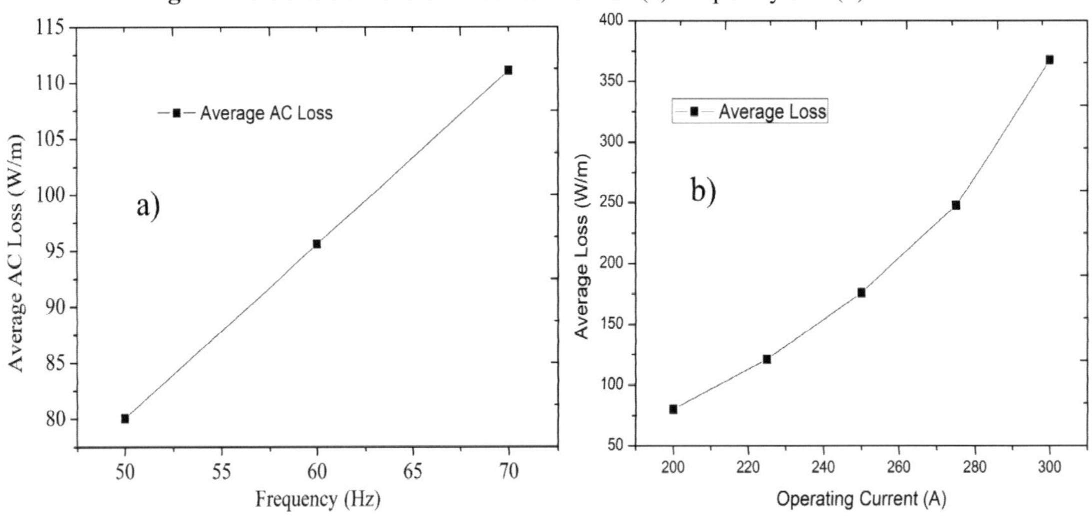

Figure 4. Average AC Losses variation with (a) frequency and (b) current.

5. Conclusions

For a current of 200 A and 60 turns on a single pancake coil, instantaneous losses are found to increase with frequency and the average losses are varying linearly with the frequency, however, the instantaneous losses are found to increase with increase in operating current and the rate at which these losses are increasing is more near the critical current. The average losses for different operating currents are increasing exponentially compared to the variation due to frequency change. So, it is beneficial if current should reach its steady state slowly means charging duration must be longer.

6. References

[1] S. Ma, S. Wang, C. Zhang, and S. Zhang, "A method to improve the efficiency of an electric aircraft propulsion system," *Energy*, vol. 140, pp. 436–443, 2017.

[2] S. C. Clarke, "Aircraft Electric Propulsion Systems: Applied Research at NASA," pp. 1–37, 2015.

[3] T. J. Haugan, "Development of Superconducting and Cryogenic Power Systems and Impact for Aircraft Propulsion," *Air Force Res. Lab.*, no. April, 2015.

[4] A. Kumar, A. Agrawal, and J. M. L. Jeyan, "A Numerical Model Comprising the Effect of Number of Turns on AC Losses in 2G HTS Coated Conductor at 77K using H-formulations," in *2019 IEEE 2nd International Conference on Power and Energy Applications (ICPEA)*, 2019, pp. 115–118.

[5] A. Kumar, J. V. M. L. Jeyan, and A. Agarwal, "Electromagnetic Analysis on 2.5 MJ High Temperature Superconducting Magnetic Energy Storage (SMES) Coil to be used in Uninterruptible Power Applications," *Mater. Today Proc.*, vol. 21, pp. 1755–1762, 2020.

[6] A. Kumar and R. Kaur, "Electromagnetic analysis of 1MJ class of high temperature superconducting magnetic energy storage (SMES) coil to be used in power applications," in *AIP Conference Proceedings*, 2018, vol. 2005, no. 1, p. 50003.

[7] A. Kumar, J. V. M. L. Jeyan, and A. Agarwal, "Numerical analysis on 10 MJ solenoidal high temperature superconducting magnetic energy storage system to evaluate magnetic flux and Lorentz force distribution," *Phys. C Supercond. its Appl.*, vol. 558, pp. 17–24, 2019.

[8] A. Agrawal and A. Kumar, "Effect of Self-field and Loosening of Stacked Superconducting Tapes on Critical Current of a Single Pancake Coil," *Int. J. Eng. Adv. Technol.*, vol. 9, no. 1S3, pp. 155–158, 2019.

[9] A. Kumar and A. Agrawal, "Effect of Substrate Layer Thickness on the AC Losses in Stacked Superconducting Pancake Coils using Direct H-formulations," *Int. J. Eng. Adv. Technol.*, vol. 9, no. 1S3, pp. 150–154, 2019.

[10] V. Lombardo, E. Barzi, D. Turrioni, and A. V Zlobin, "Critical Currents of YBCO Tapes and BSCCO Wires at Different Temperatures and Magnetic Fields," *IEEE Trans. Appl. Supercond.*, vol. 21, no. 3, pp. 3247–3250, 2010.

[11] R. Brambilla, F. Grilli, and L. Martini, "Development of an edge-element model for AC loss computation of high-temperature superconductors," vol. 16, 2007.

[12] V. M. R. Zermeno, F. Grilli, and F. Sirois, "A full 3D time-dependent electromagnetic model for Roebel cables," *Supercond. Sci. Technol.*, vol. 26, no. 5, 2013.

Homopolar superconducting AC machines, with HTS dynamo driven field coils, for aerospace applications

S Kalsi[1], R A Badcock[2], K Hamilton[2] and J G Storey[2]

[1]Kalsi Green Power Systems, LLC, Princeton, NJ 08540
[2] Robinson Research Institute, Victoria University of Wellington, Lower Hutt 5046, New Zealand

rod.badcock@vuw.ac.nz

Abstract. There is worldwide interest in high-speed motors and generators with characteristics of compactness, light weight and high efficiency for aerospace applications. Several options are under consideration. However, machines employing high temperature superconductors (HTS) look promising for enabling machines with the desired characteristics. Machines employing excitation field windings on the rotor are constrained by the stress limit of rotor teeth and mechanisms for holding the winding at very high speed. Homopolar AC synchronous machines characteristically employ both the DC field excitation winding and AC armature windings in the stator. The rotor is merely a magnetic iron forging with salient pole lumps, which could be rotated at very high speeds up to the stress limit of the rotor materials. Rotational speeds of 50,000 RPM and higher are achievable. The high rotational speed enables more compact lightweight machines.

This paper describes a 2 MW 25,000 RPM concept designs for machines employing HTS field excitation windings. The AC armature winding is made of actively cooled copper Litz conductor. The field winding consists of a small turn-count HTS coil that could be ramped up or down with a contactless HTS dynamo. This eliminates current leads spanning room-temperature and cryogenic regions and are major source for thermal conduction into the cryogenic region and thereby increase thermal load to be removed with refrigerators. For early adaption of this technology for the aerospace applications, this 2 MW machine weighing 380 kg with an efficiency > 99% represents an attractive option.

1. Introduction

The aviation industry is working diligently to develop an all-electric aircraft [1], with liquid hydrogen and fuel cells being considered as the prime generation source for aircraft propulsion [2]. Compact, lightweight, high efficiency motors and generators are essential for such applications [3] [4]. General Electric's Homopolar Inductor Alternator (HIA) is a good example of a prototype for an airborne generator; achieving a power density of ~ 9 kW/kg for a 5 MVA, 35,000 RPM machine [5].

Such AC homopolar synchronous machines are an ideal choice for near term aerospace applications. These machines include both AC armature and Direct-Current (DC) excitation windings within the stationary part of the machine. The stationary excitation winding magnetizes a solid steel rotor, enabling operating speeds limited only by the mechanical stress limit of the rotor steel. The operating speeds are many multiples of conventional power 50/60 Hz machines. Significant cooling requirements limit machines of this type utilizing copper excitation windings to only a few kilowatts. However, megawatt ratings become possible when superconducting excitation coils are used. The superconducting excitation coil is circular in shape and is accommodated in the stator. Its major thermal load is via conduction through current leads that span room-temperature and the cryogenic region of the coil. This thermal load must be removed with a refrigerator, which adds mass to the motor system and lowers overall efficiency. An alternative is to charge the field coil wirelessly using a DC dynamo [6, 7, 8] being developed by Victoria University of Wellington. In DC dynamos, the HTS stator is subjected to a time-varying magnetic field provided by a rotor housing permanent magnets [9, 10]. A time-averaged non-zero DC voltage is developed in the dynamo stator [11]. This voltage is utilized for exciting the field coil [12]. Dynamos capable of delivering kiloamp class currents have recently been demonstrated [13, 14]. Since there are no warm components directly connected to the field coil the cryogenic heat-leak is minimized, enabling the use of high current coil design methodology and quasi-persistent operation. This paper describes 2 MW, 25000 RPM concept designs for machines employing an HTS field excitation winding energized by a DC dynamo.

2. Aircraft Motor Specifications

Table 1 summarises the specifications for the 2 MW 25,000 RPM motor excited with a DC dynamo.

Table 1. Specifications for a 2 MW, 25,000-RPM Aircraft Propulsion Motor

Parameter	Value
Motor Rating	2 MW
Motor Speed	25,000 RPM
Line Voltage	~1000 V
Rated power factor	0.9 lag
Rotor diameter	< 500 mm
Axial length	< 800 mm
Field excitation winding	REBCO
Operating temperature	50 K

This motor design is based on the following assumptions;

a) Synchronous machine with 6-poles
b) Both DC field excitation and AC armature windings are located in the stator
c) Armature 3-phase winding employs a suitable Litz copper wire cable
d) Current density in the Litz copper strands is 6 A/mm^2
e) Armature coils are liquid cooled (water or oil)
f) Superconductor field winding uses Rare-Earth-Barium-Copper-Oxide (REBCO) conductor
g) REBCO windings operate at 50K
h) Field winding is cooled with a cryocooler available off-the-shelf
i) Field winding is charged wirelessly with a DC dynamo
j) Rotor is made of high permeability Carpenter Steel's Aermet 310 magnetic steel
k) Stator laminations are 0.1 mm thick Japanese JNEX-Core (model 10JNEX900)

The design and analysis for such machines is based on reference [15]. A similar machine was previously designed for a flywheel energy storage system [16] but did not incorporate excitation by a dynamo.

3. Motor Configuration

A cutaway diagram of the AC homopolar motor/generator is shown in Figure 1. The shaft and rotor have not been sectioned so as to show the 6-pole layout, with 60° rotationally offset poles. The three armature coil colors illustrate the three phases winding scheme. The HTS coil cryostat and its surrounding insulation have been omitted for clarity, as has an electromagnetic (EM) shield located between the armatures and the superconducting coils. The ferromagnetic poles are shaped as in any salient pole machine so as to reduce the harmonic and torque ripple to acceptable levels. The motor housing is maintained in a partial vacuum for reducing windage friction drag at high rotational speeds. Frictionless non-contact magnetic bearings are required since mechanical bearings would not be able to operate continuously due to frictional losses and wear. Further configurational details are available in reference [16].

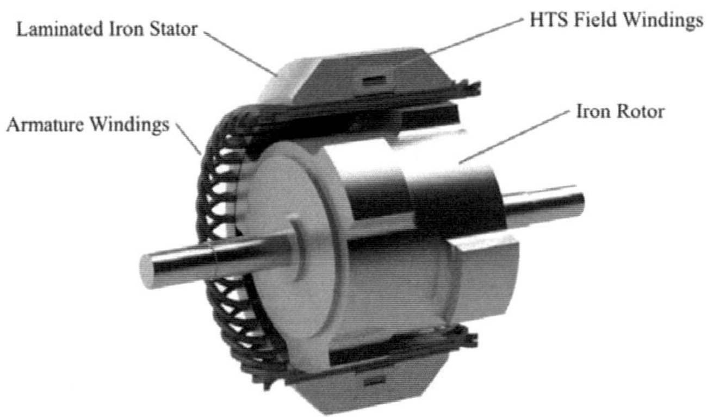

Figure 1. Sectioned view of the AC homopolar motor/generator

Initial sizing is conducted using 2D finite-element analysis (FEA) code. In this approximation the rotor poles are not offset by 90 elect.-degrees as the behavior of interest is the capability of the machine to effectively route flux within the rotor and stator. Figure 2 shows a 2D FEA model cross-section with armature coils housed between iron stator teeth. Note that the slots in the iron core (as shown in the figure) are for quantifying the effect of the iron teeth. In a real machine, the slots and teeth run parallel to the rotational axis of the machine. A double-layer, 3-phase AC winding configuration is selected. The dimensions of this cross-section suit the specifications of Table 1. The location of the field excitation coil is also shown in the figure.

The 2D model of Figure 2 was calibrated with a 3D Opera FEA model shown in Figure 3. Empirical corrections were applied to the 2D model to match the results of 3D model. The 2D model is preferred because of its simplicity and ease of rapidly comparing different designs. The 3D calculated field experienced by the full-pitched stator windings is decomposed into fundamental and harmonics that are listed in Table 2. Most harmonics of concern (5th, 7th, 11th and 13th) are quite small and are not expected to be problematic in creating excessive eddy-current heating in the stator coil.

Preliminary design of the machine is summarized in Table 3. The machine is sized for 2 MW at a 3-phase line voltage of 1292 V. The table also lists preliminary component weights for the machine. The power density of this machine is 5.4 kW/kg, which is comparable with 9.2 kW/kg 5 MW @35,000 RPM machine presented by GE [2]; i.e. 2 MW @25,000 RPM scaled to 35,000 RPM (5.4*35000/25000 = 7.6 kW/kg. Since the power density and efficiency of such machines are highly size dependent, this 2 MW machine could achieve power densities even higher than GE projection in 5 MW rating.

The field coil is designed using SuperPower 3 mm wide HTS tapes with critical current (I_c) enhanced by a factor of 3.3 (to account for thick film conductor being proposed by University of Houston). The self-field I_c for this conductor is 165 A @ 77 K and 690 A @ 50 K [17, 18]. However, the cross-section of HTS coil is very small compared to the allocated cavity size being controlled by cryostat walls. Thus, a higher I_c conductor is likely to have little influence on this machine size or mass. The maximum perpendicular field experienced by the HTS field coil is 0.16 T (at full-load). This field coil creates 1.7 T field in the magnetic iron. It is not desirable to operate the iron at higher field in such machines because they operate on basis of difference in reluctance of magnetic path under a pole and in inter-pole region. Higher magnetic field in the iron will reduce this difference casing increase in machine size and mass. The field coil carries 364 A at full-load. If current leads are employed the thermal load conducted to cryogenic environment would be about 36 W. Estimated thermal conduction through coil cryostat is about 30 W. Using an HTS dynamo, the thermal conduction will be reduced by 90% of that due to current leads (~ 4 W). Thus, total thermal load with and without the dynamo would be 34 W and 66 W, respectively. Table 4 compares the two systems; with current leads and with dynamo.

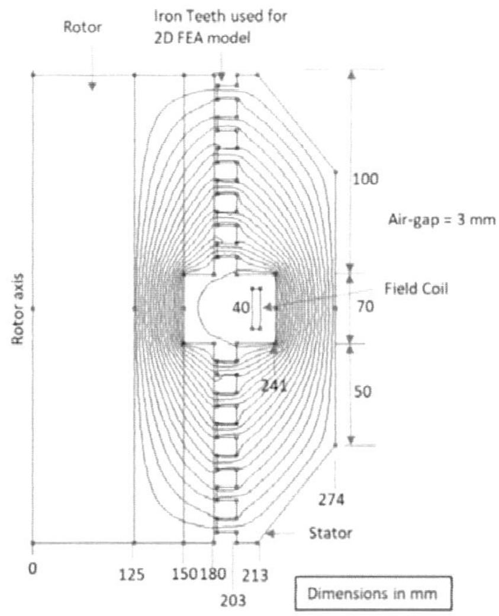

Figure 2. Representative cross-sections of the machine (slots in iron core are intentionally oriented as shown to assess effect of teeth with 2D FEA modelling).

Table 2. Field harmonics experienced by the full-pitched stator coils.

Harmonic	Field(G)	Fraction of fundamental
1	3512	1
3	71.9	0.02
5	162.8	0.046
7	78.4	0.022
9	118.6	0.034
11	39.3	0.011
13	40.6	0.012
15	33.4	0.009

4. Field Winding Details

The field coil is circular in shape with a cross-section of 40 mm x 8 mm and has a mean radius of 212 mm. It has 60 turns of 3 mm wide REBCO conductor – 12 turns/layer and 5 layers. The overall dimensions of the coil cryostat are 70 mm x 33 mm. Inductance of the field coil is 1.19 mH. Field currents at no-load and rated full-load are 188 A and 364 A, respectively. The dynamo to be integrated with this coil is shown in Figure 4. and will be capable of managing field current over this range.

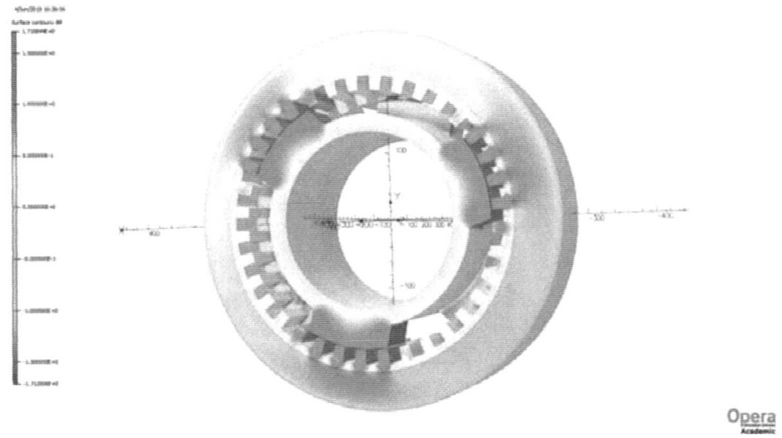

Figure 3. Opera 3-D finite-element model of the machine. Colour scale illustrates the radial field component B_r between ± 1.72 T.

Table 3. Preliminary design summary for the machine with dynamo exciter.

Parameter	Double Layer
Power Rating, kVA	2020
Output power at full-load, kW	2000
Line voltage, V-rms	1292
Phase current, A-rms	922
Overall axial length, m	0.45
Overall diameter, m	0.56
Mass of the machine alone, kg	381
Mass of cryo-cooling system, kg	3
Total mass, kg	383
Efficiency at full-load, %	99.1
Cryocooler load, W	38[a]
Other parameters of interest	
Rated speed, RPM	25000
Number of poles	6
Frequency, Hz	833
Field winding details	
Number of turns	60

Field winding critical current- no-load, A	621
Field winding critical current – rated-load, A	560
Field winding current at rated load, A	364
HTS wire width, mm	3
HTS wire length, m	84
Operating temperature, K	50
Stator winding details	
Active length under each pole, mm	100
Number of armature turns/ph	16
Number of armature circuits	6
Number of coils in armature	36
Number of turns/coil	8
Field coil inductance, mH	1.19
Machine component weight summary	
- Shaft, kg	5
- Rotor yoke, kg	76
- Poles, kg	21
- Stator case, kg	50
- Cooling system, kg	2[a]
Total machine mass, kg	380
Total system mass, kg	380
Torque density, N*m/kg	2.06
Power density, kW/kg	5.4

[a] Calculated using Ref: Ray Radebaugh, "Ray Radebaugh Cryocoolers for Aircraft Superconducting Generators and Motors", NIST, AIP Conference Proceedings 1434. 171 (2012): doi: 10.1063/1.4706918

Table 4. Comparison of field winding thermal load with and without a dynamo.

Description	Current Leads	Dynamo
Thermal conduction through cryostat, W	30	30
Thermal conduction through exciter, W	36	4
Total thermal load, W	66	34
Power input to refrigerator, kW	1.94	1.05
Weight of refrigerator, kg	4	2

5. Conclusions

The AC homopolar concept currently represents least risky option for aircraft applications in near term. It could be built by integrating components already prototyped and tested. With an efficiency > 99% and power density > 5.4 kW/kg might be acceptable to early adopters, while research continues to seek alternatives to achieve higher efficiencies and power densities. It must be noted that the power density and efficiency of this class of machines are highly rating dependent. For example, a 5 MW, 35,000 RPM machine could achieve power density > 12 kW/kg – exceeding the 9.2 kW/kg reported for the GE machine [5].

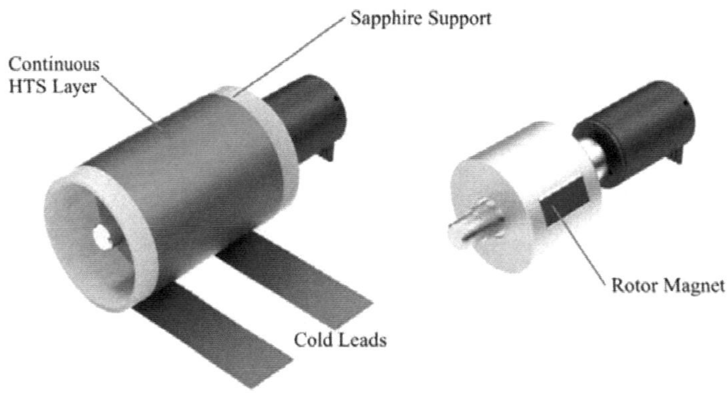

Figure 4. DC HTS dynamo concept to be integrated with the field coil. The right image shows the device with the HTS and sapphire support removed to illustrate the rotating magnet.

6. References

[1] Brelje B J and Martins J R R A 2019 Electric, hybrid, and turboelectric fixed-wing aircraft: A review of concepts, models, and design approaches Progress in Aerospace Sciences 104 1–19
[2] Ashcraft S W, Padron A S, Pascioni K A, Stout G W and Huff D L 2011 Review of Propulsion Technologies for N+3 Subsonic Vehicle Concepts 38
[3] Haran K S, Kalsi S, Arndt T, Karmaker H, Badcock R, Buckley B, Haugan T, Izumi M, Loder D and Bray J W 2017 High power density superconducting rotating machines – development status and technology roadmap Superconductor Science and Technology 30 123002
[4] Brown G 2011 Weights and Efficiencies of Electric Components of a Turboelectric Aircraft Propulsion System 49th AIAA Aerospace Sciences Meeting including the New Horizons Forum and Aerospace Exposition 49th AIAA Aerospace Sciences Meeting including the New Horizons Forum and Aerospace Exposition (Orlando, Florida: American Institute of Aeronautics and Astronautics)
[5] Sivasubramaniam K, Zhang T, Lokhandwalla M, Laskaris E T, Bray J W, Gerstler B, Shah M R and Alexander J P 2009 Development of a high speed HTS generator for airborne applications IEEE Transactions on Applied Superconductivity 19 1656–61
[6] Bumby C W, Jiang Z, Storey J G, Pantoja A E and Badcock R A 2016 Anomalous open-circuit voltage from a high-Tc superconducting dynamo Appl. Phys. Lett. 108 122601
[7] Bumby C W, Badcock R A, Sung H-J, Kim K-M, Jiang Z, Pantoja A E, Bernado P, Park M and Buckley R G 2016 Development of a brushless HTS exciter for a 10 kW HTS synchronous generator Superconductor Science and Technology 29 024008
[8] Hoffmann C, Pooke D and Caplin A D 2011 Flux Pump for HTS Magnets IEEE Trans. Appl. Supercond. 21 1628–31

[9] Ma J, Geng J, Gawith J, Zhang H, Li C, Shen B, Dong Q, Yang J, Chen J, Li Z and Coombs T A 2019 Rotating Permanent Magnets Based Flux Pump for HTS No-Insulation Coil IEEE Trans. Appl. Supercond. 8663420

[10] Hoffmann C, Walsh R, Karrer-Mueller E and Pooke D 2012 Design Parameters for an HTS Flux Pump Phys. Procedia 36 1324–9

[11] Mataira R C, Ainslie M D, Badcock R A and Bumby C W 2019 Origin of the dc output voltage from a high-Tc superconducting dynamo Appl. Phys. Lett. 114 162601

[12] Jeon H, Lee J, Han S, Kim J H, Hyeon C J, Kim H M, Park D, Chung Y D, Ko T K and Yoon Y S 2018 Methods for Increasing the Saturation Current and Charging Speed of a Rotary HTS Flux-Pump to Charge the Field Coil of a Synchronous Motor IEEE Trans. Appl. Supercond. 28 1–5

[13] Hamilton K, Pantoja A E, Storey J G, Jiang Z, Badcock R A and Bumby C W 2019 Asynchronous magnet–stator topologies in a squirrel-cage superconducting dynamo IEEE Transactions on Applied Superconductivity 29 5200705

[14] Walsh R M, Bumby C W, Badcock R A, Slade R A, Jiang Z, Hamilton K A and Fee M G 2018 Superconducting Current Pump U.S. Patent 9,972,429

[15] Kalsi S S 2011 Applications of High Temperature Superconductors to Electric Power Equipment (Hoboken: John Wiley & Sons, Inc.)

[16] Kalsi S, Hamilton K, Buckley R and Badcock R 2018 Superconducting ac homopolar machines for high-speed applications Energies 12 86

[17] Wimbush S C and Strickland N M 2016 A public database of high-temperature superconductor critical current data IEEE Trans. Appl. Supercond. 27 1–5

[18] Wimbush S and Strickland N 2016 A high-temperature superconducting (HTS) wire critical current database https://doi.org/10.6084/m9.figshare.c.2861821.v1

IOP Conf. Series: Materials Science and Engineering **756** (2020) 012029 doi:10.1088/1757-899X/756/1/012029

Coolant transfer coupling with integrated dynamo for rotor with HTS windings

S Kalsi[1], R A Badcock[2] and K A Hamilton[2]

[1]Kalsi Green Power Systems, LLC, Princeton, NJ 08540
[2] Robinson Research Institute, Victoria University of Wellington, Lower Hutt 5046, New Zealand

skalsi@KalsiGPS.com

Abstract. Most superconducting synchronous machines employ HTS field excitation windings on the rotor operating at cryogenic temperature. These windings are usually cooled by coolant supplied from a stationary source to the rotor with rotary couplings. Closed loop gaseous helium couplings have been employed on mega-watt size machines operating at both low speed and high speed. However, these couplings were marred with leaking of cryogen out of the closed loop and needed periodic replenishment. This undesirable problem has been recognized long ago but no suitable solution has emerged in the open literature. Currently, the HTS machines are being considered for the aerospace applications, wherein leakage of cryogen from closed loop is highly undesirable. This paper presents a concept that prevents the cryogen leakage and/or its collections it for returning to the closed cooling loop. An option is also explored to include an HTS dynamo for providing the field excitation without current leads. This concept would need de-risking before using in the motors and generators for the aerospace applications. Possible cryogens for cooling include gaseous helium, Neon, H2 and N2.

1. Introduction

Superconducting machines have long been associated with characteristics like compactness, lightweight and high efficiency [1]. Most machines built to date have been synchronous type with DC excitation field windings on the rotor. In most cases, resistive losses in the field winding and stator AC winding are comparable. Superconducting field windings have been demonstrated to increase the magnetic field experienced by the stator winding, while reducing the field winding losses by nearly 80% as compared to copper equivalents. This has reduced the size and mass of machines used for industrial as well as ship propulsion applications [2,3]. Many such machines have been built, and all have been maintained at cryogenic temperatures during operation by transferring coolant from stationary coolers to the rotor through a rotary coupling [4,5,6,7].

Superconducting motors with further reduced mass may be suitable as propulsion motors for All Electric Aircraft. Electrically propelled aircraft have the potential to be more efficient, have reduced emissions, and operate with less noise both on the ground and in the cabin, as well as many other potential advantages as described in [3].

Figure 1 shows a concept AC synchronous machine [8] where both the DC rotor field winding and the AC stator armature winding are superconducting. The rotating magnetic field produced by physical rotation of a superconducting DC field winding is typically twice that produced in a machine built with conventional copper field windings.

The machine described in this paper has both stator and rotor windings requiring cryogenic cooling. Each winding has individual vacuum space for ease of construction, testing and maintenance during operation. In this concept the cooling for each winding is supplied from a non-rotating source. This paper presents a concept design for the transfer coupling to supply coolant to the rotor in a closed loop where the coolant warmed by the rotor losses is recovered. The recovered coolant may then be re-cooled and reused.

An HTS dynamo [9] is also integrated with the transfer coupling for wireless excitation of the field windings. This is intended to increase the efficiency of the motor by reducing the thermal conduction into the rotor cryogenic environment, and removing resistive components from the field winding supply circuit.

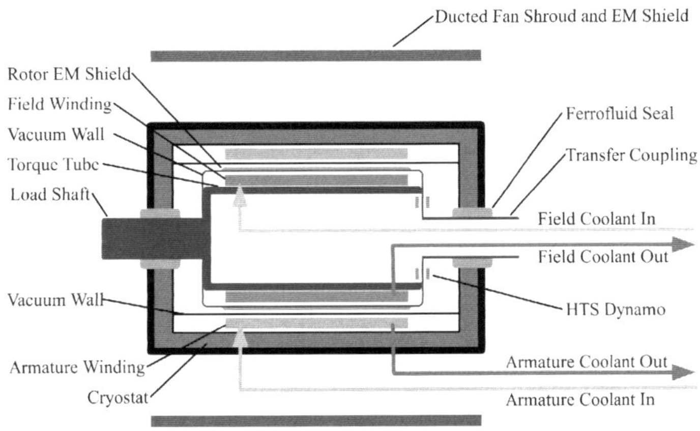

Figure 1. Superconducting motor concept with HTS field winding.

2. Transfer Coupling Concept Description

A coupling concept shown in Figure 2 uses a double walled bayonet to transfer cold coolant to the rotor and return warm coolant out of the rotor. This bayonet mates with the end of a second single walled bayonet fixed to the rotor. A Teflon cold slip seal separates the cryocooled coolant in stream from the warmer coolant out stream.

The warmed returning coolant enters the outer rotating tube through holes in the tube wall. This rotating tube is attached to the room temperature motor shaft. Thus, this coolant tube is cold at the slip seal, and at room temperature at the shaft end. On the warm end, stationary and rotary parts are supported with a room temperature ferrofluid seal. This seal can hold a small pressure difference across it, but the sealing action is not totally leak free. The bayonets serve as an interface between the cold rotating parts and stationary cooling system. Since a partial vacuum is required in the air gap of the high-speed machine, the motor shaft is supported with another set of ferrofluid seals at the load shaft end of the motor.

Rotary seal concepts like this have been used by others [10]. However, the smaller ferrofluid seal between the rotating and stationary bayonets is not fully capable of stopping leakage of coolants such as gaseous helium to the environment. Closed loop cooling systems have been built [6] with room-temperature ferrofluidic seals. Currently no suitable solution is available in the published literature that prevents coolant leakage through the ferrofluid seals. The loss of coolant, however, is not acceptable in machines for airplanes and other applications.

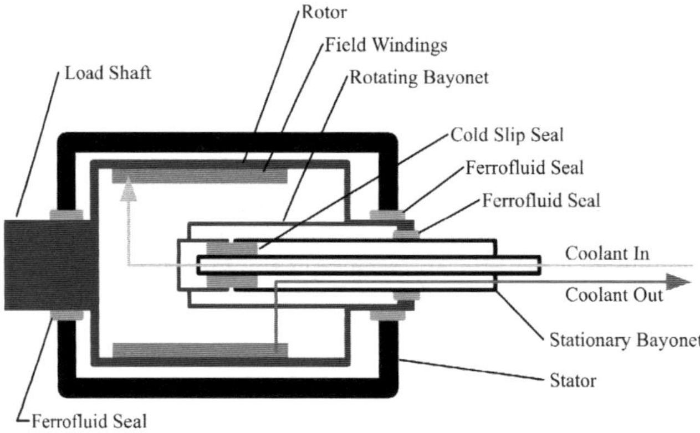

Figure 2. Coolant transfer coupling concept.

In order to fix the coolant loss, this paper presents a glove box type structure shown in Figure 3. This arrangement is expected to prevent leakage of the coolant from the machine by recovering the coolant that passes through the inter-bayonet ferrofluid seal. The glove box type structure encases all rotating parts against the motor wall and attaches to the stationary bayonet with a bolted flange. The pump out port in the glove box may be used for collecting the leaking coolant, or providing positive pressure to prevent the leakage.

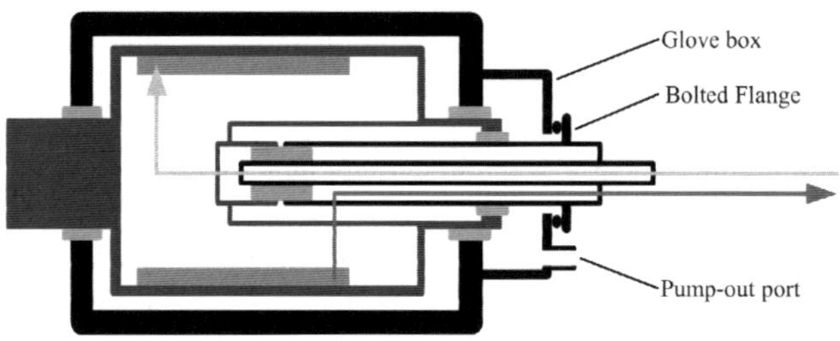

Figure 3. Leak-free coolant transfer coupling concept

Figure 4 shows the interface between the rotor and stationary cooling system. Coolant recovered via the pump out port is added to the closed loop gas circulating system. The coolant in the closed loop is circulated with a gas pump. Although this configuration has not yet been tried, almost all components needed for this cooling system are available off-the-shelf.

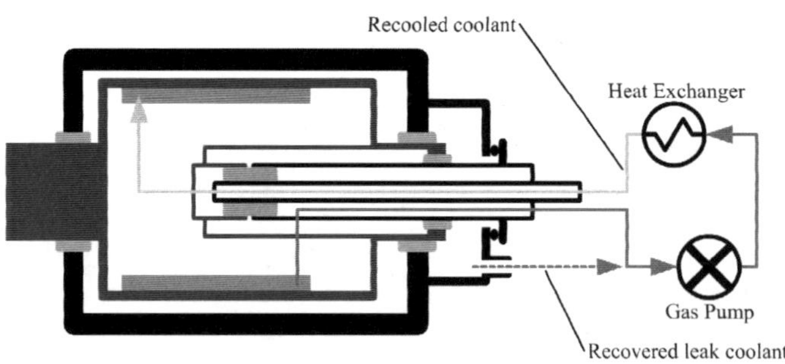

Figure 4. Rotor cooling system with stationary cooler

3. Integration of Dynamo for Field Winding Excitation

Field poles must be excited wirelessly without need for slip rings and brushes for avoiding reliability and maintenance issues for high-speed machines. In the past, brushless concepts based on magnetic iron core were used that employed current leads spanning room-temperature and cryogenic regions. These current leads created a large thermal load for the cryocooler. Moreover, such an exciter is too heavy for aerospace applications where high specific power is required. A possible option is to employ the HTS dynamo [9,11] developed by Victoria University of Wellington (VUW). Its weight is expected to be less than 10% of the weight of magnetic core exciter. Figure 5 illustrates the HTS dynamo concept that is capable of supplying more than a kA of current [12]. Figure 6 shows how such a dynamo may be integrated into the motor as a passive device for energising the field winding without current leads spanning crossing the cryostat wall.

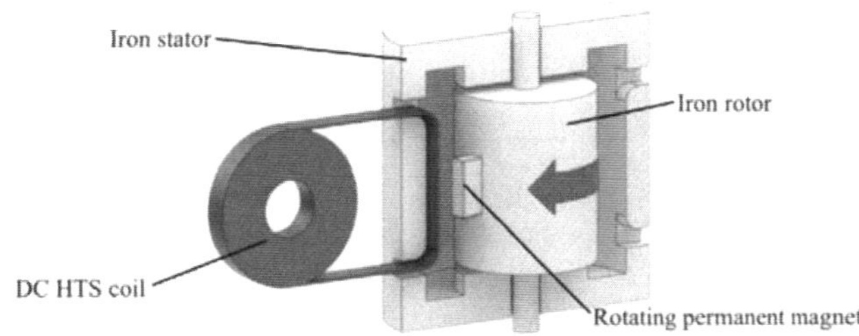

Figure 5. Brushless HTS dynamo concept.

The permanent magnets (PM) for excitation are attached to the outer wall of the stationary bayonet within the air core of the motor. The superconducting elements in which DC voltage is induced is attached to the outer wall of the rotating bayonet. To prevent the rotating magnets acting as a magnetic brake the rotating bayonet must be manufactured from a non-metallic material such as carbon fiber composite.

The field windings may be energized to high DC currents over many rotations of the magnets relative to the superconducting elements. Energization will occur continuously provided there is relative movement between these components, thereby providing the current to make up any losses caused by losses due to non-superconducting joints in the HTS circuit, or AC effects. The HTS dynamo enables all components of the field winding circuit to be superconducting, and high currents may be supported with little loss. A high circuit current within the field windings allows large magnetic output coils to be produced with relatively few windings. In addition to being lighter and cheaper to produce, low turn coils have smaller inductance and are faster to charge.

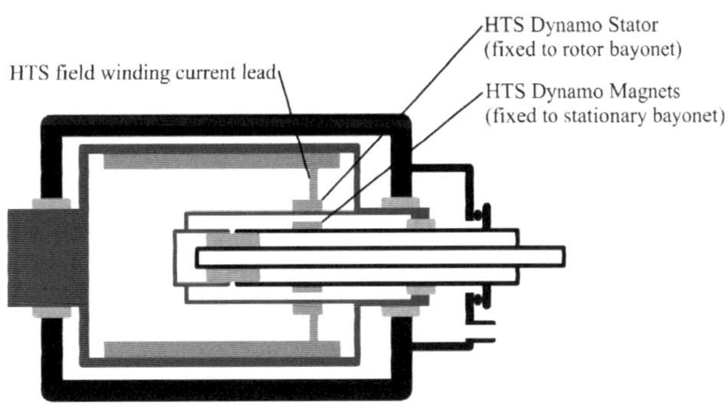

Figure 6. HTS Dynamo integrated with transfer coupling for rotor field excitation

It may be necessary in some cases to charge the rotor poles before a synchronous machine is started. This would require the PMs to rotate when the rotor is stationary. This could be accomplished by rotating the outer tube of bayonet using external means on the room-temperature end. However, Figure 7 shows an additional tube carrying PMs that could be rotated independent of the rotor. This will require an additional ferrofluid seal on the room temperature end. This tube is rotated with a small motor located within the glove box. Once the field windings are energized, the small motor may be disengaged so the dynamo may operate in passive mode as in figure 6. Alternatively, the motor may be used to vary the current in the field windings throughout their operation as desired.

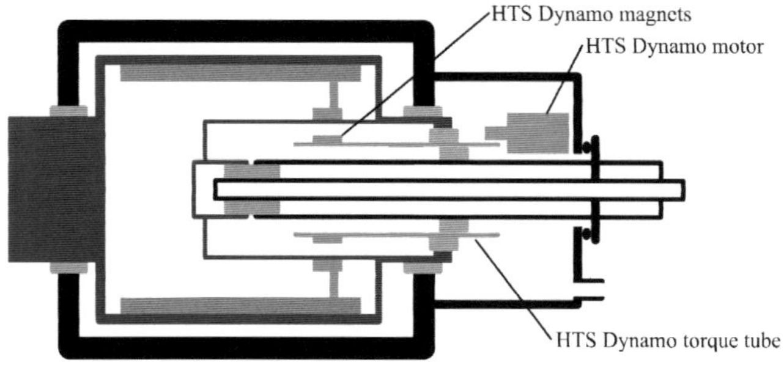

Figure 7. HTS Dynamo integrated with independent HTS dynamo drive tube

4. Conclusions

A rotary seal concept is presented to facilitate transfer of coolant from stationary source to the rotor for cooling the field winding poles. This rotary seal is expected to capture coolant leaking through the room-temperature Ferro-Magnetic seals. The rotary coupling also incorporates DC dynamo for charging the field coils without current leads spanning room-temperatures and cold environment. Leak-free transfer coupling with dynamo is highly essential for achieving highest possible power densities (kW/kg) for airplane applications.

Acknowledgments

This work was supported by the New Zealand Ministry of Business, Innovation and Employment (MBIE) under contract no. RTVU1707.

References

[1] Kalsi S, Weeber K, Takesue H, Lewis C, Neumueller H, Blaugher R. Development status of rotating machines employing superconducting field windings. Proceedings of the IEEE. 2004;92(10):1688-1704. doi: 10.1109/JPROC.2004.833676
[2] Kalsi S. Applications of high temperature superconductors to electric power equipment. Piscataway, NJ: IEEE Press; 2011, ISBN 978-0-470-16768-7
[3] Haran K, Kalsi S, Arndt T, Karmaker H, Badcock R, Buckley R, Haugan T, Izumi M, Loder D, Bray J, Masson P, Stautner W. "High Power Density Superconducting Rotating Machines – Development Status and Technology Roadmap", Supercond. Sci. Technol. 30, (2017) 123002. doi: 10.1088/1361-6668/aa833e
[4] Lizon J. Rotating feed-though for cryogenic fluid, Cryogenics 38 (1998) 853-854
[5] Gamble B. Development of a helium transfer coupling for a superconducting generator rotor, Advances in Cryogenic Engineering, pp. 125-131, Vol. 23, 1978. doi: 10.1007/978-1-4613-4039-3_15
[6] Maguire J, Winn P, Sidi-Yekhlef A and Yan J. Cooling System for HTS Machines, US Patent # 6,625,992 B2, September 30, 2003
[7] Gamble B, Sidi-Yekhlef A, Schwall B, Driscoll D, Shoykhet B. Superconductor rotor cooling system, U.S. Patent 6,812,601, Nov. 2004

[8] Kalsi S, Hamilton K and Badcock R. Superconducting rotating machines for aerospace applications, AIAA/IEEE Electric Aircraft Technologies Symposium, Cincinnati, OH on July 12-13, 2018. doi: 10.2514/6.2018-4796
[9] Bumby C, Badcock R, Sung H-J, Kim K-M, Jiang Z, Pantoja A, Bernardo P, Park M and Buckley R. Development of a brushless HTS exciter for a 10kW HTS synchronous generator, Supercond. Sci. Technol. 29 (2016) 024008 (11pp). doi:10.1088/0953-2048/29/2/024008
[10] Felder B, Miki M, Tsuzuki K, Sato R, Hayakawa H and Izumi M. Cryogenic rotary joints applied to the cooling of superconducting rotating machinery, IEEE Transactions on Applied Superconductivity, 23(3), 5201204-5201204. (2013). doi: 10.1109/TASC.2013.2241382
[11] Walsh R, Bumby C, Badcock R, Slade R, Jiang Z, Hamilton K and Fee M. Victoria Link Ltd, 2018. Superconducting current pump. U.S. Patent 9,972,429
[12] Hamilton K, Pantoja A, Storey J, Jiang Z, Badcock A, and Bumby C. Asynchronous Magnet–Stator Topologies in a Squirrel-Cage Superconducting Dynamo. IEEE Transactions on Applied Superconductivity 29, no. 5 (2019): 1-5. doi: 10.1109/TASC.2019.2892773

IOP Conf. Series: Materials Science and Engineering **756** (2020) 012030 doi:10.1088/1757-899X/756/1/012030

A fully superconducting air-core machine for aircraft propulsion

T Balachandran [1], D Lee [1], N Salk [1], K S Haran [1]

[1] Department of Electrical and Computer Engineering, University of Illinois at Urbana-Champaign, 306 N. Wright Street, Urbana, IL 61801, United States of America

E-mail: kharan@illinois.edu

Abstract. Partial and fully superconducting (SC) machines promise high power density capabilities required for electric propulsion. These machines need to achieve high power densities while reducing electrical heat losses to minimize the required cryogenic power and subsequent additional weight. Hydrogen powered all-electric planes provide a design space where ac losses are manageable. However, the high electrical frequencies in high-speed fully superconducting machines pose a significant challenge to reducing armature ac losses. In high-speed applications, coupling loss in the SC armature coils dominates and becomes a barrier for practical application of these machines. In this paper a fully superconducting machine is proposed for a hydrogen powered regional all-electric plane. An air core design is considered utilizing low ac loss MgB_2 wires. The design is targeted to achieve 50 kW/kg specific power while requiring ac losses to be less than 3 kW. This study explores the possibility of replacing a passive iron shield with active shielding coils to contain the magnetic flux inside the machine while reducing weight and increasing power density. The study focuses on minimizing weight as well as ac losses in the armature coils. An optimization algorithm is used to determine the trade-offs between iron shield and active shield coil designs. Results show that optimal designs for electric propulsion eliminate the passive shield in favor of active shielding coils - increasing the power density of the machine while maintaining the outside flux density below standard safety limits.

1. Introduction

Commercial electric aviation targets demand higher power density (>20 kW/kg) motors with higher efficiency (>92%) to leverage the weight benefits of electric propulsion [1] - [2]. Fully superconducting (SC) machines have the potential to achieve these high-power density targets by increasing electric and magnetic loading without losing efficiency to ohmic heat losses. However, due to the high-speed requirement of electric propulsion, armature winding ac losses become a significant challenge for fully superconducting machines. This paper investigates the feasibility of a hydrogen fuel-cell powered electric propulsion scheme for a regional commercial all-electric aircraft. A fully superconducting machine is proposed considering 30% of the total available cooling capacity of liquid hydrogen's latent heat for motor cooling. To increase the power density, an inside-out synchronous machine is proposed where traditional passive shielding is replaced with active SC shield coils. The additional coils are placed radially outward to contain the flux within the machine. Since core loss and saturation are eliminated, higher flux density designs can be achieved with this topology. Initial studies have shown that flux densities up to 3 to 5 times that of conventional iron core designs are possible using superconductors, resulting in a more compact machine. Previous studies of partially SC machines indicated that with a

5 T air gap flux density, 35% volume reduction can be attained with only a 17% increase in superconductor usage [3]. Converting this machine to a fully SC design can potentially offer further performance improvements. However, armature winding ac losses in fully SC machine designs introduce a limitation on the achievable air-gap field strength and frequency. This paper explores fully superconducting, actively shielded machine designs for electric aircraft propulsion. A 2.5 MW, 4,500 rpm MgB_2 based fully SC machine design is investigated. A feasibility study of this topology for electric propulsion is conducted considering efficiency, weight, and cost. Special attention is given to modeling and reducing ac losses in the machine. An optimal design is targeted which reduces the SC usage and weight while constraining the total ac loss in the machine below a practical limit of 3 kW. A pole count study is conducted to identify the optimal operating electrical frequency which balances tolerable ac losses with high power density. The results are compared against a traditional iron core machine to evaluate the benefits.

2. Fully Superconducting Machines for Electric Propulsion

Several design studies have been conducted on high-speed fully SC machines for electric propulsion. A compilation and analysis of previously proposed high-speed SC machines is reported in [4]. The authors point out that ac losses are greatly affected by machine topology and that, consequently, an optimization needs to be performed to identify the best topology for minimizing these losses. Most previous design studies demonstrate that fully SC machines have the potential to meet the power density required for electric propulsion. However, minimizing ac losses to reach necessary efficiency levels and reduce cryogenic power and weight is still the roadblock for most applications.

2.1. Conventional Fully SC Designs

Conventional SC machine designs use iron in their stator and rotor, or as the back iron to shield magnetic flux inside the machine. A 1 MW, 12,000 rpm, fully SC machine using MgB_2 (Hyper Tech multi-filament wire with 0.18 mm filament diameter, 0.36 mm wire diameter and stainless steel sheath) windings is proposed in [5]. A design with an air-core stator and an iron core rotor is estimated to have 5,378 W of ac losses with a 1 T air-gap flux density. This design resulted in an 850.7 kW machine with a power density of 11.7 kW/kg. An iron core stator design is estimated to have 304 W of ac losses and 3,314 W of core losses. This design resulted in a power density of approximately 6.5 kW/kg. These losses and power densities need to be improved to surpass conventional electric machine power densities and to meet the specifications required for commercial electric propulsion [1] - [2]. Another design study of a 10 MW fully SC machine using REBCO tape ($REBa_2Cu_3O_y$) is presented in [6]. The performance of an air-core stator, 2-pole machine design is compared at three different speeds (3,000 rpm, 6,000 rpm and 9,000 rpm). Results showed that a 25.6 kW/kg power density could be achieved at 9,000 rpm, while 99.2% efficiency can be obtained at 3,000 rpm. A design operating at 20 K with 1 T air-gap flux density at 3,000 rpm had 49.9 kW of estimated losses, and at higher frequencies this loss increased further. It is reported that half of the weight of the machine came from iron used as a passive shield, and if active shield coils are used instead, the power density could be improved significantly.

A comparison study between a partial SC machine and fully SC machine employing MgB_2 (Hyper Tech multi-filament wire with 0.8 mm wire diameter and 25 mm twist pitch) in the armature windings and YBCO tapes as field winding is presented in [7]. EM FEA was carried out on a 5,000 rpm motor at 1.5, 3 and 5 MW; different pole counts were considered. The results showed that fully superconducting motors have the potential to reach a power density of 16 kg/kW. Terao *et al.* showed that for fully superconducting machines whose air-gap flux density is 1 T, ac losses are over 10% of the rated power - rendering them infeasible for use in electric propulsion. An EM study of 3 MW and 5 MW fully SC machines employing MgB_2

(multi-filament wire) cables in the armature coils and YBCO tapes for the field coils is presented in [8]. The 4-pole designs at 5,000 rpm showed that output densities of 19.4 and 25.2 kW/kg could be reached by the 3 MW and 5MW machines respectively. In this study, the impact of back iron is investigated by varying its radial depth. It is evident that the back iron component adds significant weight and also influences ac losses in the machine. For feasible designs, the ac losses need to be reduced by several kilowatts.

While these design studies provide crucial weight and ac loss information, an optimization to minimize ac losses was never carried out. Since ac losses of a given SC wire and machine weight are influenced by several geometric and operational motor parameters, a multi-objective optimization should be explored to develop a front of optimal designs that minimize both objectives.

2.2. Actively shielded design

Traditionally, a passive iron shield is used to maintain flux within the machine, limiting external flux to accepted levels. The active shield machine topology uses a set of SC coils introduced radially outward from the field coils to accomplish the same task. This increases the power density by eliminating the need for a heavy iron shield. Furthermore, iron losses are also eliminated, resulting in a reduction of associated cryogenic power and system weight. Previously conducted EM studies [3], [9] and [10] demonstrate that an active shield topology has the potential to significantly increase power density. A multi-objective optimization study showed that a 32% reduction in volume can be achieved with a 33% increase in SC usage [9]. A similar study to determine the optimal pole-count for a compact machine showed that a 74% volume reduction can be obtained with 104% more SC usage [10].

3. Cryogenic Hydrogen-Energy Electric Transport Aircraft (CHEETA) Concept

Cryogenic hydrogen-energy electric transport aircraft is a concept proposed to utilize the high specific energy content of liquid hydrogen (LH_2) through fuel cell energy conversion and an ultra-efficient distributed electric propulsion system. The concept uses (LH_2) not only as the energy storage mechanism, but also as a cryogen to enable a highly efficient SC electric system. This enables the usage of SC machines with SC power cables to efficiently convert electrical energy into mechanical energy. Letting gaseous hydrogen used in the fuel cells evaporate by heat losses in the SC motor and cables, the system provides free cryogenic cooling for SC electrical components. If the required cryogenic power to remove generated heat losses in the electrical system is maintained below the available free cooling, an ultra-efficient electrical system could be obtained using this concept. In this study, a plane which uses 920 GJ of total energy during a mission and has 20 MW of combined propulsion is explored. The weight of required LH_2 fuel is estimated to be 12,778 kg. Assuming a 0.059 kg/s boil-off rate, 26,383.4 W of total cooling power can be extracted. If 8 motors are used for distributed propulsion, the power of each motor and the available cooling power per motor can be estimated.

$$Power\ of\ each\ motor = 20/8 = 2.5\ \mathrm{MW} \tag{1}$$

$$Available\ free\ cooling\ power\ per\ motor = 26,383.4/8 = 3298\ \mathrm{W} \tag{2}$$

$$Targeted\ power\ density\ of\ each\ motor \geq 25\ \mathrm{kW/kg} \tag{3}$$

Therefore, if each 2.5 MW motor is designed to have ac losses below 3 kW, no additional cryo-cooling refrigeration equipment is necessary. However, a method of cooling that utilizes the available LH_2 needs to be developed and will add significant weight to the machine.

4. Proposed fully superconducting machine

In this study a 2.5 MW, 4,500 rpm fully SC actively shielded motor is proposed for electric propulsion. The proposed machine specifications are tabulated in Table 1. However, to demonstrate the trade-offs between active and passive shield designs, a topology which consists of both shielding methods is explored in this paper. A genetic algorithm based optimization software (GOSET) is used to determine the best design suitable for electric propulsion between the two concepts.

Multi-filament MgB_2 SC cables are chosen for the armature windings due to their low ac loss characteristics. The same wires are also chosen for the field and shield coils. MgB_2 cable specifications are tabulated in Table 2. While setting the field and shield current densities to be at 200 A/mm^2, the armature current density is allowed to vary during optimization. However, designs which have more than 50% of the critical current density evaluated at the applied field are considered infeasible as they exceed our risk margin.

Table 1. Proposed Machine Specifications

Specification	Value
Power	2.5 [MW]
Speed	4,500 [rpm]
Power density	$\geq$ 25 [kW/kg]
Air-gap length	5 [mm]
Outer diameter	max 0.5 [m]
Pole number	2, 4, 6, 8, 10, 12
Superconductor	MgB_2
Operating temperature	20 K
Operating phase voltage	750 V_{rms}
Operating current	1111 A_{rms}
Armature slots per pole	6
Number of parallel connection	4
Armature current density, RMS	max 200 [A_{rms}/mm^2]
Field current density	200 [A/mm^2]
Shield current density	200 [A/mm^2]

Table 2. MgB_2 Conductor Data (D_0=0.32/d_f=10/L=5) [mm/μm/mm]

Parameter	Symbol	Value
Critical current density at 20 K and self field 2 T [A/mm^2]	J_c	1780
SC outer diameter [mm]	D_o	0.32
Filament diameter [mm]	d_f	0.01
Number of filaments	n	114
SC fill factor	λ	0.15
Effective fill factor	λ_e	0.49
Transverse resistivity [Ω-mm]	ρ	3.65E^{-4}
Effective transverse resistivity [Ω-mm]	ρ_e	1.25E^{-4}
Twist pitch [mm]	L	5

5. Estimating AC Losses in Armature Coils

When alternating fields and currents are introduced to SC materials they generate losses in the form of heat, known as ac losses. These losses are cyclical and increase with both applied frequency and the square of the peak applied flux density. Because motors for electric propulsion operate at high speeds and therefore experience high electrical frequencies, losses in the armature coils are significant and determine the feasibility of a design. In this paper, Carr's model [11] is used to estimate ac losses in the armature windings. These losses are evaluated across a unit length of wire and integrated over the total volume to compute the total loss. This model is widely used and was tweaked to estimate ac losses at high frequencies in [12] for MgB_2 cables and YBCO tapes.

The field required to penetrate the superconductor is evaluated as

$$B_p = 0.4\mu_0 J_c d_f \tag{4}$$

where μ_0, J_c, and d_f are the magnetic permeability, critical current density, and filament diameter respectively. The following equations are used to compute hysteresis (P_h), eddy current (P_e), coupling (P_c), and transport current (P_t) losses when the peak applied flux density, B_m, is much greater than B_p.

$$P_h = \frac{4}{3} J_c d_f B_m f \lambda \tag{5}$$

$$P_e = \frac{\pi^2}{k\rho_e}(D_o B_m f)^2 \tag{6}$$

$$P_c = \frac{1}{n\rho_e}(L B_m f)^2 \tag{7}$$

$$P_t = \frac{\mu_0 f}{\pi} I_c^2 \frac{\left[(1-\frac{I_0}{I_c})\ln(1-\frac{I_0}{I_c}) + \frac{I_0}{I_c} - 0.5(\frac{I_0}{I_c})^2\right]}{\pi(\frac{D_o}{2})^2} \tag{8}$$

where I_0, I_c, f, λ, ρ_e, L, and D_0 are the operating current, critical current, applied frequency, fill factor, copper matrix effective transverse resistivity, twist pitch, and SC outer diameter respectively. The constants k and n are set to be 4 and 2 respectively for circular, multi-filament SC wires.

6. Optimal Design Scheme

A simulation-based approach is chosen to explore the optimal design. An objective function shown in Equation 9 is formulated to minimize weight and ac losses using the total SC and shield-iron usage in the machine:

$$\min_{(\theta=variables)} \left\{ F_{\text{objective}}(\theta) = [weight,\ ac\,loss] \right\} \tag{9}$$

constrained such that

$$P_{out} = 2.5\ \text{MW} \tag{10}$$

$$Machine\ Diameter \leq 0.5\ \text{m} \tag{11}$$

$$Outside\ Flux\ Density\ at\ 200\ mm\ from\ OD \leq 1\ \text{mT} \tag{12}$$

$$Active\ Length \leq 1.5\ \text{m} \tag{13}$$

$$J_0 \leq \frac{J_c}{2} \tag{14}$$

In this optimization, only the cost and weight of SC wire and iron yoke are considered. SC wire weight is calculated assuming a race track winding format and is separately computed for the active winding and end-winding regions. The weight of the machine is evaluated from the superconducting volume (V_{SC}) and shield iron volume (V_{SI}).

$$V_{SC} = A_a(l_s + l_a) + A_f(l_s + l_f) + A_{sh}(l_s + l_{sh}) \tag{15}$$

$$V_{SI} = A_{SI} l_s \tag{16}$$

$$weight = \rho_{SC} V_{SC} + \rho_{SI} V_{SI} \tag{17}$$

where A_a, A_f, and A_{sh} are the SC cross-sectional areas for the armature, field, and shield respectively and l_a, l_f, and l_{sh} are their corresponding end-winding lengths. The stack length and shield iron cross-sectional area are represented by l_s and A_{SI} respectively, while ρ_{SC} and ρ_{SI} are the respective SC and shield iron densities considered as $8.25 \times 10^3 \text{kg/m}^3$. Cost of MgB_2 is considered as \$2/m and iron cost is considered as \$2/kg. AC losses are evaluated only for the armature, and radial variations of flux density within the coils are considered by subdividing the armature cross-section into three paths and computing the average field in each section.

6.1. Optimization Algorithm

A genetic algorithm based optimization is chosen to minimize weight and losses. Optimization is performed by changing the machine design parameters shown in Figure 1. Design boundaries for each variable are tabulated in Table 3.

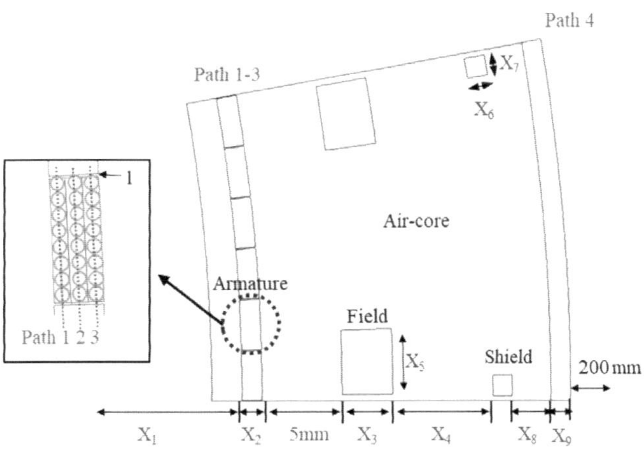

Figure 1. Design variables of the machine

6.2. Optimization procedure

The open source MATLAB tool box, GOSET, is a genetic algorithm (GA) used for the multi-objective optimization of this machine. A flow chart of the optimization procedure is shown in Figure 2. Due to the non-linear magnetic characteristics of iron, finite element analysis (FEA) is used to determine the flux density at various points in the machine. First, GOSET randomly generates an initial population with variables within the given boundaries. These design parameters are then passed to the FEA tool for modeling and EM analysis of the motor. FEA analysis is initially performed assuming a 1 m stack length and the air-gap flux density and torque production are evaluated. Based on the resulting power output, the stack length is modified to reach 2.5 MW and additional torque production from the end windings is accounted

Table 3. Design Space

Parameter	Minimum	Maximum
Armature slot inner radius (X_1) [mm]	70	200
Armature slot radial height (X_2) [mm]	1	20
Radial distance between field coils and shield coils (X_3) [mm]	3	50
Field slot radial height (X_4) [mm]	1	20
Field slot circumferential width (X_5) [degrees]	1	*f (# of poles)*
Shield slot radial height (X_6) [mm]	0.5	10
Shield slot circumferential width (X_7) [mm]	0.5	*f (# of poles)*
Radial distance between shield coil and iron shield (X_8) [mm]	0.1	2
Radial iron shield height (X_9) [mm]	0.1	30
Armature urrent density (X_{10}) [A/mm^2]	50	200

for with a 30% increase in torque. These results are then passed back to MATLAB and fed into analytical models to evaluate the losses and weight of the machine. A function then assigns fitness values to the resulting weight and ac losses for use by the GA in developing a new generation of designs. This procedure continues for a specified number of generations. After the optimization completes, an optimal front is extracted from the best results.

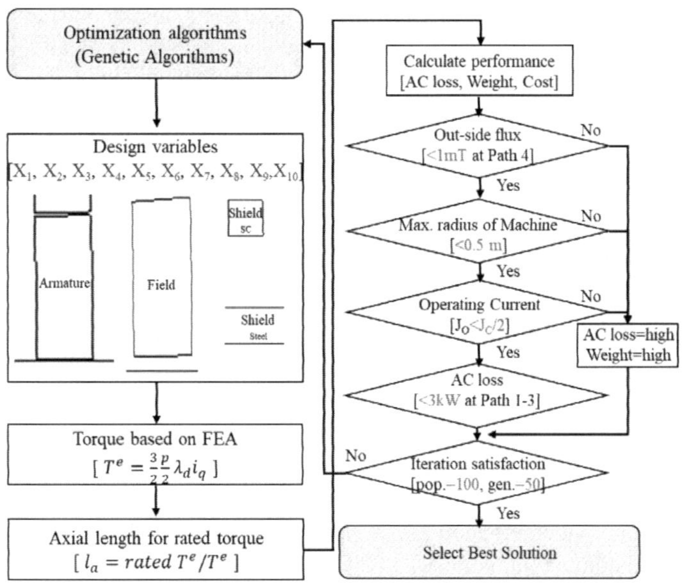

Figure 2. Optimization flowchart

7. Preliminary Results

A Pareto-optimal front for the entire design space is shown in Figure 3. Every design in the Pareto-front is a 2.5 MW machine that satisfies the diameter, axial length, and external flux density constraints. Figure 4 shows a Pareto-front of feasible designs for the CHEETA concept. All feasible designs have total losses below 3,000 W. From the results, a clear trade-off between weight and ac losses is evident. From the preliminary results it is evident that pole count significantly impacts weight and ac losses. Higher pole counts result in lower minimum weight designs; however, the minimum weight is achieved at the expense of higher losses. 8-pole machines showed the best results within the feasible design space. Therefore, 8-pole designs were further explored by opening up the boundaries and increasing the number of generation. The resultant Pareto-optimal front of an 8-pole design is shown in Figure 5. Low ac loss designs use more iron shield and are heavier as a result, while minimum weight designs eliminate the iron shield entirely as can be seen in Figure 6 and Figure 7. Best designs which have both optimal weight and losses also converge towards active shielding alone, shown in Figure 8.

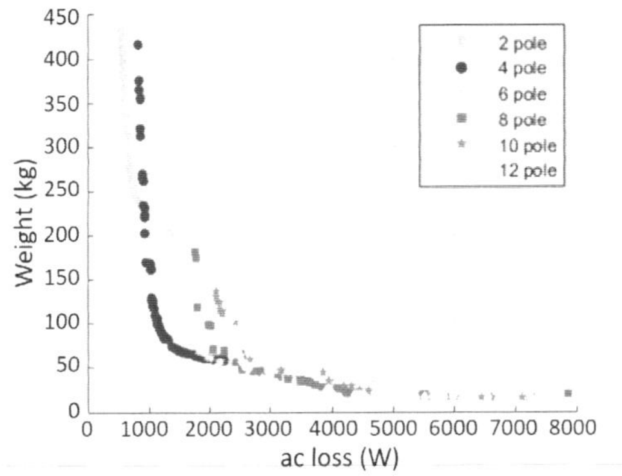

Figure 3. Optimal front for entire design space

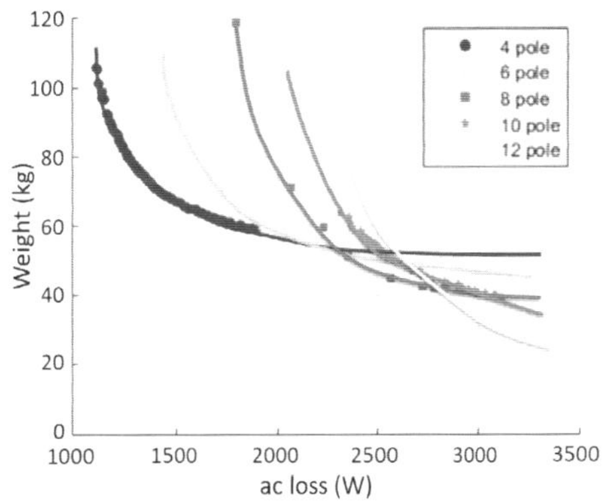

Figure 4. Optimal front for feasible design space for CHEETA concept

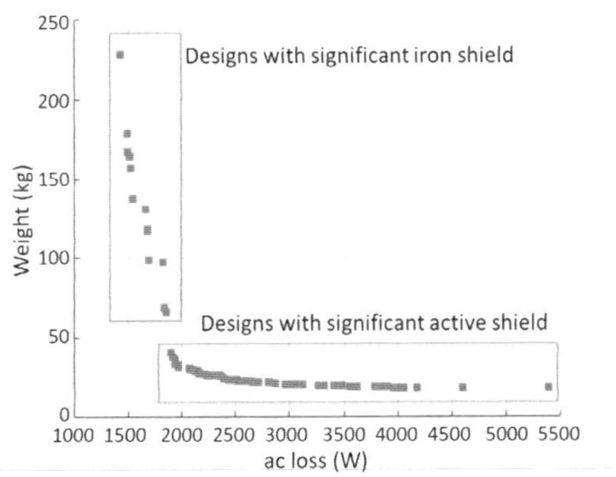

Figure 5. Optimal-Pareto front for 8-pole machine

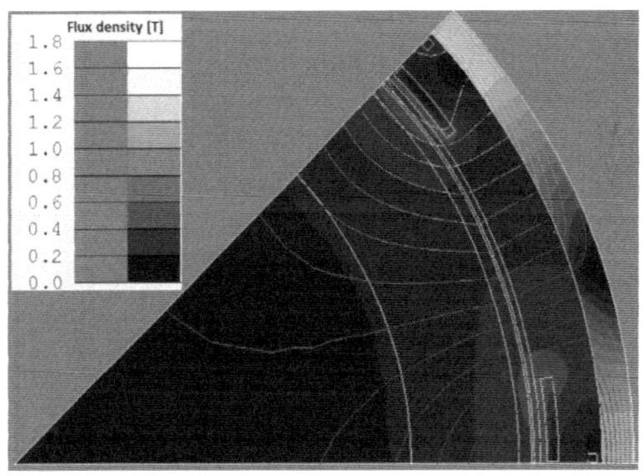

Figure 6. Minimum ac loss design for 8-pole machine

8. Mechanical Design

An inside-out synchronous motor topology will be adopted for this machine. Field and shield coils are placed on the outer rotor while armature coils are placed on the stator. This results in more compact volumetric designs. Figure 9 shows the proposed mechanical design of an optimal motor. The overall machine design is a layered vacuum structure to maximize thermal paths to the windings, and LH_2 coolant will be dispersed to the stator and rotor through couplers connected to a fixed shaft. The LH_2 will be sent through metal pipes and the heat will be extracted from the winding using conduction cooling. Field and shield excitation will be accomplished using a superconducting brushless exciter as proposed in [13].

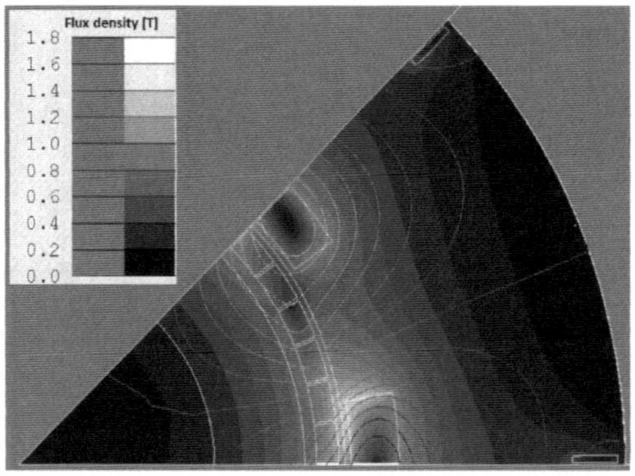

Figure 7. Minimum weight design for 8-pole machine

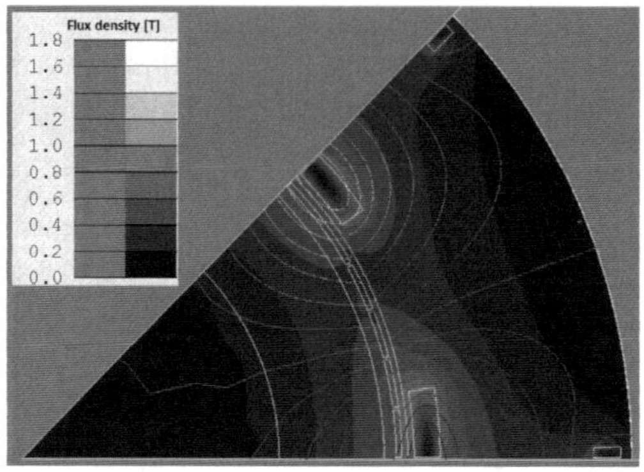

Figure 8. Optimized weight and ac loss design for 8-pole machine

Table 4. Comparison of Motors

Parameter	Best Weight Design	Best AC Loss Design	Best Overall Design
Outer diameter [m]	0.249	0.249	0.249
Axial length [m]	0.215	1.277	0.67
Average torque [Nm]	24708	4154	7917
Air-gap flux density [T]	0.814	0.33	0.464
Outside flux density [mT]	0.953	0.001	0.1
Armature SC length	10.921	11.554	11.224
Field SC length [km]	13.435	23.640	21.186
Shield SC length [km]	2.425	3.094	4.337
Total SC length [km]	26.782	38.288	36.74
Iron shield weight [kg]	0.277	253	0.924
Total loss [W]	4158	1502	2255
Weight (iron and SC) [kg]	18.01	278	25.3
Cost (iron and SC) [k$]	53.56	77.08	73.5

9. Conclusion and Future Work

In this paper, a fully SC machine is proposed for electric aircraft propulsion. A genetic algorithm based optimization is performed to minimize the machine weight and ac losses. Results show that actively shielded designs result in significant power density improvement compared to their passive iron shield counterparts. A comparison between different pole counts reveals that 8-pole machines best suit the design constraints. A 25 kg (SC and iron only) motor could be achieved with 2,255 W of ac losses. Assuming another 100% of additional weight for additional components a 50 kw/kg specific power motor could be achieved. Utilizing the CHEETA concept, a 99.9% efficient electric propulsion motor could be obtained with a fully SC machine. Nearly all feasible designs have air-gap flux densities between 0.3 T and 0.8 T. As these flux densities are achievable with permanent magnets (PM), future studies will explore the potential of partially SC PM machines. Future work will also focus on experimentally validating the ac loss model used in this study, further developing a mechanical design including a cooling scheme for better weight estimates, and adapting the air-core topology to SC coil-formers. Furthermore, a 3D flux

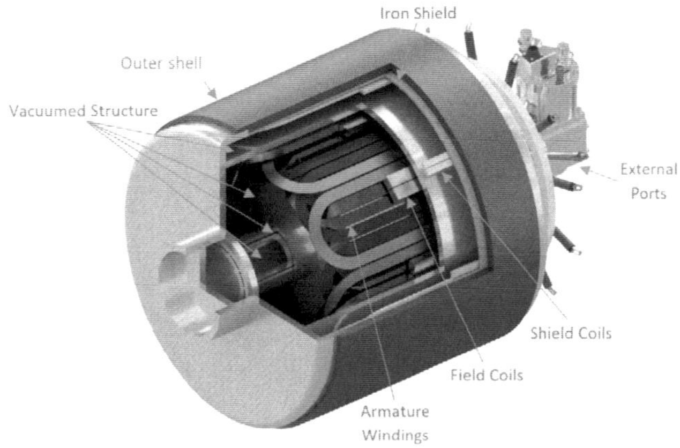

Figure 9. Model of an 8-pole machine

analysis will investigate the effects of end-windings on torque production.

References

[1] Jansen R, Brown G V, Felder J L,Duffy K P 2015 Turboelectric Aircraft Drive Key Performance Parameters and Functional Requirements *51st AIAA/SAE/ASEE Joint Propulsion Conference*

[2] National Academies of Sciences Engineering & Medicine 2016 *Commercial Aircraft Propulsion and Energy Systems Research: Reducing Global Carbon Emissions* First Edition (The National Academies Press).

[3] Haran K S, Lodder D, Deppen T O, Zheng L 2016 Active shielded high-field air-core superconducting machines *IEEE Trans. Applied Superconductivity* **26** 98-105

[4] Patel A, Climente-Alarcon C, Baskys A, Glowacki B A, Reis T 2018 Design consideration for fully superconducting synchronous motors at future electric aircraft *Proc. International Conference on Electrical System for Aircraft, Railways, Ship Propulsion and Road Vehicles & International Transportation Electrification Conference* (Nottingham, UK, 7-9 November, 2018)

[5] Manolopoulous D C, Iacchetti M F, Smith A C, Tuohy P M 2018 Design of superconducting AC propulsion motors for hybrid electric aerospace *Proc. AIAA\IEEE Electric Aircraft Technology Symposium* (Cincinnati, Ohio, USA, 9-11 July, 2018)

[6] Komiya M, Aikawa T, Sasa H, Miura S, Iwakuma M, Yoshida T, Sasayama T, Tomioka A, Konno M, Izumi T 2019 Design Study of 10MW REBECO fully Superconducting Synchronous Generator for Electric Aircraft *IEEE Trans. Applied Superconductivity* **29** 1-6

[7] Terao Y, Kong W, Ohsaki H, 2018 Electromagnetic Design of Superconducting Synchronous Motors for Electric Aircraft Propulsion *IEEE Trans. Applied Superconductivity* **28** 1-5

[8] Terao Y,Seta A, Ohsaki H, Oyori H, Morioka N 2019 Lightweight design of fully supercounducting motors for electrical aircraft propulsion *IEEE Trans. Applied Superconductivity* **29** 1-5

[9] Lodder D C, Haran K S 2015 Multi-objective optimization of an actively shielded superconducting field winding *Proc. IEEE Power and Energy Conference at Illinois (PECI)* (Champaign, Illinois, USA,2015)

[10] Lodder D C Haran K S 2015 Multi-objective optimization of an actively shielded superconducting field winding: Pole-count study *Proc. IEEE International Electric Machines & Drive Conference (IEMDC)* (Coeur d'Alene, Idaho, USA, 2015)

[11] Carr W J 2001 *AC Loss and Macroscopic Theory of Superconductors* Second Edition (Taylor and Francis Inc).

[12] Sumption M D 2018 AC loss of superconducting materials in motors and generators for very high density motors and generators for hybrid-electric aircraft 2018 AIAA/IEEE Electric Aircraft Technologies Symp. AIAA Propulsion and Energy Forum.

[13] H J Sung, B S Go, H Park, R A Badcock, M Park, I K Yu, 2017 Design, Fabrication, and Analysis of HTS Coils for a 10-kW Wind Power Generator Employing a Brushless Exciter, *IEEE Trans. Applied Superconductivity* **27** 1-5

IOP Conf. Series: Materials Science and Engineering **756** (2020) 012031 doi:10.1088/1757-899X/756/1/012031

2-layer Cable-in-Conduit for Hybrid-Coil Magnets

J Breitschopf[1], D Chavez[2], T Elliott[3], J Kellams[4], P McIntyre[5], and A Sattarov[2]

[1] Department of Physics, Colorado State University, Fort Collins, CO 80523 USA
[2] Institute de Physica, Universidad Guanajuato, Leon, GTO 37150 Mexico
[3] Accelerator Technology Corporation, 9701 Valley View Drive, College Station, TX 77845 USA
[4] Lockheed Martin Aeronautics, 1011 Lockheed Way, B601, Palmdale, CA. 93599 USA

[5] corresponding author: Peter McIntyre, Department of Physics, Texas A&M University, College Station, TX 77845 USA p-mcintyre@tamu.edu

Abstract. Recent results are reported in the development of 2-layer cable-in-conduit (SuperCIC) that is designed for use in hybrid-coil magnets. SuperCIC preserves the full performance of the individual wires, and can be formed into flared-end windings for dipoles into layer-wound toroids and solenoids for hybrid windings for tokamaks. The structure of the SuperCIC windings is designed to accommodate winding and heat-treating sub-windings of Bi-2212, Nb3Sn, and NbTi separately and then assembling them and preloading in the magnet.

1. Introduction

Ever-higher magnetic field is desired for proposed hadron colliders, tokomaks, and similar applications. Each of the superconductors that has been rendered into practical wire has an upper limit to the magnetic field that immerses the winding: NbTi is limited to ~8 T, Nb_3Sn is limited to ~15 T, and only the Bi-2212 and REBCO and Bi-2223 can operate at higher fields. At the same time the superconductor cost increases dramatically for each succeeding step in operating field: ~$1/kA-m for NbTi [1], ~$5/kA-m for Nb_3Sn, and ~$100/kA-m for Bi-2212 and Bi-2223, and considerable more for REBCO [2]. There is therefore a strong motivation to develop hybrid magnet technology, in which the winding is segmented into sub-windings of the respective superconductors in which the field at each superconductor conductor is within its field range.

Hybrid windings have long been used successfully for superconducting solenoids, Nb_3Sn/NbTi assemblies for many years [] and recently for NMR spectrometers [3]. The geometry of a solenoid accommodates hybrid windings naturally, since the sub-windings are configured as concentric cylindrical shells, each with its own stress shell to support Lorentz forces.

The cable technology used in all collider dipoles is Rutherford cable [4], shown in Fig. 1a. The high-current windings for the toroids and solenoids of tokamaks and most windings for superconducting magnetic energy storage (SMES) utilize the 'cable-of-ropes' CIC shown in Fig. 1b.

Hybrid windings with both Rutherford cable and CIC pose major challenges for fabrication, assembly, stress management, and magnetization effects on field homogeneity. NbTi and REBCO sub-windings do not require heat treatment, but Nb_3Sn sub-windings require heat treatment in final form to 650 C in inert gas, and Bi-2212 sub-windings require heat treatment in final form to 890 C in a gas mixture of 49 bar N_2 +1 bar O_2. Nb_3Sn and Bi-2212 sub-windings are fragile after heat treatment and so assembly of sub-windings is challenging. Lorentz forces within the winding increase ~B^2 and would cause strain damage in Nb_3Sn or Bi-2212 with B > 15 T unless the Lorenz stresses on inner windings were intercepted and bypassed past the outer windings. Lastly REBCO tapes have much larger effective filament size than NbTi or Nb_3Sn, yet they must be used in the highest field region of the winding which is closest to the high-field region of a collider dipole. Persistent currents develop in the filaments and would generate significant multipole errors for high-field collider dipoles. In light of the above challenges, no hybrid windings have yet been built for either dipole or toroid applications.

In recent years CERN has examined the possibility of doubling the energy of LHC by replacing the 8 T magnet ring by a >16 T ring [5,6,7] – HE-LHC, and the larger vision of a 100 TeV hadron collider – FCC-hh [8]. Superconducting dipole R&D efforts are progressing at a number of laboratories worldwide [9]. All of the designs under development utilize Rutherford cable in winding geometries of cos θ, block-coil, canted cos θ, and common-coil. All designs face a number of daunting challenges: how to manage Lorentz stress within the thick windings; how to configure the ends of each turn to accommodate the beam tube yet provide a compact stress support; how to integrate hybrid windings that would minimize the quantity of the most expensive superconductors.

Also in recent years there has been growing interest in developing fusion tokamaks with stronger toroidal field at the plasma radius and more compact geometry than that of ITER [10]. The interplay of parameters that govern the potential of a tokamak to reach practical breakeven as an energy source has led to designs for a compact spherical tokamak with ~1 m plasma radius and aspect ratio ~2:1. Of particular importance is the overall winding power density J_{WP} (including CIC and structure) [11]. Adopting hybrid windings for a tokamak would require that each sector winding be fabricated as a layer-wound configuration. In the ~40 kA Nb_3Sn windings of the ITER toroid there is significant degradation of critical current in the wires of the 'cable-of-ropes' configuration of cable-in-conduit (CIC) [12]. Along the length of any wire in the ITER CIC, there are unsupported regions and regions where adjacent wires are crushed onto one another to make a divot. Both effects significantly reduce critical current in the wires so that J_{WP} is only ~14 A/mm^2. The analysis of Ref. 11 shows that J_{WP} > 70 A/mm^2 will be required for optimum electric power generation and breakeven for power generation.

A collaboration among Texas A&M University (TAMU), Accelerator Technology Corp. (ATC), and HyperTech Research has developed a novel ordered-structure cable-in-conduit (SuperCIC), shown in 5 1c, that specifically address the above challenges [13]. The ordered-structure CIC provides uniform stress support for all wires within the cable, and stress management at the cable level throughout a winding, including the flared ends. It accommodates separate fabrication and heat treatment of sub-windings, and assembly and preload of assembled windings.

The initial development of the CIC was done using one layer of NbTi wire for use in a 3 T large-aperture dipole [14] for the arcs of the proposed electron-ion collider JLEIC [15]. Then a 2-layer NbTi CIC (Figure 1d) was developed for use in a 6 T dipole [16], shown in Figure 4, to double the ion beam energy for JLEIC [16].

This paper reports recent development of 2-layer SuperCIC using Nb_3Sn and Bi-2212 wire to provide a basis for hybrid-coil dipoles and tokamaks with bore fields >16 T.

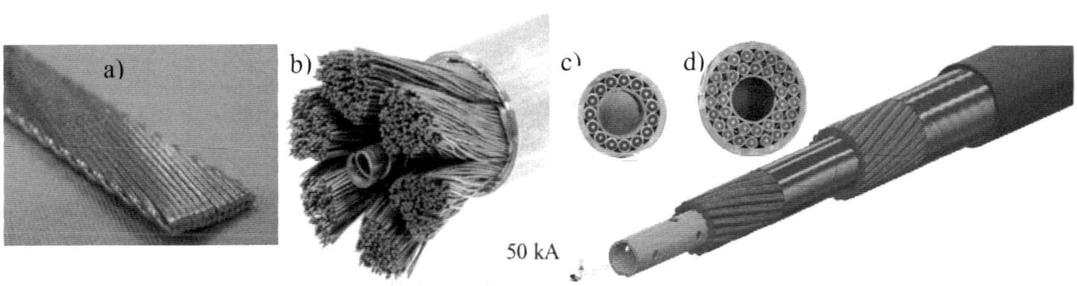

Figure 1. Cables used in superconducting magnets (to same scale): a) Rutherford cable is used in superconducting dipoles; b) 'cable-of-ropes' CIC is used in toroids and solenoids for tokamaks and energy storage; c) single-layer SuperCIC was developed the arc dipoles of JLEIC; d) 2-layer SuperCIC is being developed for use in high-field hybrid-coil dipoles, toroids, and solenoids. The 2-layer SuperCIC and ITER CIC are shown to same scale, with same operating current.

Figure 2. Fabrication procedure for SuperCIC.

2. Two-layer NbTi cable-in-conduit

The single-layer NbTi CIC used in the 3 T JLEIC dipole is formed by cabling wires onto a thin-wall perforated stainless steel (316SS) center tube Fig. 2a) with a twist pitch. The cabling is done using a 24-spool stranding machine, shown in Fig. 2b. Next a spiral over-wrap of SS foil tape is applied, and the cable is inserted in a seamless CuNi sheath tube (Fig. 2c), and the sheath tube is drawn onto the cable (Fig. 2d) to compress all wires against the center tube and immobilize them.
Several innovations were important to the success of the 3 T CIC dipole, each has been adapted for the 2-layer NbTi Super CIC for a 6 T JLEIC dipole, and each is being utilized in developing 2-layer Super-CIC containing Nb_3Sn and Bi-2212 for high-field applications:

2.1. Robotic benders to form the flared ends for dipole windings

The ends of each turn in a dipole must be flared up/down to accommodate the beam tube. The flared ends have presented a focal challenge in most high-field dipole developments. A set of robotic benders has been developed that form a constant-radius bend of the SuperCIC while maintaining the sheath tube to be locally round in transverse profile as it is bent into an arc. This is accomplished using conformal die sets and robotic bend tools shown in Fig. 3a,b. The sheath tube is actually inelastically deformed throughout the bend to maintain its round contour, so that the interior geometry remains benign to the registration of the enclosed superconducting wires.

Figure 3. Robotic bending tools used to form a) the 180° bend and b) the 90° saddle bend that comprise the end geometry of each winding turn for a collider dipole; c) a completed 24-turn end region of the 3 T JLEIC dipole.

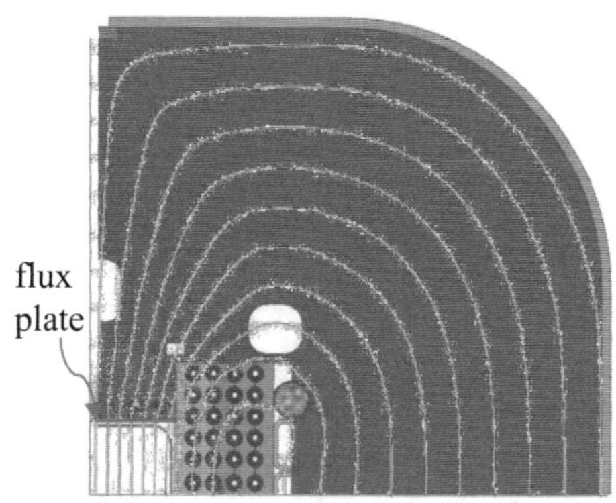

Figure 4. Quadrant cross-section of the 6 T CIC dipole: green and gray– G-11 structure elements, blue – steel flux return. Note the steel flux plate.

2.2. Eliminating differential strain in the bends of flared ends

As an end is formed, wires located on the outside of the bend must transit a longer catenary length than those on the inside. The twist pitch of the wires in each layer of cable is chosen to be exactly equal to the mean bend arc length used to form the end windings around a 90° arc. With that choice all wires have exactly the same catenary length around each bend segment, so that no strain is propagated into the straight regions of cable flanking the bend.

With that provision, the wires still must locally shift transversely as each 90° bend is formed as the each wire spirals from inside of the bend to the outside or *vice versa*. To facilitate this transverse shift without residual strain in the wires, it was found necessary to provide a spiral-wound over-wrap of 316 stainless steel (316SS) which provides a slip-surface on the outside of each wire layer. Samples of the cable were formed and bent to a 180° arc with the bend radius required for the dipole winding. Samples were dissected, wires were etched to determine filament breakage, and wires were tested for short-sample current and compared with witness samples. After optimization of the bend tools and of a functionalized multi-layer over-wrap, there was no filament breakage and extracted wires in the bend-formed SuperCIC retained the same short-sample current as witness wires.

In order to achieve the larger cable current required for the 6 T NbTi dipole for JLEIC, it was necessary to add a second layer of wires to the SuperCIC cable. This was accomplished using the same stranding machine (Fig. 1b), but significant development was required to achieve the bending properties for the larger and stiffer SuperCIC structure.

A problem arose in tight-radius bending of the 2-layer SuperCIC. The spiral over-wrap was applied on the inner layer, leaving a 1 mm-wide gap between foil turns to provide for liquid helium flow during magnet operation. When segments of the 2-layer SuperCIC were formed into a U-bend and dissected, it was found that the foil of the over-wrap had carved divots into the wires of the inner layer in the region of the bend (Figure 5a). The divots were formed by the edge of each spiral wrap of 316SS foil digging into the wires as the bend was formed. A multi-laminar inter-layer over-wrap was applied between the first and second layers. The inner wires are cabled with a clockwise (CW) pitch, the inner 316SS over-wrap is applied with a counter-clockwise (CCW) pitch and 1 mm gap. Two layers of Cu foil tape are applied with opposite signs of pitch and 1 mm gap: the outer 316SS over-wrap is applied with a CW pitch and 1 mm gap, and the outer wires are cabled with a CCW pitch. Samples of the SuperCIC were formed and bent and then dissected as before. No surface damage was observed on either layer (Figure 5b). Similar procedures were used to validate that wire performance is preserved in the finished 2-layer NbTi CIC.

3. Support structure for CIC winding for a dipole

A support structure was developed to support the inner SuperCIC turns in precise geometry, and to similarly position succeeding SuperCIC layers precisely. It is an epoxy-impregnated assembly of a fiber-reinforced polymer (FRP) support beam with the 10 cm x 6 cm 316SS beam tube, and a set of FRP side plates that support the succeeding layers of the winding as shown in Figure 4.

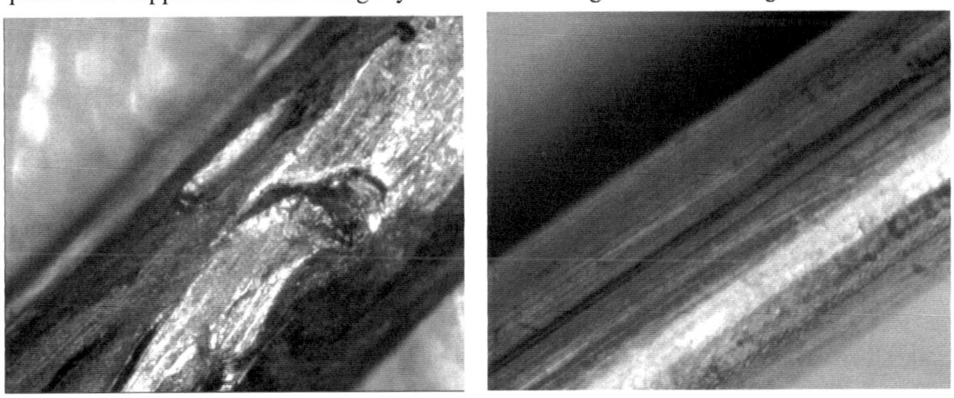

Figure 5. a) single-layer over-wrap carved divots in the wires of the inner layer; b).a multi-layer over-wrap was applied between inner and outer wire layers, there was no surface damage.

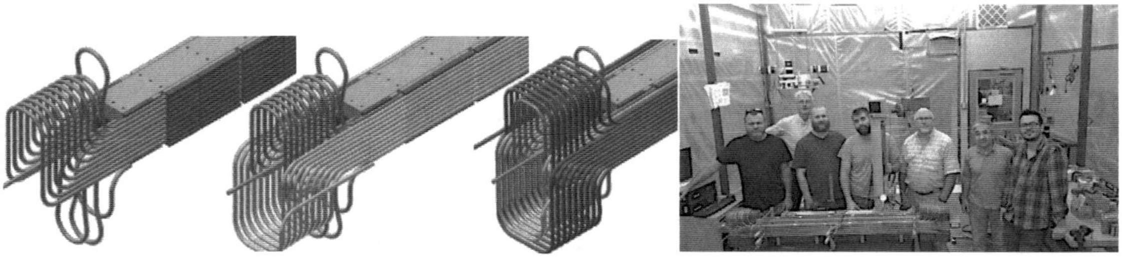

Figure 6. Winding the layers of a CIC block-geometry dipole: a) completed winding of 3 T JLEIC dipole; b) winding strategy for the 6 T CIC winding, showing the stages when successive layers are complete.

The pattern of channels that support and position all SuperCIC turns was precision-machined in the support beam and side plates. The SuperCIC turns for a 1.2 m model dipole were wound, and metrology was performed to measure the r.m.s. precision with which the SuperCIC turns are confined in the positions defined in the magnetic design to produce field homogeneity within the bore tube. The measured r.m.s. position error of individual SuperCIC turns was $<.05$ mm in the global frame of the dipole, corresponding to random field multipoles a_n, $b_n < 1$ unit (10^{-4} with reference radius 2/3 of the bore) over the field range from injection to collision energy [14].

Figure 6a summarizes the sequence for barrel-winding the SuperCIC layers of a block-coil dipole. Figure 6b shows a completed winding for the 3-T JLEIC dipole, still mounted on the winding table. The robotic bending tools can be seen suspended above. The total length of the flared-end winding on the 6 T JLEIC dipole is 300 m.

4. Development of 2-layer Nb_3Sn SuperCIC

A collaboration of Texas A&M, Accelerator Technology Corp., and HyperTech Research developed a single-layer Nb_3Sn-based CIC [17]. The development was done using HyperTech's fine-filament tube-process Nb_3Sn wire. In developing the 2-layer NbTi CIC and the Nb_3Sn SuperCIC, a proprietary sensor method was used to detect any filament breakage within strands during the forming of the bends. The method was very useful in optimizing the material choices and pre-heat treatments for the perforated center tube, the foil over-wraps, and the sheath.

Figure 7 shows a 2-layer Nb_3Sn SuperCIC: the cross-section of the tube-process Nb_3Sn wire, the succession of over-wraps, and the completed SuperCIC. U-bend samples have been evaluated by dissection (Figure 8), etching, and short-sample measurement, and there is no filament breakage from the optimized bending process. Testing of sample segments of 2-layer Nb_3Sn in background field is planned.

Figure 7. 2-layer Nb_3Sn SuperCIC: a) HyperTech's tube-process Nb_3Sn wire, b) over-wraps of inner layer 316SS and Cu foil tapes applied; c) outer layers of Cu foil and 316SS tapes applied; d) finished SuperCIC.

5. Development of 2-layer Bi-2212 SuperCIC

Bi-2212/Ag wire is soft and has limited strength. Also the cable must be heat-treated in final shape through a partial melt, during which the metal atoms in the liquid phase are highly mobile in the silver matrix and susceptible to formation of parasitic phases. Also metal atoms from surfaces that contact the wires can diffuse through the silver matrix and form intermetallics within the superconducting cores. Considerable effort was made to develop a combination of under- and over-wrap foils that provide two distinct functions:

- a diffusion barrier foil provides a soft cushion to protect the wires from deformation from the alloy foil layers during bending of flared ends, and prevents metal migration between the wires and the center tube and sheath tube;
- slip-surface foil over-wrap on the outer layer to enable wires to re-arrange configuration as the SuperCIC is bent.

An optimum choice of materials and surface treatments was found that provides those functions, and sample segments of 2-layer Bi-2212 SuperCIC were fabricated and bent with the required 5 cm bend radius. Figure 7a shows the completed SuperCIC. Figure 7b shows an enlargement of the interlayer region showing the under- and over-wrap layers. A specimen of the SuperCIC was carefully dissected to examine all of its components, shown in Fig. 7c. There was slight deformation of the cross-section of the wires, but less than is occasioned in forming of Rutherford cable. There are no divots or other damage to any wires.

The 2-layer Bi-2212/Ag SuperCIC will be heat-treated using overpressure processing, in which 50 atm pressure is maintained during heat treatment (including 1 atm partial pressure of O_2). The center tube and sheath tube have been made from Haynes 230 alloy, which is not damaged by exposure to oxygen for periods of hours. The Haynes 230 sheath tube has sufficient strength at 870 C to serve as a 50 atm pressure barrier. Flow of the high-pressure buffer gas is provided through the center tube, and the perforations in the it and the gaps in the over-wraps enable ready access of oxygen during the periods of partial melt and annealing. This eliminates the need for a high-pressure furnace that puts an entire winding in the 50 atm buffer gas.

6. Cryogenic considerations

Cryogen flow is supported through the center tube. It is anticipated to operate the SuperCIC using supercritical helium (SCHe), so that single-phase flow is assured during operation. Simulations of quench propagation have been made for the high-windings in the dipole and toroid magnets described in the following section. When quench protection heaters are fired at both ends of a 150 m-long

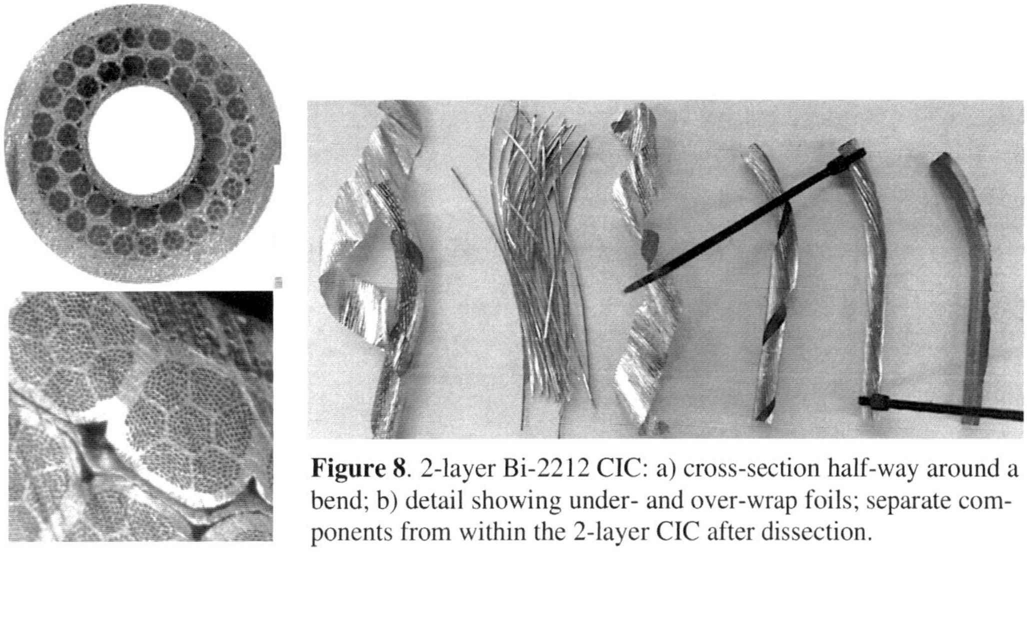

Figure 8. 2-layer Bi-2212 CIC: a) cross-section half-way around a bend; b) detail showing under- and over-wrap foils; separate components from within the 2-layer CIC after dissection.

SuperCIC winding, a maximum temperature of ~100 K is reached and the maximum pressure within the sheath tube is ~60 atm, well within the strength of the Haynes 230 sheath tube.

7. Application of Super CIC for high-field hybrid-coil dipoles

The 2-layer SuperCIC offers an attractive option for the hybrid windings for high-field dipoles [18]. Figure 9 shows a design for a 19 Tesla dipole suitable for use in HE-LHC. It contains sub-windings of Bi-2212, Nb_3Sn, and NbTi, each configured as barrel-wound sub-windings with planar vertical boundaries. The divisions are made so that the magnetic field at conductor for each sub-winding is within the limits for that superconductor. The sub-windings are assembled in Russian doll fashion, connected in series, and the size and number of wires in each 2-layer CIC are chosen so that all sub-windings operate at the same fraction of critical current at maximum excitation. Parameters are summarized in Table 2.

Table 1. Main parameters for hybrid-winding 16 T toroid.

R_0	1.2	m
B @R_0	6.7	T
B @coil	17.4	T
A	2.0	
CIC	# strands, wire dia.	
Bi-2212	42 strands, 0.97 mm	
Nb_3Sn	42 strands, 0.97 mm	
J_{WP}	140	MA/m^2
I_{op}	28.7	kA
I_{op}/I_c @4.2K	0.7	
# layers	11	
SC(Layers)	Bi-2212 – 5 layers Nb_3Sn – 6 layers	

Table 2. Main parameters of a hybrid-winding 16 T collider dipole.

Coil current				28	kA
Operating temp range				4.2-4.6	K
2-layer CIC sub-windings:					
	NbTi:	# layers, turns/bore	1	14	
		# wires, wire dia.	16+22	1.2	mm
		B_{max} in sub-winding		7.3	T
	Nb_3Sn:	# layers, turns/bore	4	80	
		# wires, wire dia.	17+23	0.88	mm
		B_{max} in sub-winding		12.4	T
	Bi-2212	# layers, turns/bore	4	68	
		# wires, wire dia.	17+23	0.88	mm
		B_{max} in sub-winding		19.4	T

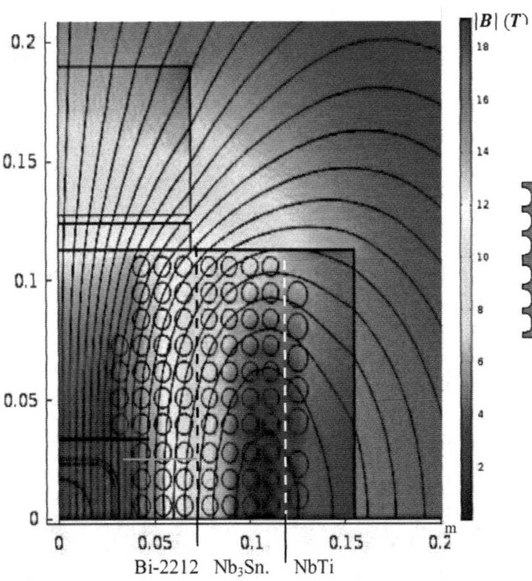

Figure 9. 19 T hybrid dual dipole. Lines of force and color-code of $|B|$ are shown at full excitation (27 kA).

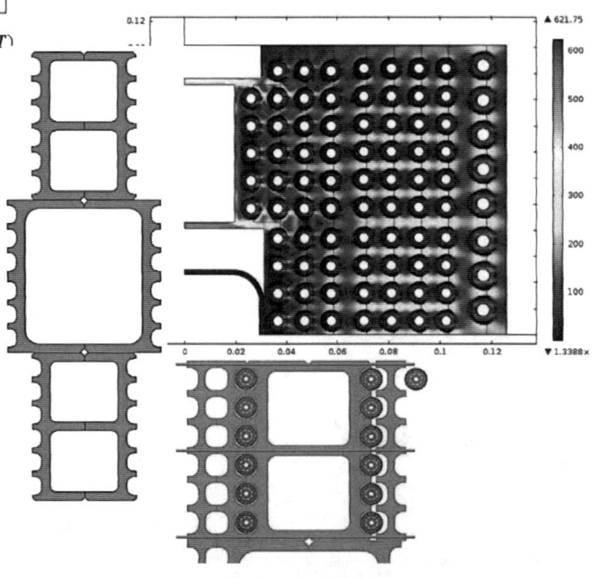

Figure 10. a) Cross-section of the central support structure of the hybrid dipole; b) simulated von Mises stress (MPa) in the windings, c) detail showing first stages of winding the Bi-2212 sub-winding.

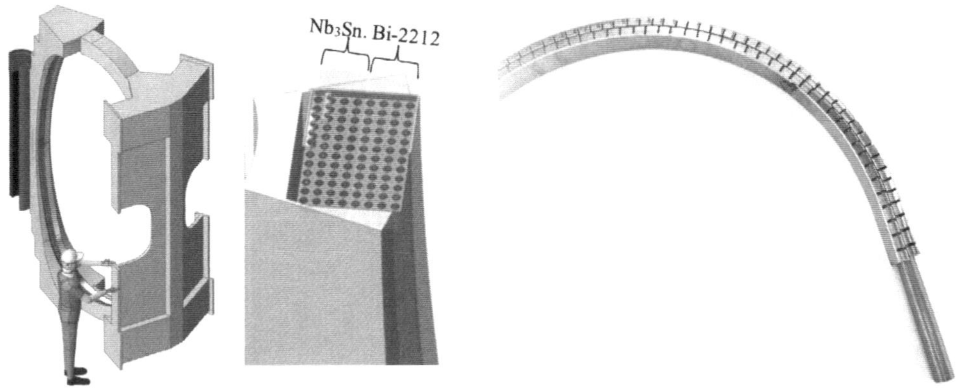

Figure 11. a) 16 T compact tokomak currently under design at PPPL using hybrid 2-layer CIC; b) detail of one hybrid winding showing the layer-wound sub-windings of Nb_3 Sn and Bi-2212; c) armor shell for the 2-layer CIC, co-wound with the CIC to form the toroid winding.

The structural elements are made of Ti-6Al-4V alloy, precision-machined to define the positions of all SuperCIC turns, as shown in Fig. 10a. Figure 10b shows a simulation of the von Mises stress distribution within the winding at 18 T bore field. The maximum stress in the sheath tubes is 219 MPa, 38 MPa, and 338 MPa, respectively. The maximum stress in the wires within each turn of SuperCIC appears to be ~half of those values (and so within the limits to avoid strain degradation), but multi-scale modeling of a twisted cable within a conduit with spring-loaded center tube in a winding is extremely difficult and must ultimately be validated in actual model dipoles.

8. Application of SuperCIC for high-field hybrid-coil toroids

Figure 11a shows a conceptual design of an 18 T toroid for a compact spherical tokamak designed by Menard and Brown [11]. It utilizes a layer-wound hybrid winding, containing sub-windings of Bi-2212 (for the inner layers) and Nb_3Sn (for the outer layers), shown in Fig. 11b. The solenoidal winding is made entirely of Nb_3Sn SuperCIC. Each 18 T toroid winding is a layer-wound hybrid in which the outer 4-layer sub-winding (in lowest field) is wound with Nb_3Sn SuperCIC and the inner 4-layer Bi-2212 sub-winding is wound with Bi-2212 CIC. By using the expensive Bi-2212 only in regions where the field in conductor is >14 T, the superconductor cost is minimized and the winding current density is maximized. SuperCIC windings uniquely make it possible to use hybrid-coil toroids, to make separate heat-treats for each sub-winding, and to assemble and preload the windings with precise geometry.

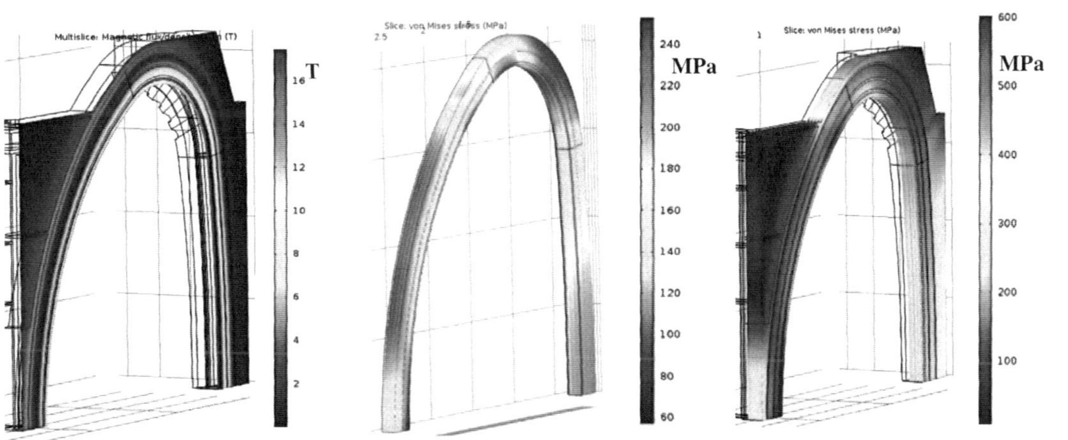

Figure 12. a) Calculated field distribution in a winding module of a 16 T toroid; b) von Mises stress distribution in the co-wound armor of the winding.

The accumulating stress in a large-scale toroid winding must be managed by enclosing the SuperCIC in an armor shell made of high-strength alloy. We have made an innovation for the winding structure, in which the armor shell is configured as two half-shells that enclose the SuperCIC. The three elements - inner half-shell, SuperCIC, and outer half-shell – are co-wound to form the sub-winding of each toroid segment. The armor half-shells are slit as shown in Fig. 11c with a row of kerf-cuts. The half-shells can be bent readily to the required curvature radius without deforming its bulk metal or the SuperCIC. This uses the same principal as a carpenter uses when he saws a row of kerf-cuts part-way through the wooden molding for a stair-case. The hybrid SuperCIC and the kerf-cut co-wound armor are key to achieving $J_{WP} \sim 140$ A/mm^2.

Table 1 summarizes the main parameters of a 16 T toroid for the compact spherical tokamak of Ref. 11 that provides 6.7 T at the plasma radius 1.2 m. Each toroid winding is configured as hybrid sub-windings of Nb_3Sn and Bi-2212. shows the calculated stress distribution in the overall structure and in the armor sheaths of the SuperCIC windings. As with the dipole, the maximum stress in the wires is estimated to be <half that in the armor sheath, but accurate simulation requires multi-scale modeling (in progress) and will need to be validated in model windings.

9. Conclusions

A novel approach to superconducting cable-in-conduit is being developed for applications requiring hybrid windings operating at high magnetic field. SuperCIC supports all wires within the CIC so that strain degradation can be prevented even at very high fields. The robust sheath tube provides stress management within each sub-winding. SuperCIC is now available as a product in lengths up to 150 m.

The internal structure permits bending of SuperCIC on a radius ~8x cable radius, which accommodates applications for collider dipoles. A co-wound armor shell can be applied for toroid applications to provide high-strength structure without degradation of the cable during winding. Designs for SuperCIC have been made in prototype for up to 50 kA operating current. SuperCIC hybrid windings appear to offer promising performance in high-field magnets for particle accelerators and for fusion magnetics.

10. References

[1] N. Peray, 'NbTi *vs*. Nb_3Sn', https://indico.cern.ch/event/624538/contributions/2626568/attachments/1477868/2290192/Nb3Sn__D2-N_Peray.pdf

[2] L. Cooley and I. Pong, 'Cost drivers for very high energy p-p collider magnet conductors', https://indico.cern.ch/event/438866/contributions/1085142/attachments/1257973/1858756/Cost_drivers_for_VHEPP_magnet_conductors-v2.pdf

[3] M. Breschi *et al*., 'Analysis of quench in the NHMFL REBCO prototype coils for the 32 T Magnet Project', Superconduct. Sci. and Tech., **29**, 5, 055002 (2016).

[4] G. Willering, 'Stability of superconducting Rutherford cables', dissertation at Universiteit Twente, 1978. http://inspirehep.net/record/887091/files/CERN-THESIS-2010-011.pdf

[5] P. McIntyre *et al*., '20 T dipoles and Bi-2212: the path to the LHC energy upgrade', Proc. EuCARD Workshop on High-Energy LHC, Malta, Oct. 14-16, 2010. CERN-2011-003. http://arxiv.org/abs/1108.1640

[6] P. McIntyre and A. Sattarov, 'On the feasibility of a tripler upgrade for LHC', Proc. 21st Particle Accelerator Conference, Knoxville, pp. 611-613 (2006). doi: 10.1109/PAC.2005.1590511

[7] J.L. Abelleira *et al*., 'High-Energy LHC design', J. Phys.: Conf. Ser., vol. 1067, 022009 (2018). doi 10.1088/1742-6596/1067/2/022009

[8] M. Benedikt *et al*., 'Future Circular Collider v3: FCC-hh', CERN-ACC-2018-0058 https://cds.cern.ch/record/2651300

[9] D. Tommassini *et al*., 'Status of the 16 T dipole development program for a future hadron collider', IEEE Trans. Appl. Superconduct., vol. 28, no. 3, 4001305 (2018). doi 10.1109/TASC.2017.2780045

[10] L. Muzzi *et al*., 'Cable-in-conduit conductor: lessons from the recent past for future developments of low and high temperature superconductors', Supercond. Sci. Technol. **28**, 053001 (2015). doi 10.1088/0953-2048/28/5/053001

[11] J. Menard, 'Compact steady-state tokamak performance dependence on magnet and core physics limits', Phil. Trans. **A377**, 20170440 (2018). doi 10.1098/rsta.2017.0440

[12] J. Qin *et al.*, 'New design of cable-in-conduit conductor for application in future fusion reactors', Superconduct. Sci. and Tech. **30**, 115012 (2017). DOI: 10.1088/1361-6668/aa8900

[13] D. Chavez *et al.*, 'Cable-in-conduit using MgB_2, Nb_3Sn and Bi-2212 for wind-and-react coils', Proc. Appl. Superconduct. Conf., Seattle, Oct. 28-Nov.2, 2018.

[14] P. McIntyre *et al.*, 'Cable-in-Conduit dipoles for the Ion Ring of JLEIC', Proc. Appl. Superconduct. Conf, Seattle, 28 Oct - 2 Nov 2018.

[15] C. Montag, 'eRHIC design status', Proc. Int'l Part. Accel. Conf., Copenhagen, May 14, 2017. http://accel-conf.web.cern.ch/AccelConf/ipac2018/papers/tuygbd3.pdf

[16] P. McIntyre *et al.*, '6 T Cable-in-Conduit dipole to double the ion energy for JLEIC', Proc. Int'l Part. Accel. Conf., Melbourne, May 15-22, 2019.

[17] D. Chavez *et al.*, 'Cable-in-conduit using MgB_2, Nb_3Sn and Bi-2212 for wind-and-react coils', Proc. Appl. Superconduct. Conf., Seattle, Oct. 28-Nov.2, 2018.

[18] P. McIntyre *et al.*, 'LHC Doubler: CIC dipole technology may make it feasible and affordable', Proc. Int'l Part. Accel. Conf., Melbourne, May 15-22, 2019.

Development of HTS Current Leads for the ITER Project

P Bauer[1*], A Ballarino[3], A Devred[3], K Ding[2], Y Dong[2], E Niu[4], Q Du[2], CY Gung[1], Q Han[2], R Heller[5], X Huang[2], T Ichihara[6], S Lee[1], J Li[2], C Liu[2], Cl Liu[2], K Lu[2], N Mitchell[1], Y Song[2], T Spina[3], Q Ran[2], T Taylor[7], S Yamada[8], Y Yang[9], T Zhou[1]

[1]ITER Organization, CS 90 046, 13067 St. Paul Lez Durance Cedex, France
[2]Institute of Plasma Physics, 350 Shushanhu Rd., 230031 Hefei, Anhui, China
[3]CERN, Route de Meyrin, 1211 Geneva, Switzerland
[4] China Fusion Domestic Agency, 15 Fuxing Rd, 100862, Beijing, China
[5] Karlsruhe Institute of Technology, PO box 3640, 76021 Karlsruhe, Germany
[6] Fusac Technologies, Osaka, Japan
[7] AT Scientific LLC, Geneva, Switzerland
[8] National Institute of Fusion Science, 322-6 Oroshi-cho, Toki, 509-5292 Japan
[9] Institute of Cryogenics, Univ of Southampton, University Road, SO17 1BJ, UK

E-mail: pierre.bauer@iter.org

Abstract. The HTS current leads for the ITER project will be the largest ever operated, with unprecedented currents, up to 68 kA and voltages, up to 14 kV. According to the ITER agreement they will be provided in-kind by China. After an extensive development program at the Hefei Institute of Plasma Physics (ASIPP), the ITER current leads were designed and qualified. The following discusses the main highlights of this development, with particular emphasis on the description of the design of the different types of ITER current leads and their final qualification in dedicated cold tests in nominal conditions.

1. Introduction

The ITER project to demonstrate the feasibility of power generation by nuclear fusion technology, now under construction in the south of France, requires very large superconducting magnets for confinement and control of the burning plasma. These large coils - 18 Toroidal Field Coils (TFC), six Central Solenoid (CS) modules, six Poloidal Field Coils (PFC) and 18 Correction Coils (CC) – are supplied with 60 HTS current leads, ranging from very large (68 kA for the TF) to medium (10 kA for the CC), transferring up to 2.6 MA into and out of the cryogenic environment (table 1). The largest (TF-type) leads are over 3 m long and weigh 600 kg. The ITER current leads are mounted horizontally into the so-called Coil Terminal Boxes (CTB) of the feeders, of which they are one of the key components [ref 1].

* To whom any correspondence should be addressed.

Table 1. Nominal operation parameters of the three types of ITER HTS current leads.

Parameter	TF	PF/CS	CC
Number of pairs	9	12	9
Design/operational max DC current (kA)	68/68	55/52	10/10
Operational max voltage/fault voltage to ground (kV)	7/19	14/29	0.3/3
Nominal operating temperature GHe supply to heat exchanger (K)		50	
Nominal operating pressure GHe supply to heat exchanger (MPa)		0.4	
Nominal operating temperature room temperature terminal (K)		~300	
Nominal operating temperature HTS top/warm end (K)		65	
Operating temperature HTS bottom/cold end (and LTS section) (K)		<7	
Maximum ramp-rate during pulse (kA/s)	-	20	10

Following CERN's successful development of HTS current leads for the LHC [ref 2, 3], ITER also launched R&D for HTS leads. Among the trials was the 70 kA "demonstrator", built by Germany's KIT and tested successfully at 70 kA in 2004 [ref 4-8]. Key technical design choices from the demonstrator survived into the final design, such as the counter-flow cooling of the resistive section with 50 K[1] helium gas from an intermediate cooling stage of the cryo-plant and the conduction cooling of the HTS section from the cold end. In 2004 the decision was made that the Chinese ITER partner supplies these HTS leads. ASIPP, which had recently completed the commissioning of the 16.5 kA HTS leads for the EAST Tokamak [ref 9], soon launched the manufacturing of several pre-prototypes for ITER [ref 10-17]. A 70 kA pre-prototype ("trial lead") was tested at up to 100 kA in December 2008 using the EAST power and cryogenic infrastructure, followed by a 52 kA PF-type lead and a 10 kA CC-type lead. These pre-prototype leads explored different resistive Heat eXchanger (HX) designs and jointing technologies. The final HTS shunt design incorporates many features of the ASIPP trial leads. The main technological design choices were finally fixed end of 2010. CERN's HX design was adopted, as well as its application of e-beam welding for precision and low heat input technique for joining the different lead sub-sections – HX, terminal and shunt [ref 3]. Generally, the ITER project chose to employ proven techniques and designs to minimize technical risk. The final design was jointly developed by ASIPP and ITER, under guidance from the HTS working group, a panel of experts. The following describes this development.

2. ITER Current Lead Design

The details of the design of the ITER HTS current leads can be found in [ref 19]. The main operational parameters are listed in table 1. As shown in figure 1 the ITER HTS current leads are assembled from: -1- the HTS shunt / LTS-linker and joint assembly, -2- the resistive Cu Heat eXchanger (HX), and -3- the Room Temperature (RT) terminal block.

The HTS-shunt, LTS linker and joint assembly is the component containing the superconductor, HTS and LTS. The complexity of this assembly comes from the multiple electrical splices - from the HX Cu to the HTS and from the HTS to the LTS. Located at the far end of the ITER feeders, the leads operate in a much lower magnetic field than the coils, such that the LTS linker is essentially a one meter piece of feeder cable in conduit (CIC) NbTi busbar [ref 20], terminated in a standard "Twin-Box" (TB) type joint [ref 21]. The HTS material is BiSCCO2223 (or Bi2223) tape with a specially alloyed Ag matrix containing 3 at% of Au to reduce the matrix thermal conductivity.

The resistive HX is machined from thick Cu stock, to form a central axial conductor with radial cooling fins. It is inserted into a thin-walled steel jacket, with very tight tolerance to limit bypass gas flow. The zigzag flow of GHe cooling is obtained by clipping the radial fins on alternating sides, maximizing the heat exchange surface. Its design is based on extensive modelling [ref 22]. Independent verifications with simplified models were also conducted [ref 23].

The RT terminal is designed for low current density, also in the contacts to the room temperature

[1] As shown in [ref 18] the ITER leads are operated close to the optimum cost with GHe injected at 50 K, implying a temperature of the HTS warm (top) end of 65 K.

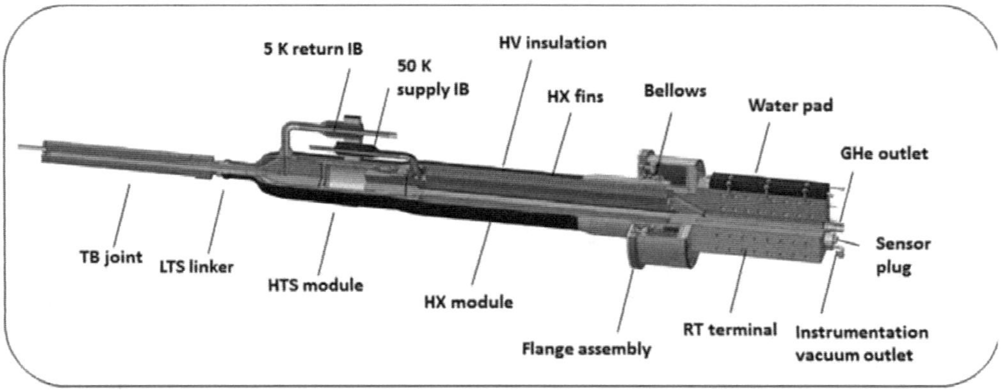

Figure 1. ITER HTS current lead design (TF-type).

busbar system. The HV insulation is designed to resist 30 kV with a large safety factor (~3-4). It is made of a number of layers of polyimide combined with prepreg fiberglass sheets, adding up to six mm as described in [ref 24].

3. Qualification

The general methodology for the qualification of ASIPP and its two main sub-suppliers Juneng and Keye is discussed in [ref 25]. The complete supply chain had to be qualified, from material procurement, through targeted trials of critical manufacturing steps to the fabrication of technology mock-ups, and finally prototype leads of each type (PF, TF and CC).

First, all materials used require certification according to the applicable codes. The stainless steel is to conform to AISI 316L (level 3.1 certification according to EN 10204). OFE C10200 Cu must conform to standard ASTM B152 with an additional specification for the RRR value (40-200). Rigorous test schedules apply to the HTS tape (supplied by SEI-Japan and Innovia-China), including a minimum critical current I_c of 100 A @ 77 K in self-field (at 1 μV/cm according to IEC 61788-3) and many more parameters. Continuous I_c and geometrical measurements are performed for the entire tape length. The insulation system is qualified for low void fraction (<2%), minimum laminate shear strength (>50 MPa) and minimum ultimate tensile strength (500 MPa @ 0°, 200 MPa @ 90° to fiber), all after thermal cycling ([ref 26, 27]). With hundreds of box-type joints inside the ITER feeder system, the box joints were separately qualified in the framework of the feeder qualification. Further details of this are reported in [ref 21]. A major related effort was the development of suitable 316L steel – OFE Cu clad plates used for machining of the bimetal joint boxes [ref 28]. All the TIG welds were qualified according to EN ISO 15614-1:2004 (acceptance levels ISO5817 level B), including the additional requirement of five thermal cycles to liquid nitrogen (LN2) temperature. The brazing of the two major brazed subassemblies, HTS shunt and RT terminal, were qualified according to EN 13134:2000, with full-scale mock-ups, which were assessed through visual inspection (EN970), leak tested after 5 thermal shocks and destructively evaluated (metallography and tensile test according to ASME section IX). Imperfections were within the limits set by ISO 13919-1 level B. In the tensile test, failure occurred, as expected, in the copper (at ~180 MPa), not in the braze joint. Electron-Beam Welding (EBW) was qualified according to EN ISO 15614-11:2004. A special mock-up was made to develop the welding procedure. In addition to the ISO 13919-1 level B acceptance requirement, special requirements were also defined in this case including vacuum leak tests after thermal shocks and the evaluation of the length of the heat affected zone. The mock-up was also destructively tested to verify the quality and strength of the weld.

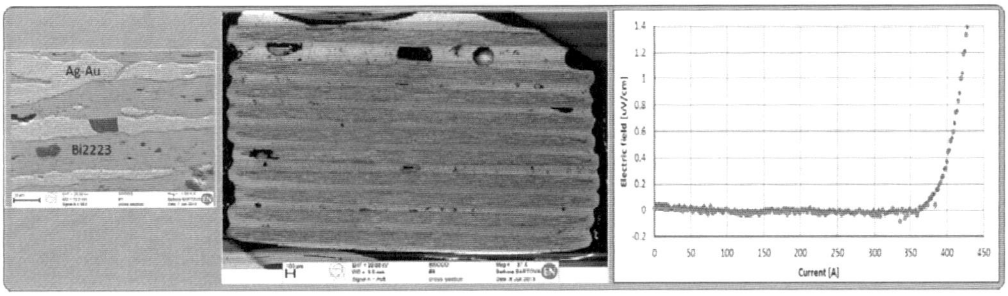

Figure 2. Left: micrograph of Innovia Bi-2223 tape, Middle: stack made by CERN from Innovia tape, Right: critical current measurement of two 5-stacks (CC-type) in self-field @ 77 K.

3.1 Technological Mock-ups

Technological mock-ups of different critical current lead subassemblies were built and tested for the different types of current leads by the respective suppliers [ref 29, 30], discussed in further detail below.

To qualify the stack-shunt soldering joints a 1/30 sample of a TF type shunt module was manufactured and tested. It consisted of three Bi-2223 HTS stacks soldered to a Cu-steel brazed plate, with a LTS sub-cable soldered into it. A major goal of this effort was to measure the joint resistance. Tape samples were also investigated at CERN and made into stacks (figure 2). Porosity found in the Ag-Au matrix of earlier tapes from Innovia were not found in subsequent samples.

The current lead sub-components requiring the most complex assembly is the low temperature end of the lead, which includes the joint box and the LTS section with several soldered electrical contacts. The most critical step is the soldering process between the LTS sub-cables and the shunt (figure 3). A specification for the electrical resistance must be met, requiring 100% wetting by solder of the bare strands, a tight fit between sub-cable and hole and complete filling of voids with solder. A resistance <1 nΩ (for TF type lead) was found as required. After reverse plating, the other (joint box side) end was assembled into the joint-box together with a 0.3 mm Sn63Pb37 solder foil to improve the electrical contact to the Cu-sole. Microscopic, x-ray and mechanical examination of the mock-up joints demonstrated good bonding (peeling tests).

Following an early HX mock-up made at CERN to demonstrate the general feasibility and manufacturing technology, full size HX mock-ups of the TF and CC types were manufactured at Juneng and Keye. The tight H7g6 manufacturing tolerances between the core and honed tube were verified with coordinate measurement machine measurements of the HX core straightness (<0.07 mm) and OD (<0.03 mm tolerance) and with a feeler gage inserted into the honed tube with a special tooling. The HX mock-ups were submitted to a series of pressure drop tests with GN2 (nitrogen gas) and GHe at room temperature to establish a set of reference data for the future prototyping and series manufacturing. Due to their importance the measurements were performed at ASIPP and then repeated at CERN. The comparison of the experimental data to the model also helped the calibration of the HX model, as discussed in [ref 31]. A special measurement was performed on the TF HX mock-up in which capillaries were inserted along the HX to measure the pressure drop along four equally long sections. As expected the pressure drops of the four sections agreed, except for a slight increase from upstream to downstream resulting from the inlet pressure (into the specific segment), which is smaller for each segment as a result of the pressure drop in the previous segment(s). Finally X-ray tomography was used at CERN to verify that there were no bent fins or other defects in the HX honed tube assembly, which could not be detected by the pressure drop method (none was found).

Stringent qualification requirements were defined for the insulation system, as discussed above. To these are added particular requirements for the resin curing temperature (≤80°C) and for the maximum boron content (zero, to prevent nuclear activation). The test requirements also include partial discharge

Figure 3. Preparing the sub-cables (TF-type): Left: chemical Ni removal, Middle: pre-tinning, Right: soldering into the shunt blind holes.

and Paschen testing. The development of a 30 kV class, combined polyimide / impregnated glass fiber ("GK") system required significant R&D (it was completed only at the very end of the lead development). The ASIPP efforts are discussed in [ref 32]. Partial discharge levels found in the mock-up tests are consistent with those of small, flat samples for which the void fraction was measured to be within specification. Extensive R&D was also conducted for the insulated heaters and room temperature Insulating Breaks (IB) for the RT terminals [ref 33, 34], special developments for the HTS leads.

3.2 Prototypes

A pair of each type of current lead was tested as the final step of the qualification process. More details on CC, PF and TF prototype manufacture and testing are discussed in: [ref 35, 36, 37, 38 for CC] and [ref 39, PF] and [ref 40, TF]. The test data are summarized in table 2. The tests were conducted in a new ASIPP test facility including a 900 W @ 4.5 K He plant and a 80 kA power supply system as described in [ref 41,42]. It also includes a control and interlock system supplied by ITER. Figure 4 shows the TF lead prototypes mounted in series inside a prototype CTB, shorted by a removable superconducting U-bend with clamped TB joints. The leads were fully instrumented with voltage taps for quench detection and measurement of the different joint resistances and with temperature sensors in key locations. The 50 K and 5 K hydraulic circuits were instrumented with pressure gages and temperature sensors at the inlet/outlet and flow-meters at the outlet. The two HX were supplied in parallel with 50 K GHe with the flow being regulated with a valve downstream of the 300 K GHe exhaust of each lead. At the 5 K ends the leads were supplied with 5 K SHe, entering through the TB joint of one lead, then flowing through the CIC- type U-bend and exiting after the LTS linker of the other lead. The SHe could be heated to room temperature in a large heat exchanger downstream of the CTB, to allow a precise flow measurement with a flow-controller. As will be done in ITER, large water-cooled Al busbars transferred the currents from the power supply to the HTS current lead test-station, where air-cooled flexibles connected them to the current lead terminals.

Steady state tests were performed at different currents (80, 90, 100 and 110% of the nominal). Figure 5 shows the performance parameters of the HX (voltage drop, steady state mass-flow) for the different currents. In the case of the TF prototypes the agreement of these parameters with the CERN model is good, as can be seen from the figure. Also for the CC mass-flow rates are predicted well by the model from [ref 38]. The joint resistances found during steady operation were below specification (table 2).

Following the steady state operation test, the HTS was quenched by a LOFA (Loss Of Flow Acc) event, i.e. by stopping the 50 K GHe cooling flow while keeping the full current. The quench is defined

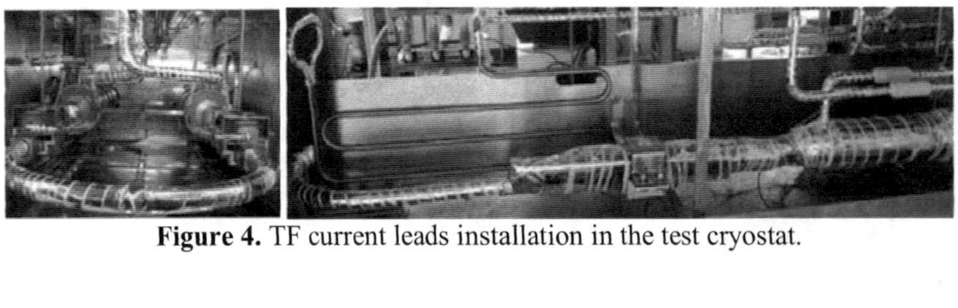

Figure 4. TF current leads installation in the test cryostat.

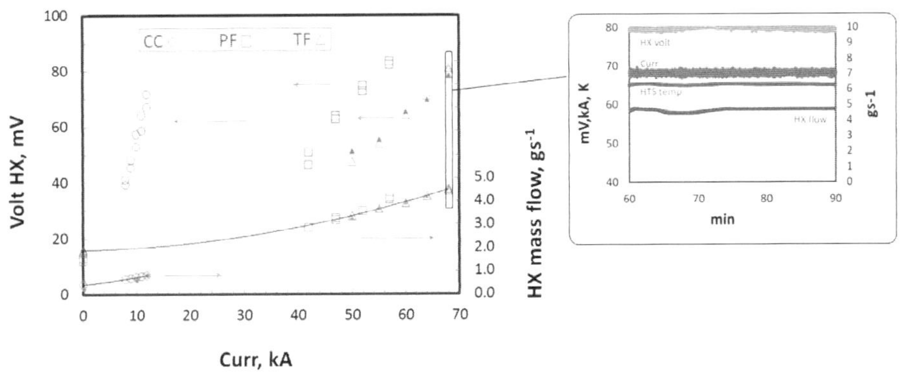

Figure 5. Prototype steady state HX voltage and mass-flow rate compared to model (full markers - CERN model for TF, [ref 38] model for CC). Insert: steady state in the TF prototype at 68 kA.

as occurring when the HTS voltage exceeds 2 mV. The temperature at the top of the HTS first rises slowly to ~90 K when the quench occurs and the temperature rises fast with the current being switched off when the hot-spot reaches 200 K. The hot-spot is located a bit below the top (warm) end of the HTS, where the shunt steel cross-section underneath the HTS is reduced and the heat diffusion and current sharing therefore less effective. Thermal-cycling and multiple LOFA quenches did not impact on the performance of the leads.

The heat conduction of the current lead was measured calorimetrically at zero current. After stabilizing the 5 K SHe cooling of the LTS loop (including the U-bend) for a given HTS top temperature in steady state (65, 80 or 100 K), the heat load was computed from the enthalpy difference (obtained from SHe temperature and pressure at the inlet/outlet) multiplied by the mass-flow. This technique applies to the entire 5 K loop and does not differentiate between the two leads (the total number is therefore divided by 2 in table 2). To check that the heat conduction from the two leads was balanced, the heat load was also derived from the shunt temperature measurements. This method consists of plotting the recorded temperature readings along the shunt and tuning the conduction heat load in the calculated temperature profile until it matches. Although this method was less precise (with only three temperature measurement points), its consistency with the heat load value obtained from halving the 5 K circuit enthalpy validated the calorimetric approach (and the material properties models).

A key parameter, which was modified after completion of the current lead design, is the 50 K GHe supply condition for the resistive heat exchanger. Originally defined as 50 K (±2 K), it was changed to a looser 40 K - 52 K. Special studies were performed to explore the impact of lower inlet temperatures (and pressures) for the HX operation. It emerged from this study that, provided the difference between GHe inlet temperature and the HTS top temperature remains 15 K, there is no difference in HX operation between 50 K and 40 K GHe supply temperature. The mass-flow is close to nominal and the lead operates stably. However, if this temperature interval is allowed to increase, the HX becomes unstable. For example, the HX voltage cannot stabilize for the condition $T_{He,in}$ = 40 K and $T_{HTS\text{-}top}$ = 65 K, as the mass-flow for this temperature at the cold end of the HX is insufficient to cool its warm end. This result implies that operation in ITER with "looser" temperature tolerances at the inlet will require also changing the $T_{HTS\text{-}top}$ set-points for controlling mass-flow (e.g. for 40 K supply $T_{HTS\text{-}top}$ = 55 K). While the effect of lower inlet temperatures can be managed with an improved control algorithm, the effect of inlet pressure is more serious. The pressure drop in the HX increases when the inlet pressure is reduced, leading to a choking effect, i.e. at lower supply pressure the mass-flow through the HX is reduced, possibly leading to a thermal runaway. Lower pressure at the outlet also affects the electrical breakdown strength of the helium in the RTIBs. It is recommended that for operation the GHe supply pressure remains above ~3.8 bar (and below ~5 bar, as the safety valves for the current leads open at ~5.5 bar).

While the ITER TF type current leads operate in steady state, the PF-type and CC-type leads operate

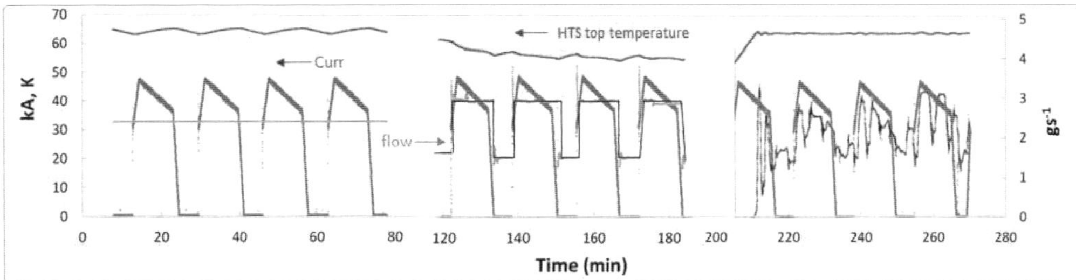

Figure 6. Three different flow control modes in the PF type current lead 1. Left: single flow (manual), Middle: "two-flow" (manual), Right: PID control. Temperature profiles shown are for steady state.

in pulsed mode. The prototypes were therefore charged with a typical ITER pulse current pattern to investigate their response. For the PF-type leads, the simulated pulse was rectangular with fast ramps (figure 6). Three different current lead control modes were explored. The simplest consisted of applying a constant flow in the HX calculated from the average current in the pulse (note that the zero current flow rate is the "stand-by" flow rate and not zero). The so-called "two-flow" mode oscillates between the stand-by flow at zero-current and the max-flow at the pulse maximum current. Finally a PID control mode was also tested, with the PID parameters set according to the results of open loop response tests. In all three modes the HTS top temperature is reasonably well controlled. There are, however, qualitative differences: the GHe control valve (located at the 300 K exhaust) works much harder in the automatic control mode. It shall also be considered to correlate the required flow-rate with the square of the current.

The HX pressure drops were usually obtained from the difference between the inlet and outlet absolute pressure measurements. This is not as accurate as the use of properly chosen differential pressure gages, but was good enough, at least for the TF and PF-type leads, which have a significant p-drop in nominal conditions (~1 bar). As shown in figure 5 they also compare well to the prediction by the CERN model. The pressure drop in the CC leads, however, is very small, only a few kPa and not measurable with the absolute gauges used. In this case pressure drop measurements were conducted with GHe at room temperature and compared to those obtained in similar conditions in the CC HX mock-up. These comparisons showed no evidence of by-pass flow, further corroborating the fact that geometrical data obtained on the HX cores and honed tubes during manufacturing were within specifications. One of the TF and CC prototypes were subjected to a full suite of HV measurements, including hi-pot to 30 kV, Partial Discharge (PD) to 10 kV and Paschen tests (the other leads did not have the finally developed insulation scheme). These leads passed these tests, see table 2.

Table 2. Test results for the ITER prototype HTS current leads.

Qualification Parameter	TF I	TF II	PF I	PF II	CC I	CC II
Leak rate 50 K loop at test pressure (10^{-9}Pa-m^3/s)	0.49	0.49	0.12	0.25	0.4	0.4
Leak rate 5 K loop at test pressure (10^{-9}Pa-m^3/s)	0.83	0.83	0.27	0.27	0.28	0.28
Cold end heat load with HTS top at 65 K (W)	20/2	20/2	20/2	20/2	2*	N/A
Pressure drop in HX section at max DC curr (MPa)	0.13	0.125	0.066	0.065	0.005	0.007
50 K GHe steady state flow-rate at max DC curr (g/s)	4.6	4.7	3.6	3.6	0.7	0.6
50 K GHe steady state flow-rate at zero curr** (g/s)	1.9	1.9	1.55	1.45	0.3	0.3
Nominal operating voltage HX at max DC curr (mV)	77	76	74	73	60	63
Temp margin (2 mV crit) in ITER condition (K)***	27	27	31	31	30	30
LOFA time at max DC operation current (2mV) (s)***	480	480	690	690	1020	1020
HTS over-heating time (2 mV criterion) (s)***	18	15	37	27	28	26
RT terminal and interface BB contact resistance (nΩ)	400	200	200	100	N/A	N/A
HX to HTS joint resistance (nΩ)	5.5	5.4	5.1	4.8	15.8	9.4
HTS to LTS joint resistance (nΩ)	0.1	0.4	0.2	0.3	0.4	0.4
HTS-CL to feeder busbar TB joint resistance (nΩ)	0.5	0.5	0.6	0.6	0.4	0.2
Resistance to ground in hi-pot test (GΩ)	3.3	N/A	30	N/A	50	4.5
Equivalent charge during PD test (nC)	5.0	N/A	2.5	N/A	N/A	0.1
Paschen test (@ 15 kV)	pass	N/A	N/A	N/A	N/A	pass

* obtained with the temperature profile method only; ** with 65 K at the HTS top; ***does not take into account the external stray field of ITER;

4. Summary

HTS current leads are one of the enabling technologies for fusion Tokamaks, as they reduce the input power requirement for plant operation. At 68 kA the ITER TF type HTS current leads will be the largest ever operated. The considerable effort to design them and qualify their manufacturing is described here.

5. Disclaimer

The views and opinions expressed herein do not necessarily reflect those of the ITER organization.

6. References

[1] Bauer P et al, 2012 *IEEE Trans Appl SC* **22/3** 4800504
[2] Taylor T, 1999 *IEEE Trans Appl SC* **9** 412
[3] Ballarino A, 2002 *Physica C* **372-376** 1413
[4] Heller R et al, 2004 *IEEE Trans Appl SC* **14/2** 1774
[5] Heller R et al, 2005 *IEEE Trans Appl SC* **15/2** 1496
[6] Heller R, Fietz WH, Lietzow R, Tanna VL,Vostner A, Wesche R, Zahn GR, 2006 *IEEE Trans Appl SC* **16/2** 823
[7] Fietz W, Heller R, Lietzow R, Tanna VL, Zahn G, Wesche R, Salpietro E, Vostner A, 2005 *21st IEEE/NPS Symposium on Fusion Engineering*
[8] Heller R, 2009 *IEEE Trans Appl SC* **19/3** 1504
[9] Bi Y, Chen X, Ma D, Liu X, Wu S, Li J, 2005 *Proceedings of ICEC 20* 681
[10] Bauer P et al, 2009 *IEEE Trans Appl SC* **19/3** 1500
[11] Bi Y et al, 2010 *IEEE Trans Appl SC* **20/3** 1718
[12] Ding K, Bi Y, Yu J, Zhou T, Liu CL, Lin X, Shen G, Song Y, 2010 *IEEE Trans Appl SC* **20/3** 1729
[13] Huang X, Bi Y, Cheng Y, Lu L, Wang C, Shen G, Song Y, 2010 *IEEE Trans Appl SC* **20/3** 1722
[14] Bauer P et al, 2011 *IEEE Trans Appl SC* **21/3** 1074
[15] Liu C et al, 2012 *IEEE Trans Appl SC* **22/3** 4800204
[16] Ding K et al, 2012 *Physics Procedia* **36** 931
[17] Zhou T, Ding K, Liu CL, Bi Y, Song Y, 2015 *IEEE Trans Appl SC* **25/2** 4802704
[18] Wesche R, Heller R, Bruzzone P, Fietz WH, Lietzow R, Vostner A, 2007 *Fusion Eng and Design* **82** 1385
[19] Taylor T, 2012 *IEEE Trans Appl SC* **22/3** 4800304
[20] Calvi M, Bauer P, Bessette D, Cau F, Marinucci C, Bruzzone PL, 2010 *IEEE Trans Appl SC* Vol **20** No 3 p 407
[21] Ilin Y et al, 2016 *IEEE Trans Appl SC* **26/4** 4800905
[22] Sitko M, Bordini B, Ballarino A, Bauer P, Devred A, 2013 *IEEE Trans Appl SC* **23/3** 4900805
[23] Rizzo E, Heller R, Savoldi Richard L, Zanino R, 2012 *IEEE Trans Appl SC* **22/3** 4801104
[24] Knaster J, Rajainmaki H, Evans D, Losasso M, 2013 *IEEE Trans Appl SC* **23/3** 3800505
[25] Bauer P et al, 2013 *IEEE Trans Appl SC* **23/3** 4801104
[26] Clayton N, Crouchen M, Devred A, Evans D, Gung CY, Lathwell I, 2017 *Cryogenics* **83** 64
[27] Huang X et al, 2017 *Fusion Eng and Design* **114** 13
[28] Langeslag S A E, Sgobba S, Libeyre P, Gung C Y, 2015 *Physics Procedia* **67** 1036
[29] Zhou T, Lu K, Ran Q, Ding K, Feng HS, Wu H, Liu CL, Song Y, Niu E, Bauer P, 2014 *IEEE Trans Appl SC* **24/3** 4801205
[30] Zhou T, Lu K, Ran Q, Ding K, Feng HS, Wu H, Liu CL, Song Y, Niu E, Bauer P, Devred A, 2015 *Fusion Eng and Design* **96–97** 388
[31] Bauer P, Ballarino A, Bordini B, Devred A, Ding K, Niu E, Sitko M, Taylor T, Yang Y, Zhou T, 2015 *IOP Conf. Series: Materials Science and Engineering* **101** 012119
[32] Huang X et al, 2019 *Fusion Eng and Design*, **146** 955
[33] Fietz WH, Fink S, Kraft G, Scheller H, Urbach E, Zwecker V, 2013 *IEEE Trans Appl SC* **23/3** 4200604
[34] Pan W, Bauer P, Carrillo D, Journeaux J, Zhu Y, Wu C, Cao Y, Chen X, 2019 *Fusion Eng and Design* **143** 154
[35] Zhou T et al, 2016 *IEEE Trans Appl SC* **26/6** 4802505
[36] Zhou T et al, 2015 2015 *IEEE 26th Symposium on Fusion Engineering (SOFE)*
[37] Ding K et al, 2016 *IEEE Trans Appl SC* **26/4** 4803404
[38] Zappatore A, Heller R, Bauer P, Savoldi L, Zanino R, 2016 *Cryogenics* **80/3** 325
[39] Liu C, Ding K, Zhou T, Du Q, Dong Y, Han Q, Lu K, Song Y, Niu E, Bauer P, 2018 *IEEE Trans Plasma Sci* **46/9** 3219
[40] Ding K et al, 2019 *IEEE Trans Appl SC* **29/5** 4802605
[41] Ding K, Liu CL, Bi Y, Feng HS, Du Q, Zhou T, Lu K, Song Y, Xu L, 2016 *IEEE Trans Appl SC* **26/3** 9500604
[42] Liu C, Ding K, Du Q, Dong Y, Wang J, Lu K, Song Y, Niu E, Bauer P, 2019 *IEEE Transactions on Plasma Sci* **47/1** 795

IOP Conf. Series: Materials Science and Engineering **756** (2020) 012033 doi:10.1088/1757-899X/756/1/012033

Liquid Nitrogen Cooled Superconducting Power Cable with No Solid Insulation

A Al-Taie[1,2,3], S Telikapalli[1,2], P Cheetham[1,2], C H Kim[2], S V Pamidi[1,2]

[1] Florida A&M University - Florida State University College of Engineering, Tallahassee, FL 32310, USA

[2] Center for Advanced Power Systems, Florida State University, Tallahassee, FL 32310, USA

[3] University of Technology, Baghdad, 10066, Iraq

Email: aha16b@my.fsu.edu

Abstract. In an attempt to eliminate solid insulation challenges in cryogenic superconducting power cables, a new design concept for liquid cryogen cooled superconducting power cable was investigated. The design is based on superconducting gas insulated line (S-GIL). The design used liquid cryogen as the sole insulation medium. The suitability of the design for medium voltage power cables is discussed and the benefits of eliminating a solid insulation were identified. Experiments on 1-m long model cables with insulator tubes as spacers showed that the design is suitable for cables at 50 kV or higher. The actual limits could not be identified because of the experimental limitations originated from limited standoff distances in the measurement setup used. On a fundamental level, the investigations presented in the study showed a direct correlation between the intrinsic dielectric strength of the cryogen used and the maximum tolerated voltage for a given diameter of the cable system. The results show the promise for liquid nitrogen (LN_2) and liquid hydrogen (LH_2) cooled cables for various medium voltage applications, including electric aviation and electric ships.

1. Introduction

When designing high temperature superconducting (HTS) power cables for medium voltage applications, it is necessary to consider the interdependencies between the electrical and thermal aspects. For HTS power cables the temperature gradient along the cable route reduces both the voltage and current ratings. Superconductivity is a function of temperature, current and applied magnetic field [1], therefore, the maximum current which the HTS cable can carry before quenching is a function of the highest temperature along the cable route. By minimizing the temperature gradient along the cable route, higher power rating of an HTS cable can be achieved for a given design. When a liquid cryogen is used, the temperature gradient along the cable route needs to be maintained to prevent a phase change. The presence of bubbles within a liquid cryogen significantly reduces the dielectric strength of the cryogen and increases the potential of an in-service electrical fault occurring. The temperature rise which exists along an HTS cable is proportional to the losses generated within the cable as well as the cooling capacity of the cryogenic system. For DC applications, the heat loads within an HTS cable come from the terminations and the heat leak through the cryostat. Terminations provide the necessary interfaces between the power devices at the room temperature and the HTS cables operating at cryogenic temperatures. The heat load produced by a termination depends on the design and rated current of the HTS cables and are typically a few hundred watts [2]. The heat load from the cryostat is approximately 1 W/m. For AC applications an additional heat load exists from the AC losses which is typically 2 W/m for 60 Hz operation. From the heat loads described above, for cable lengths less than 100 m the dominant heat load is from the terminations. However, for AC cables longer than 100 m, the AC losses generated in the HTS conductor become the dominant heat load.

There are several potential applications for medium voltage HTS power cables which vary in cable length. These applications include the electrical power grid, industrial applications, data centers, electric ships, and electric aircraft. Long length HTS power cables have been envisioned to enable the integration of renewable energy sources into the electrical power grid [3]. Liquid nitrogen is the cryogen used for the majority of HTS electrical power grid applications due to its low cost and large heat capacity. LN_2 cooled AC and DC HTS cables have been demonstrated in electrical power grids in a point-to-point configurations for short distances, and a few of them are currently in operation [4–6]. Lapped-tape insulation has been used with LN_2 as the dielectric medium in all the designs and the HTS cable covered with the insulation sits directly on the cryostat at ground potential. Some LN_2 cooled HTS cables used a more compact design that utilizes one cryogenic envelope for all the poles/phases. Such configurations are used for medium and low voltage applications. Solid insulation was used between the phases and on the whole cable [5–7].

For shipboard applications helium gas is the preferred cryogen for HTS power cables due to its larger operating temperature range and reduced risk of asphyxiation in the event of a cryogen leak [8]. Our recent work has focused on developing helium gas cooled power cables for medium voltage DC applications. This includes designing a superconducting gas insulated transmission line (S-GIL) which utilizes helium gas as both the cryogen and dielectric medium for the HTS power cable [9]. It has been demonstrated that for a given geometry the dielectric strength of the S-GIL is a function of the dielectric strength of the cryogen. This promising result led us to investigate the applicability of liquid cooled HTS power cables with no solid insulation.

For aircraft application such as NASA N3-X program, liquid hydrogen (LH_2) is being considered as a potential cryogen. Utilizing LH_2 as the cryogen would allow it to be stored within a fuel tank which eliminates or reduces the need for cryocoolers. LH_2 absorbs the heat loads from HTS devices and evaporates. The evaporated gas is used as fuel in the engines [10]. The designs developed here for LN_2 cooled HTS power cable with no solid insulation are directly applicable for LH_2 or LN_2 cooled HTS power cables. The changes in the intrinsic properties of the cryogen which include operating temperature range, dielectric strength, heat capacity, and mass flow rate have to be adjusted depending on the liquid cryogen used.

This paper discusses the design consideration and potential system level benefits by developing LN_2 cooled HTS power cables with no solid insulation. Several 1-m long model cables were fabricated and their performance under high voltage was measured in pressurized, subcooled and boiling LN_2.

2. Design Considerations of LN_2 Cooled HTS Power Cables with No Solid Insulation

There are several benefits of implementing LN_2 cooled HTS power cables with no solid insulation for medium voltage applications compared to the traditionally utilized lapped-tape insulation design. These benefits include simpler and quicker manufacturing process for long length HTS power cables, potential to reduce the size of the cryogenic cooling plant required for the HTS cable route, ability to utilize prefabricated HTS conductor, retention of dielectric integrity after electrical fault, and reduced hotspots/thermal gradients in HTS section due to the enhanced heat transfer.

Lapped-tape insulation requires multiple layers of the lapped-tape to be helically wrapped on top of one another along the entire length of the HTS conductor. This process needs to ensure that each helical layer is offset to one another to ensure the butt gaps, which are part of the design, do not overlap from one layer to the next. This process also needs to prevent any damage to the lapped-tape or the HTS conductor to ensure optimal performance of the HTS power cable. Any defects during the HTS cable manufacturing process can only be identified once the HTS power cable is installed and cooled to the operating temperature. Replacing the cable at that stage is costly if defects are identified and the dielectric integrity is not as expected. Eliminating the solid lapped-tape insulation and using the cryogen as the sole dielectric medium simplifies the HTS cable manufacturing process and enables the possibility of factory acceptance testing. As the performance of no solid insulation design is a function of the dielectric strength of the cryogen surrounding the HTS conductor, it may be possible to perform factory acceptance testing at room temperature using a gas with similar dielectric properties as liquid nitrogen. This allows for greater consumer confidence for HTS cable technology.

While there has been development with HTS tape efficiency, there has also been development with HTS conductor technology such as CORC® HTS conductors, which are capable of operating at 5-10 kA whilst only having an outer diameter between 3-7 mm [11]. The size of these conductors may allow for smaller cryostats to be utilized which reduces the overall volume of cryogen needed for the cable route. The ability to utilize commercially available HTS conductors as part of the HTS power cable design is advantageous to reduce the manufacturing and lead time of HTS power cables. This should help to make HTS power cables a more attractive and feasible option to industry.

When designing an LN_2 cooled HTS power cables with no solid insulation, the achievable voltage rating is a function of the physical and electrical properties of the HTS conductor, cryostat, insulator spacers and LN_2. For a single phase LN_2 cooled HTS power cables with no solid insulation, the electric field within the cryostat can be derived from the coaxial electric field equation [12]. To ensure the HTS conductor remains concentric to the cryostat at ground potential, insulator spacers are required. As part of our previous work on the S-GIL, various insulator spacer configurations have been investigated along with the relative diameters of the HTS conductor and the cryostat selected. Based on our previous studies, a 12.7 mm diameter HTS conductor and a cryostat (at ground potential) with inner diameter of 39 mm were selected for model cables. These parameters ensure the optimized electric field based on the coaxial equation [13]. Our previous work on gaseous helium cooled HTS power cables demonstrated that a simple design that used tubular insulators ensure the conductor remains concentric to the cryostat. The use of tubular spacers also promotes large openings to facilitate the circulation of the cryogen. 1-m long model cables were fabricated to measure the dielectric response of the HTS conductor and cryostat configuration with and without insulator spacers (Figure 1).

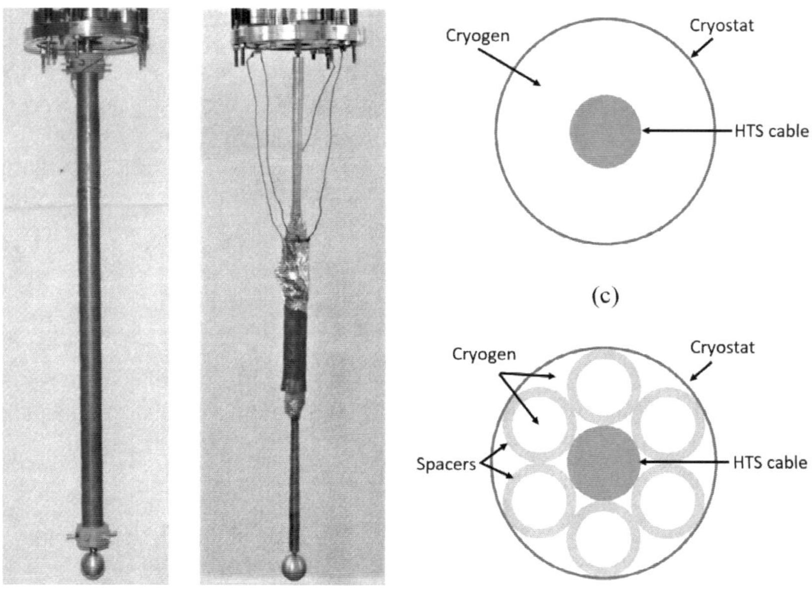

(a) (b) (d)

Figure 1. (a) No-spacer HTS cable experimental setup (b) bundled spacers HTS cable experimental setup (c) no-spacer HTS cable configuration cross-section (d) bundled spacers HTS cable configuration cross-section

For the HTS power cables design shown in Figure 1, the dielectric behavior is a function of the dielectric strength of LN_2. For the design shown in Figure 1 (a & c) the maximum withstand voltage is a function of the intrinsic dielectric strength of LN2. The introduction of the insulator spacers (Figure 1(b & d)) results in a reduction of the maximum withstand voltage as the mechanism of electrical breakdown changes from intrinsic breakdown through the LN2 to surface flashover occurring along the outer wall of the insulator spacer. For a LN_2 cooled HTS power cables with no solid insulation, the limiting voltage rating is expected to be as a result of surface flashover occurring along the insulator spacers. The insulator spacer design shown in Figure 1 is one potential design so it is also important to measure the intrinsic case to know the maximum possible voltage for given dimensions of the HTS conductor and the cryostat.

LN_2 power cables that have been previously demonstrated for the electrical power grid applications used pressurized and subcooled liquid nitrogen. Pressurizing and subcooling the LN_2 allows for a larger permissible temperature gradient (63-80 K) without the occurrence of bubbles. The dielectric strength of liquid nitrogen has been reported to be between 12 and 50 kV/mm at pressure levels between 200 kPa and 500 kPa) at 77 K when bubble formation has been prevented [14–17]. However, the dielectric integrity is disrupted with the presence of bubbles, and the dielectric strength is significantly reduced when bubbles are present [15]. At atmospheric pressure, liquid nitrogen boils at 77 K, and pressurizing and subcooling suppresses bubble formation and increases the dielectric strength of the LN_2. For example, the boiling point increases by 20 K when pressurized to 0.5 MPa [18]. Based on these facts, the experimental setup to characterize the dielectric strength of the 1-m long model cables in both atmospheric pressure and pressurized LN_2 was established.

3. Measurements in Pressurized Liquid Nitrogen

3.1. Experimental Setup and Results

For testing the dielectric performance of the solid-insulation-free LN_2 cooled HTS cable in pressurized LN_2, a stainless-steel piping system for the pressure vessel was arranged. The pressure vessel utilizes a 10 inch Conflat flange with six pipe connections for: (i) LN_2 inlet, (ii) LN_2 outlet, (iii) gas inlet, (iv) gas outlet, (v) gas pressure gauge, and (vi) extra outlet that can be used for safety if needed. In addition, a relief valve was connected to the gas in/outlet pipe to release excessive pressure when built up. Furthermore, a needle valve was inserted in the gas in/outlet connection to stop the liquid to go up in the gas pipe. A schematic of the experimental set up is shown in Figure 2, along with the associated pipe work required to achieve pressurization.

Regarding test cable preparation, a cable was prepared with 6 polyethylene (PE) bundled tubes, with outer diameter of 12.7 mm, were wrapped around the conductor, with outer diameter of 12.7 mm. Kapton tape was used at discrete locations along the polyethylene tubes to ensure contacts between the conductor and polyethylene tubes to reduce the potential of electric field enhancement at the insulator/conductor interface. The conductor with attached polyethylene tubes was installed within a 39 mm copper tube, which acts as the cable cryostat at ground potential, as shown in Figure 1(b). The tubes surrounding the conductor ensure the position of the HTS conductor at the center of the cryostat. The cable was attached to the power lead which makes the high voltage connection through the bushing, and the copper tube was connected to the lower side of the top plate flange to make the ground connection. Afterward, the vessel was sealed tightly. Due to concerns that the air inside the vessel could contain some moisture, which turns into ice when cooled down and causes surface flashover at a lower voltage levels, the vessel was filled with dry gaseous nitrogen (GN_2) at 2 MPa at room temperature. GN_2 helped to remove the moisture and checked if there are any leaks. Then, the GN_2 was flushed out and evacuated, which helped also in removing any other impurities.

Before the high voltage experiments at 77 K, room temperature experiments were performed with GN_2 at 0.5 MPa. The S-GIL design has not been tested with GN_2 before, and the experiments were conducted to generate the baseline data to compare with the results of LN_2 measurements. Accordingly, the results for LN_2 can be judged against GN_2 as the insulation medium. Three measurements were taken with GN_2 to minimize sample degradation before the measurements in LN_2. The results on GN_2 insulated cable showed that for the 6 polyethylene (PE) bundled tube design, the maximum withstand voltage before surface flashover occurred is 35.9 kV with an average of 32.3 kV (rms, 60 Hz). The results are as expected based on our previous measurements on GHe cooled cables and the relative dielectric strengths of GHe and GN_2.

For the experiments on model cables in LN_2, the pressure vessel needed to be filled with LN_2 without any contamination. The vessel was first filled with GN_2 at 0.4 MPa, at room temperature to ensure the pressure within the vessel would be above atmospheric pressure at 77 K. Then, the pressure vessel was immersed into the external LN_2 bath in a cryostat. The slight positive pressure ensures that the air does not enter the vessel during the LN_2 filling process. The hose connected to the LN_2 dewar was also flushed of any air. The pressure vessel was then filled with LN2 until LN2 was coming out from the LN2 outlet port. This condition replicated what would be achieved in a boiling LN2 bath

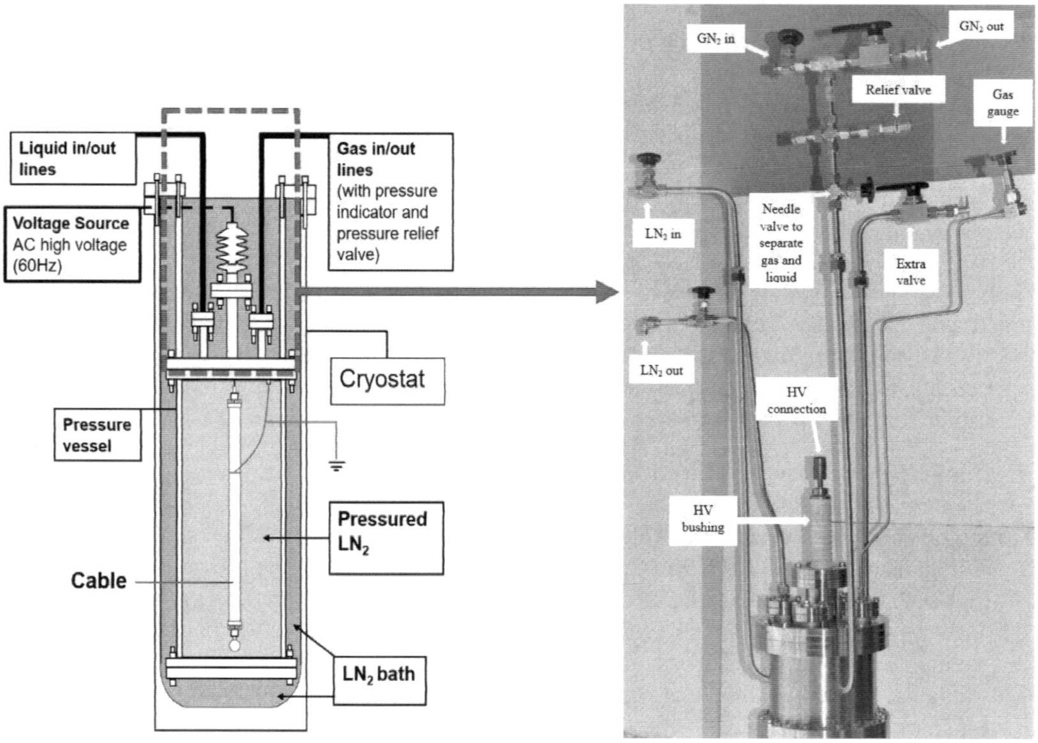

(a) (b)

Figure 2. (a) A schematic of the pressure vessel, external LN_2 bath, the test cable, and the high voltage connection used for performing the high voltage characterization of test cables (b) gas/liquid connections for the pressure vessel.

The pressure vessel was connected to a 100 kV AC (rms), 5 kVA transformer and the voltage was ramped up at approximately at 600 V/s (rms) until breakdown occurred. For the experiments in boiling LN_2 (at 0.1 MPa, 77 K), the 5 measurements performed yielded a maximum of 45.2 kV (rms) and an average of 39.8 kV (rms). Preventing bubble formation was achieved in the pressurized LN_2 (to 0.2 MPa) experiments by immersing the pressure vessel/experimental chamber in the outer open LN_2 cryostat at 77 K. The LN_2 in the pressure vessel has an elevated boiling point of ~80 K but is kept at 77 K by the LN_2 in the outer cryostat, thus maintaining the LN_2 in the inner vessel in pressurized and subcooled state to prevent bubbles.

In the experiments performed in pressurized LN_2 (0.2 MPa, 77 K), the breakdown occurred external to the pressure vessel due to the limited standoff/clearances among the HV connection on the top of the bushing and metal gas lines. The highest voltage reached breakdown outside the pressure vessel was 55 kV (rms, 60 Hz), suggesting that the cable would have reached higher voltages if the experimental limitations were not encountered. This result also implies that the withstand voltage of the model cable shown in Figure 1(a) would also exceed 55 kV in pressurized LN2. As breakdown values were recorded on the model cable when the experiment setup had boiling LN2, there was the potential to perform measurements on model cable without insulator spacers in a LN2 bath. As pressurization of the experimental setup was not required, a simpler experimental setup was utilized which consisted of the model cable being connected to the top plate via the HV bushing.

The same procedure of installing the experiment setup within the cryostat and applying the voltage to the model cable was followed as previously described. When the model cable without any insulator spacer was measured, the breakdown occurred externally to the experimental setup at approximately 55 kV. This result was as expected as this model cable should have superior performance than the model cables with insulator spacers. For verification the experiment with the model cable with insulator spacers was performed with the same experiment setup. Surprisingly, the breakdown also occurred externally from the experiment setup at approximately 55 kV. This result was approximately 10-15 kV higher than what was recorded with the original experimental setup.

4. Discussion

After completing the tests of the 6 PE bundled HTS cable sample in boiling and pressurized LN_2 using the first experimental setup, the sample was disassembled to analyze the problem that caused the variation in experimental data between the experimental setups. It was observed that the cable was clean without any charring or burn marks indicating the absence of surface flashover on the PE tubes. Removal of the bushing from the top plate showed pitting on the rod of the high voltage feedthrough (Figure 3). This suggests the limitation of the first experiment setup was due to the nitrogen gas forming near the bushing and intrinsic breakdown was occurring between the bushing feedthrough and the top plate.

For the case of 0.1 MPa, LN_2 vapor was trapped under the top plate and caused the breakdown to occur under the bushing where the dielectric strength is less than the cable dielectric strength. When the pressure was increased for the 0.2 MPa, the vapor/gas dielectric strength increased. The breakdown did not happen at that location nor on the cable, but it happened outside the vessel due to that now becoming the path of least resistance.

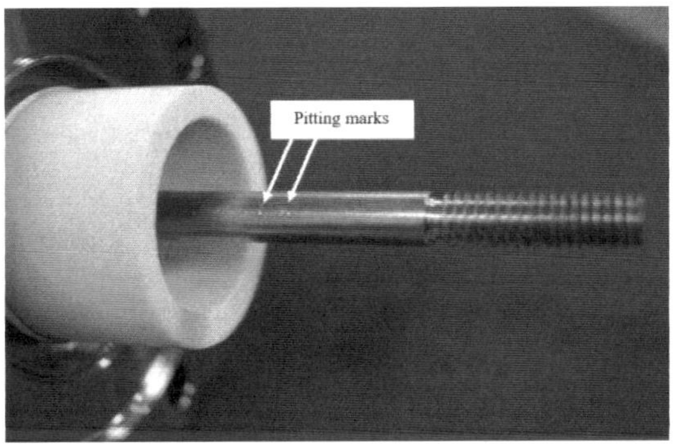

Figure 3. Pitting on the current lead part extended under the bushing

Given the configuration of the experimental setup and the associated problems, there are potential modification which can be made to improve the voltage withstand capability of the circuit to enable characterization of the model cables at higher voltages. Firstly, the limiting factor is caused from installing the 10 inch diameter pressure vessel within a cryostat with inner diameter of 12 inches. The 6 metal support bars which suspend the pressure vessel in the cryostat transition between LN2 and air. As air has relatively poor dielectric strength the breakdown occurs between the high voltage bushing and the support rods in air. Increasing the gap distance between the support rods by utilizing a larger cryostat would be the simplest method to increase the voltage rating of the experimental setup.

Regarding the first experimental setup shown in Figure 2(a) the nitrogen bubble formation could be resolved by changing the locations of the filling lines. One potential solution is to install the pressure port horizontally instead of vertically which would ensure if bubble formation was to occur it would not happen at the bushing. Horizontally installing the pressure vessel would require modification of how the model cables are installed within the pressure to ensure the cable is still suspended in the center of the cryostat and the stress sphere (at voltage potential) does not come in contact with the pressure vessel wall at ground potential.

While ultimately the experiments showed that improvements are required in the experimental setup, it demonstrated that voltage withstand capability in excess of 55 kV is expected for the given cable and cryostat parameters selected. This in turn demonstrates the potential of LN_2 cooled HTS cables with no solid insulation to be utilized for low to medium voltage applications.

5. Conclusion

LN_2 was used as the cryogen and insulation medium in S-GIL, instead of gas, to examine the dielectric strength limits for the insulation system of the bundled spacer tubes S-GIL design. The results showed that there exists a direct relationship between the dielectric strength of the insulation medium and the performance of the HTS cable. S-GIL performance increases with the increase of dielectric strength of the insulation medium which is the cryogen used. The solid-insulation-free liquid cooled HTS design has the potential to be used for applications that use LN_2 or LH_2. Electric transportation applications will benefit from such cables. The absence of solid insulation offers better heat transfer, minimal degradation during a breakdown event, and easier manufacturing.

6. References

[1] Kalsi S 2011 *Applications of high temperature superconductors to electric power equipment* (Wiley)

[2] Rey C M 2015 *Superconductors in the Power Grid: Materials and Applications* (Elsevier)

[3] Al-Taie A, Graber L and Pamidi S V 2017 Technical Challenges and Potential Solutions for Cross-Country Multi-Terminal Superconducting DC Power Cables *IOP Conference Series: Materials Science and Engineering* vol 279

[4] Maguire J F, Yuan J, Schmidt F, Soika R, Bratt S, King C, McNamara J and Welsh T E 2010 Operational experience of the world's first transmission level voltage HTS power cable *AIP Conference Proceedings* vol 1218pp 437–44

[5] Sytnikov V E, Bemert S E, Kopylov S I, Romashov M A, Ryabin T V., Shakaryan Y G and Lobyntsev V V. 2015 Status of HTS Cable Link Project for St. Petersburg Grid *IEEE Trans. Appl. Supercond.* **25** 1–4

[6] Justification P 2012 Project fact sheet 1–2

[7] Kopylov S, Sytnikov V, Bemert S, Ivanov Y, Krivetskiy I, Romashov M, Shakaryan Y, Keilin V, Shikov A, Patrikeev V, Lobyntsev V and Shcherbakov V 2014 HTS DC transmission line for megalopolis grid development *J. Phys. Conf. Ser.* **507** 1–5

[8] Kephart J T, Fitzpatrick B K, Ferrara P, Pyryt M, Pienkos J and Michael Golda E 2011 High temperature superconducting degaussing from feasibility study to fleet adoption *IEEE Trans. Appl. Supercond.* **21** 2229–32

[9] Cheetham P, Viquez J, Graber L, Kim C H, Rodrigo H and Pamidi S 2016 Novel Design Concept and Demonstration of a Superconducting Gas-Insulated Transmission Line *IEEE Trans. Appl. Supercond.* **27** 1–5

[10] Felder J, Kim H and Brown G 2009 Turboelectric Distributed Propulsion Engine Cycle Analysis for Hybrid-Wing-Body Aircraft *47th AIAA Aerosp. Sci. Meet. Incl. New Horizons Forum Aerosp. Expo.* 1–25

[11] Weiss J D, Mulder T, ten Kate H J and van der Laan D C 2017 Introduction of CORC ® wires: highly flexible, round high-temperature superconducting wires for magnet and power transmission applications *Supercond. Sci. Technol.* **30** 14002

[12] Kuffel E, Zaengl W S and Kuffel J 2016 *High Voltage Engineering Fundamentals* (Butterworth-Heinemann/Newnes)

[13] Cheetham P, Wagner J, Stamm T, Kim C H, Graber L and Pamidi S 2018 Electric field analysis of insulating spacers for a superconducting gas-insulated transmission line *Proc. IEEE Int. Conf. Prop. Appl. Dielectr. Mater.* **2018–May** 185–9

[14] Sauers I, James R, Ellis A, Tuncer E, Polizos G and Pace M 2009 Breakdown in liquid nitrogen in the presence of thermally generated bubbles for different electrode geometries *Annu. Rep. - Conf. Electr. Insul. Dielectr. Phenomena, CEIDP* 319–22

[15] Sauers I, James R, Ellis A, Tuncer E, Polizos G and Pace M 2011 Effect of bubbles on liquid nitrogen breakdown in plane-plane electrode geometry from 100-250 kPa *IEEE Trans. Appl. Supercond.* **21** 1892–5

[16] Fink S and Noe M 2008 A facility for testing the dielectric strength of liquid nitrogen *2008 IEEE International Conference on Dielectric Liquids* (IEEE) pp 1–4

[17] Blaz M and Kurrat M 2012 Studies of breakdowns in liquid nitrogen at different pressures between Rogowski electrodes *Phys. Procedia* **36** 1330–6

[18] Chang H-M, Choi Y S, Van Sciver S W and Choi K D 2003 Cryogenic cooling system of HTS transformers by natural convection of subcooled liquid nitrogen *Cryogenics (Guildf).* **43** 589–96

AUTHOR INDEX

Institute of Physics
Dirac House, Temple Back
Bristol BS1 6BE UK

ISSN: 1757-8981
ISBN 978-1-7138-1970-7

9 781713 819707

CURRAN ASSOCIATES INC.
proceedings
.com